Tourism, Global Crises and Justice

This book gathers theoretical and empirical studies exploring the link between global crises, sustainable tourism and the justice challenges being faced by vulnerable groups, individuals and society.

While any crisis may exacerbate existing inequalities, the crises of the 21st century are compounding and complicating the ways the impacts unfold and engulf individuals, communities and, indeed, the global community. Recent crises revealed how dependent our economies and societies are on the tourism and hospitality industries. While studies of crises in tourism have proliferated, with concerns for risk management, recovery and resilience, COVID-19 has exposed the need to think more profoundly on this topic. In such circumstances, therefore, tourism actors must respond to the sustainability and justice challenges resulting from current and future crises by rethinking, redefining and reorienting tourism. The chapters in this edited volume present a discussion of pertinent themes that consider just transformations, issues of climate justice, diverse worldviews and knowledges, possibilities for solidarity through tourism and concerns with power and decolonisation.

This book will be of great interest to upper-level students, researchers and academic of tourism, development studies and sustainability, as well as professionals in the field of tourism management. The chapters in this book were originally published in the *Journal of Sustainable Tourism*.

Raymond Rastegar is Lecturer and Researcher at the Department of Tourism, Sport and Hotel management at Griffith University, Australia. He holds a PhD in Tourism Management, and his scholarly interest and expertise lie in the fields of justice, sustainability transitions and environmental conservation. His research delivered new insights into the tourism phenomenon to advocate a more just and sustainable tourism future for humans and nonhumans.

Freya Higgins-Desbiolles is Adjunct Senior Lecturer in Business Unit at the University of South Australia, Australia; Adjunct Associate Professor at the Department of Recreation and Leisure at the University of Waterloo, Canada; and Visiting Professor at the Centre for Research and Innovation in Tourism at the Taylor's University of Malaysia, Malaysia. Her work focuses on social justice, human rights and sustainability issues in tourism.

Lisa Ruhanen is Professor and Deputy Head of School at the University of Queensland Business School, Brisbane, Australia. She has been involved in almost 30 academic and consultancy research projects in Australia and overseas. Her research areas include sustainable tourism destination policy and planning, climate change and Indigenous tourism.

Tourism, Global Crises and Justice

Tourism Transition to a More Just and Sustainable Future

Edited by

Raymond Rastegar, Freya Higgins-Desbiolles and Lisa Ruhanen

Routledge
Taylor & Francis Group
LONDON AND NEW YORK

First published 2025
by Routledge
4 Park Square, Milton Park, Abingdon, Oxon, OX14 4RN

and by Routledge
605 Third Avenue, New York, NY 10158

Routledge is an imprint of the Taylor & Francis Group, an informa business

British Library Cataloguing-in-Publication Data
A catalogue record for this book is available from the British Library

ISBN13: 978-1-032-79541-6 (hbk)
ISBN13: 978-1-032-79543-0 (pbk)
ISBN13: 978-1-003-49261-0 (ebk)

DOI: 10.4324/ 9781003492610

Typeset in Myriad Pro
by codeMantra

Publisher's Note
The publisher accepts responsibility for any inconsistencies that may have arisen during the conversion of this book from journal articles to book chapters, namely the inclusion of journal terminology.

Disclaimer
Every effort has been made to contact copyright holders for their permission to reprint material in this book. The publishers would be grateful to hear from any copyright holder who is not here acknowledged and will undertake to rectify any errors or omissions in future editions of this book.

Contents

Citation Information vii
Notes on Contributors x
Foreword: Academics can change the world, if they stop talking only to their peers xiii
Jeremy Sampson

Introduction: Tourism, global crises and justice: rethinking, redefining and reorienting tourism futures 1
Raymond Rastegar, Freya Higgins-Desbiolles and Lisa Ruhanen

PART I
Activating justice **17**

1 Toward critical race tourism: valuing counter-narratives and *endarkened storywork* 19
Stefanie Benjamin and Judson Laughter

2 The potential of toxic tours: indigenous perspectives on crises, relationships, justice and resurgence in Oklahoma Indian country 36
Bobbie Chew Bigby, Earl Hatley and Rebecca Jim

3 Gender justice in global tourism: exploring tourism transformation through the lens of feminist alternative economics 58
Angela B. Kalisch and Stroma Cole

4 Justice and community citizenship behavior for the environment: small tourism business entrepreneurs' perspectives 76
Mao-Ying Wu, Xinfang Wu, Qiu-cheng Li, Jie Wang and Yi Wang

PART II
Engaging marginalized tourism stakeholders **95**

5 Reimagining children's participation: a child rights informed approach to social justice in tourism 97
Antonia Canosa and Anne Graham

6 Tourism and refugee-crisis intersections: co-creating tour guide experiences in Leeds, England 110
Elisa Burrai, Dorina-Maria Buda and Emily Stevenson

7 Seeking justice beyond the platform economy: migrant workers navigating precarious lives 128
Tyler Riordan, Richard N.S. Robinson and Gerhard Hoffstaedter

PART III
Tools and processes that build (in)justice in tourism 147

8 Do international sanctions help or inhibit justice and sustainability in tourism? 149
Siamak Seyfi, Colin Michael Hall, Jarkko Saarinen and Tan Vo-Thanh

9 Tourism policies and inclusive development: the case of Kenya and Rwanda 167
Christine N. Buzinde and Tanner Caterina-Knorr

10 Tourism policy, spatial justice and COVID-19: lessons from a tourist-historic city 186
Brendan Paddison and Jenny Hall

PART IV
Confronting crises, building justice and transitioning tourism 203

11 The poor on the road: qiongyou as a collective resistance and justice tourism 205
Yimeng Yang

12 Tourism, compounding crises, and struggles for sovereignty 222
Carter A. Hunt, María José Barragán-Paladines, Juan Carlos Izurieta and Andrés Ordóñez L

13 Decolonising tourism and development: from orphanage tourism to community empowerment in Cambodia 240
Freya Higgins-Desbiolles, Regina A. Scheyvens and Bhanu Bhatia

14 Rethinking the space of tourism, its power-geometries, and spatial justice 261
Lucia Tomassini and Ian Lamond

Index 275

Citation Information

The chapters in this book were originally published in the *Journal of Sustainable Tourism*, volume 31, issue 12 (2023), except Chapter 12, which was originally published in volume 31, issue 10 (2023). When citing this material, please use the original page numbering for each article, as follows:

Foreword

Academics can change the world, if they stop talking only to their peers
Jeremy Sampson
Journal of Sustainable Tourism, volume 31, issue 12 (2023), pp. 2857–2858

Introduction

Tourism, global crises and justice: rethinking, redefining and reorienting tourism futures
Raymond Rastegar, Freya Higgins-Desbiolles and Lisa Ruhanen
Journal of Sustainable Tourism, volume 31, issue 12 (2023), pp. 2613–2627

Chapter 1

Toward critical race tourism: valuing counter-narratives and endarkened storywork
Stefanie Benjamin & Judson Laughter
Journal of Sustainable Tourism, volume 31, issue 12 (2023), pp. 2628–2644

Chapter 2

The Potential of Toxic Tours: Indigenous Perspectives on Crises, Relationships, Justice and Resurgence in Oklahoma Indian Country
Bobbie Chew Bigby, Earl Hatley and Rebecca Jim
Journal of Sustainable Tourism, volume 31, issue 12 (2023), pp. 2645–2666

Chapter 3

Gender justice in global tourism: exploring tourism transformation through the lens of feminist alternative economics
Angela B. Kalisch and Stroma Cole
Journal of Sustainable Tourism, volume 31, issue 12 (2023), pp. 2698–2715

Chapter 4

Justice and community citizenship behavior for the environment: small tourism business entrepreneurs' perspectives
Mao-Ying Wu, Xinfang Wu, Qiu-cheng Li, Jie Wang and Yi Wang
Journal of Sustainable Tourism, volume 31, issue 12 (2023), pp. 2839–2856

Chapter 5

Reimagining children's participation: a child rights informed approach to social justice in tourism
Antonia Canosa and Anne Graham
Journal of Sustainable Tourism, volume 31, issue 12 (2023), pp. 2667–2679

Chapter 6

Tourism and refugee-crisis intersections: co-creating tour guide experiences in Leeds, England
Elisa Burrai, Dorina-Maria Buda and Emily Stevenson
Journal of Sustainable Tourism, volume 31, issue 12 (2023), pp. 2680–2697

Chapter 7

Seeking justice beyond the platform economy: migrant workers navigating precarious lives
Tyler Riordan, Richard N.S. Robinson and Gerhard Hoffstaedter
Journal of Sustainable Tourism, volume 31, issue 12 (2023), pp. 2734–2751

Chapter 8

Do international sanctions help or inhibit justice and sustainability in tourism?
Siamak Seyfi, Colin Michael Hall, Jarkko Saarinen and Tan Vo-Thanh
Journal of Sustainable Tourism, volume 31, issue 12 (2023), pp. 2716–2733

Chapter 9

Tourism policies and inclusive development: the case of Kenya and Rwanda
Christine N. Buzinde and Tanner Caterina-Knorr
Journal of Sustainable Tourism, volume 31, issue 12 (2023), pp. 2752–2770

Chapter 10

Tourism policy, spatial justice and COVID-19: lessons from a tourist-historic city
Brendan Paddison and Jenny Hall
Journal of Sustainable Tourism, volume 31, issue 12 (2023), pp. 2809–2824

Chapter 11

The poor on the road: qiongyou as a collective resistance and justice tourism
Yimeng Yang
Journal of Sustainable Tourism, volume 31, issue 12 (2023), pp. 2771–2787

Chapter 12

Tourism, compounding crises, and struggles for sovereignty
Carter A. Hunt, María José Barragán-Paladines, Juan Carlos Izurieta and Andrés Ordóñez L
Journal of Sustainable Tourism, volume 31, issue 10 (2023)
pp. 2381–2398

Chapter 13

Decolonising tourism and development: from orphanage tourism to community empowerment in Cambodia
Freya Higgins-Desbiolles, Regina A. Scheyvens and Bhanu Bhatia
Journal of Sustainable Tourism, volume 31, issue 12 (2023), pp. 2788–2808

Chapter 14

Rethinking the space of tourism, its power-geometries, and spatial justice
Lucia Tomassini and Ian Lamond
Journal of Sustainable Tourism, volume 31, issue 12 (2023), pp. 2825–2838

Notes on Contributors

María José Barragán-Paladines, Charles Darwin Research Station, Charles Darwin Foundation, Galapagos, Ecuador.

Stefanie Benjamin, Department of Retail, Hospitality & Tourism Management, University of Tennessee Knoxville, USA.

Bhanu Bhatia, Asia Pacific College of Business & Law, Charles Darwin University, Australia.

Bobbie Chew Bigby, Nulungu Research Institute, University of Notre Dame Australia, Australia.

Dorina-Maria Buda, Business School, Nottingham Trent University, UK.

Elisa Burrai, School of Events, Tourism and Hospitality Management, Leeds Beckett University, UK.

Christine N. Buzinde, School of Community Resources and Development, Arizona State University, Phoenix, USA.

Antonia Canosa, Centre for Children and Young People, Faculty of Health, Southern Cross University, Australia.

Tanner Caterina-Knorr, School of Community Resources and Development, Arizona State University, Phoenix, USA.

Stroma Cole, University of Westminster, London, UK.

Anne Graham, Centre for Children and Young People, Faculty of Health, Southern Cross University, Australia.

Colin Michael Hall, Research Unit, University of Oulu, Finland; Department of Management, Marketing and Entrepreneurship, University of Canterbury, Christchurch, New Zealand; School of Business and Economics, Linnaeus University, Kalmar, Sweden; School of Tourism & Hospitality, University of Johannesburg, South Africa.

Jenny Hall, York St John University, UK.

Earl Hatley, LEAD Agency, Inc, Quechee, USA.

Freya Higgins-Desbiolles, UniSA Business, University of South Australia, Adelaide, Australia; Recreation and Leisure Studies, University of Waterloo, Canada.

Gerhard Hoffstaedter, School of Social Science, University of Queensland, St Lucia, Australia.

Carter A. Hunt, Recreation, Park, and Tourism Management, and Anthropology, Penn State University, USA.

Juan Carlos Izurieta, Charles Darwin Research Station, Charles Darwin Foundation, Galapagos, Ecuador.

Rebecca Jim, LEAD Agency, Inc, Miami, USA.

Angela B. Kalisch, University of Westminster, London, UK.

Ian Lamond, School of Events, Tourism and Hospitality Management, Leeds Beckett University, UK.

Judson Laughter, Department of Theory and Practice in Teacher Education, University of Tennessee Knoxville, USA.

Qiu-cheng Li, School of Tourism and Urban-Rural Planning, Zhejiang Gongshang University, Hangzhou, P. R. China.

Andrés Ordóñez L, Galapagos Chamber of Tourism (CAPTURGAL), Ecuador.

Brendan Paddison, York St John University, UK.

Raymond Rastegar, Department of Tourism, Sport and Hotel Management, Griffith University Brisbane, Australia; Centre for Research and Innovation in Tourism (CRiT), Taylor's University, Subang Jaya, Malaysia.

Tyler Riordan, UQ Business School, University of Queensland, St Lucia, Australia.

Richard N.S. Robinson, UQ Business School, University of Queensland, St Lucia, Australia.

Lisa Ruhanen, Business School, University of Queensland, Brisbane, Australia.

Jarkko Saarinen, Research Unit, University of Oulu, Finland.

Jeremy Sampson, The Travel Foundation.

Regina A. Scheyvens, School of People, Environment and Planning, Massey University, Palmerston North, New Zealand.

Siamak Seyfi, Research Unit, University of Oulu, Finland.

Emily Stevenson, Independent Researcher, Leeds, UK.

Tan Vo-Thanh, Department of Marketing, Excelia Business School, La Rochelle, France.

Lucia Tomassini, Academy of International Hospitality Research, NHL Stenden University of Applied Sciences, Leeuwarden, The Netherlands; University of Groningen, Campus Fryslan, Leeuwarden, The Netherlands.

Jie Wang, School of Business, University of Queensland, Brisbane, Australia.

Yi Wang, School of Management, Zhejiang University, Hangzhou, P. R. China.

Mao-Ying Wu, School of Management, Zhejiang University, Hangzhou, P. R. China.

Xinfang Wu, School of Management, Zhejiang University, Hangzhou, P. R. China.

Yimeng Yang, Graduate Institute of Building and Planning, National Taiwan University, Taiwan.

Foreword: Academics can change the world, if they stop talking only to their peers

Jeremy Sampson

I recently read a headline that caught my eye: "Academics can change the world, if they stop talking only to their peers." This is certainly just as true within tourism as elsewhere, which means a lot of good thinking, critical analysis and important ideas simply don't get the exposure they need. In a time of multiple crises: of climate emergency, of inequity and imbalance in destinations, and of recovery from a socially and economically devastating pandemic, we need to find ways to better harness the power of academia like never before.

I see a growing recognition among my peers that fundamental change is needed to the tourism model so that destinations—and the businesses that rely on them—can thrive in the twenty first Century. But to do things differently, we need new innovative approaches that are demonstrably better than what's on offer with the status quo. The good news is there is no shortage of fresh ideas, coming from both academia and the next generation of entrepreneurs and disruptors. But the not-so-good news is that these rarely become visible, practical, applied-in-the real-world solutions that address our collective desire for better tourism.

Of course, this is not only the responsibility of academics and innovators. It is also up to the rest of us to shine a light on new ideas, stimulating debate around them, finding practical applications, and building the business case for investors. We all need to bring our different strengths to this endeavour in a way that adds value and makes a difference.

At the first ever Future of Tourism Coalition Summit in Athens this past October, I was fortunate enough to spend time with Professor Xavier Font (editor of this edition of JOST), discussing the big battles that tourism faces on the road ahead. The conference focused on understanding the different roles and strengths, or "superpowers" if you will, that various players can bring to bear to fight these battles and what can happen when those superpowers merge into one great big super group like the Avengers—that's when you get what the Coalition members like to call "radical collaboration".! At the Travel Foundation, we have a complementary mantra: "destinations are everyone's business." Collaboration across and beyond our sector will be key to positive change and part of that must include working closely with academia to bring promising new ideas to light, to help the industry learn and adapt to a new paradigm and to ensure tourism is delivering on its potential to drive positive outcomes.

We have recently Engaged with researchers and universities on several new pieces of work, including a partnership with Cornell which resulted in a new course on sustainable destination management; worki with the University of Surrey on Equity, Diversity and Inclusiveness in travel and tourism businesses,a partnership with New York University focused on risk assessment as a tool for increasing investment in priority issues like climate action, and a Living Lab with Leeds Beckett to identify what practical support is needed to enable destinations to commit to and implement Climate Action Plans in the framework of the Glasgow Declaration on Climate Action in Tourism.Over the last several months The Travel Foundation partnered with the

European Tourism Futures Institute and Breda University of Applied Sciences in the Netherlands to explore what the global travel and tourism sector must implement in order to achieve the Glasgow Declaration's stated goals—contributing significantly to the global effort to halve emissions by 2030 and reach net zero before 2050. Our secently publishedindustry report, *Envisioning Tourism in 2030*, considers the future shape of tourism and the opportunities and challenges ahead for policymakers, destinations, accommodations and tour operators. The findings will set the stage for the Travel Foundation's future too. I see our role, and perhaps our superpower too, as advocating for innovation and bringing organisations together to show how tourism can transition. Academia will only become more important as we continue with this endeavour.

And so it is encouraging to see so much good thinking and so many important ideas in this Journal of Sustainable Tourism, spanning a wide variety of key issues across gender and race, interrogating ideas for community empowerment and spatial justice, as well as bringing us a wealth of perspectives, from migrant workers to mountain guides. It is time now for your superpowers to shine, as you share your story as widely as possible and build new collaborations for action. It's time to grab your shield, put your cape on and get ready to fly.

ට OPEN ACCESS

Introduction: Tourism, global crises and justice: rethinking, redefining and reorienting tourism futures

Raymond Rastegar, Freya Higgins-Desbiolles and Lisa Ruhanen

ABSTRACT

Ours is an era of crises. Having weathered the recent COVID-19 global pandemic, we are confronted with numerous interconnected crises that challenge the global community. These crises bring with them issues of justice and injustice, as different populations are differentially impacted. Certainly, we have seen obstacles to success in Global North-South inequalities, power differentials and structural injustices. It is essential to consider how we may collaborate together to manage and transition through these multitude of problems. This is the context in which contemporary tourism must operate and play its role in seeking resolutions. This introduction to the Special Issue on "Tourism Global Crises and Justice", critiques these contemporary issues and considers how we might transition tourism for more just, sustainable and equitable futures. Drawing on contributions to this Special Issue, in this article we bring together a discussion of pertinent themes that consider just transformations, issues of climate justice, diverse worldviews and knowledges, possibilities for solidarity through tourism, and concerns with power and decolonisation. In doing so, we propose a transdisciplinary analytical framework that can inform more just and equitable practices in tourism.

Introduction

The COVID-19 pandemic gave us time to pause and reflect on the fundamental underpinnings of tourism as a phenomenon. During this extraordinary moment of managing a global pandemic, borders were shut, tourism halted, lockdowns ensued for various lengths of time and economies ground to a temporary standstill. It was immediately apparent just how interdependent the sector is, and certainly its vulnerability as a result. While crisis, risk and resilience studies in tourism are not new (e.g. Mert, 2016; Prayag, 2018; Ritchie & Jiang, 2019), the scale and pervasiveness of COVID-19 forced more profound reflection, particularly as other crises and disasters emerged concurrently.

There are diverse voices from various positionalities of expertise alerting us to a time of potential existential crisis and our need to act. Arguably, we are arriving to a time with a critical difference, the potential for these crises and injustices to bring us to "collapse". However, the experiences of crises and even collapse will not be experienced in universally the same way. As we will show with our particular focus on (in)justices for poor, marginalised and oppressed groups, crises are usually far worse and capacities to adapt more limited and are therefore worthy of greater attention. We focus particularly on the Global South as the home of the majority of the world's population but also in our attunement to powerplays and failures to act in true solidarity; "the use of the phrase Global South marks a shift from a central focus on development or cultural difference toward an emphasis on geopolitical relations of power" (Dado & Connell, 2012, p. 12).

In this introduction to our Special Issue, we examine the multiple contemporary crises human societies face and the ways they impact on, intersect with, and shape tourism futures. The position we present here is that these crises share a common theme; at their centre lie issues of justice and injustice for humans, more-than-human kin and whole ecologies. Together with the articles that comprise this volume, we engage concepts, cases and dilemmas to better know where we are and where we might go. We offer an extension to these discussions through a preliminary framework for tourism, crises and justices (see Figure 1).

Building a transdisciplinary analytical framework can help inform more just and equitable practices in tourism, leveraging the cutting edge thinking from other disciplines (e.g. environmental justice, decolonial studies and peace studies) that we have drawn into our discussion to offer new insights for managing tourism in an era of crises by moving beyond managerial, technocratic and reformist approaches. This framework resonates with the papers submitted to the special issue, and collectively (this introduction and the Special Issue papers) we hope that this may prove helpful in guiding tourism scholarship and practice to consider and advocate for more fair and sustainable futures. The sections that follow build insights that support the framework, detailing the nature of impending tourism crises, the distinct claims of ecological and environmental justice and the climate justice imperative raised by climate change challenges. Narratives and experiences of injustices bring current tourism practices into question and expose the need for engagement with diverse worldviews and knowledges and transformations in relations of tourism for solidarity and decolonisation. We bring these diverse threads together to build a solid triangulated framework for critiquing contemporary tourism and reshaping it in more equitable and sustainable ways. In this era of "polycrisis" (Tooze, 2022), we need renewed thinking that grapples with the complexities of injustices in order to avoid siloed thinking and inspire the collaborations that are vital for just, sustainable and equitable solutions.

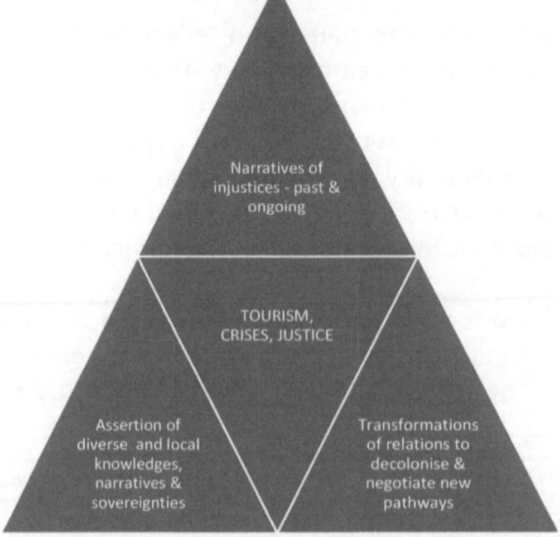

Figure 1. Framework for addressing crises and injustices in tourism towards engaged and transformative outcomes.

Tourism and justice in the face of global crises

"There's no tourism on a dead planet"- adaptation of climate activist mantra

As we write this piece, communities in the north, south, east and west are facing the devastating impacts of crises such as wars, sanctions, refugee movements, pandemics, biodiversity loss, and climate change; many of these are the direct or indirect products of capitalism and colonialism. COVID-19 was more than just a health crisis, but a crisis that has affected, and continues to affect, the most vulnerable individuals among us, including tourism, hospitality, and gig workers whose livelihoods disappeared altogether as travel and leisure halted. In other contexts, such as hospitality workers in the healthcare system, workers found themselves bearing the brunt of the crisis despite receiving relatively lower pay and benefits in many capitalistic economies.

Concurrently, biodiversity and ongoing sixth mass species extinction is threatening the well-being of many, as indicated in the recent *Living Planet Report 2022* which reported: "[...] an average decline of 69% in species populations since 1970" (WWF, 2022). Biodiversity loss will have catastrophic impacts on human lives and significantly affect many destinations and tourist experiences. Similarly, the climate change crisis disproportionately affects the environment and livelihoods of many low-income countries and poor people in high-income countries, violating their human rights and access to social justice (Levy & Patz, 2015; Venn, 2019). However, not too far in the future, no one will be immune from the devastating impacts of socio-ecological crises when for example rising temperatures fuel other crises such as "environmental degradation, natural disasters, weather extremes, food and water insecurity, economic disruption, conflict, and terrorism" (UN, 2020).

The socio-ecological crises of the present time have made it clear that failure to prevent human-induced mass extinction and climate change will affect the wellbeing of us all, but most particularly those with little "voice", including youth, future generations and nonhuman animals. The contemporary (anthropocentric) tourism system has long contributed to the environmental crises with both negative social and ecological consequences. As we well know, tourism is both vulnerable to, and a contributor to, such crises. Therefore, one key issue is to identify the human impact of tourism's role in climate ecological and environmental justice (Rastegar, 2022). Yet the meaning and importance of these distinct terms are not widely understood, and this has significant negative impacts on our ability to both prevent and respond to these multitude of crises.

These global crises suggest we must consider pathways to tourism transition. But how do we cultivate a transition mindset in tourism and what are the short-term and long-term measures that are essential to reach our goals for just, sustainable, and equitable outcomes remain disputed (see, for instance, debates on overtourism hotspot Venice in Friel, 2021; Standish, 2023). Since the experience of the global pandemic, it has been frequently claimed that tourism cannot and must not return to "business as usual" (e.g. Becken, 2021; Higgins-Desbiolles, 2022; UNWTO, 2022). However, the resumption of travel certainly casts doubt on these assessments, as consumers embrace reopened travel opportunities with a fervour; meanwhile businesses capitalise on this demand to recoup pandemic losses. We are left to wonder if the notion of a tourism transition was over before it even started. Or are tourism transitions a contested space that requires greater unpacking and critical analysis to strengthen our understanding?

Ecological justice and environmental justice: justice for human and biological diversity

As we explore the interstices of crises, justice and tourism, we must wrestle with the tensions in meeting human needs while also tending to ecological needs and the needs of the more-than-human in our shared spaces, including spaces of tourism. Additionally, humans are not a homogenous category, and there are myriad inequalities, inequities and injustices that must be recognised and addressed. In these discussions, it is vital to note the distinctions between the terms "environmental justice" and "ecological justice", despite the fact they would appear to concern similar things. Low and Gleeson explained the former addresses "the struggle for... justice of the distribution of environments among peoples" while the latter addresses "[...] justice in the relationship between humans and the rest of the natural world" (Low & Gleeson, 1998, p. 2).

The injustices of environmental racism catalysed environmental justice movements and has resulted in transformed policies, including in the US state of Massachusetts that created an *Environmental Justice Policy* in 2002 (see Agyeman & Evans, 2003, p. 36). However, this stream of environmentalism has been criticised for focusing more on social justice among humans, with less emphasis on the intrinsic value of the wilderness or non-human animals. Environmental justice advocates have countered that the fight against negative environmental impacts contributes to the overall efforts for environmental conservation and sustainability (Agyeman et al., 2002 p. 78). On the other hand, advocates for ecological justice argue that non-anthropocentric approaches of justice will be required to give full voice to the range of interests of nonhumans and futurity of all life (e.g., Treves et al., 2019). The environmental movements that solely focus on protection of the natural environment informed by experts or western approaches can however result in injustice when ignoring the rights, values, and knowledges of local and Indigenous communities. Similarly, many emerging socio-environmental conflicts are due to either ecological justice actions undermining communities' interests or promoting anthropocentric environmental justice neglecting the interests of nonhuman beings. Such development strategies are not sustainable nor just.

Combating the devastating impacts of global crises such as biodiversity loss and climate change requires addressing both environmental conservation and sociocultural wellbeing. Therefore, addressing the tension between justice among humans (social justice) and environmental protection (ecological justice) requires a collaborative and inclusive approach. Examples can be seen when the formulation of Indigenous Environmental Justice based on Indigenous philosophy, ontology, and epistemologies is suggested to address both ecological crisis and injustices being experienced by communities (McGregor et al., 2020). Similarly, Higgins-Desbiolles (2020, p. 618) advocates for "socialising tourism" by placing "tourism in the context of the society in which it occurs and to harness it for the empowerment and wellbeing of local communities" to promote social and ecological justice. This is to ensure that any tourism development is accountable to local social and ecological limits as it occurs in the local contexts that include both human and nonhuman dimensions. Identifying a social-natural sciences divide in tourism, Rastegar (2022, p. 117) proposed "a multispecies justice turn by considering a mixed moral community with human-nonhuman interaction in shared ecological, social, and political spaces".

Therefore, addressing the complex issues of justice for humans and nonhumans is essential. These efforts require interdisciplinary research involving both natural and social sciences. However, such a task requires addressing some practical challenges and barriers. For example, overcoming the communication barriers across multiple disciplines and sectors requires developing a common language, facilitating engagement, and embracing diversity in a supportive environment that values interdisciplinary research. In our considerations of tourism, crises and justice, we suggest that understanding the requirements for environmental justice, ecological justice, social sustainability and fulfilling the needs of communities are complex and interconnected. Attaining these "wisdoms" and building different, transformed approaches will be vital in facing the enormous challenges that climate change presents.

The imperative for climate justice in a climate change world

Global climate change presents one of the most pressing challenges of our time and is posing grave threats for every nation, community and sector including tourism. Devastating climate change induced impacts are already evident from recent bush fires, floods, sea level rise, extreme weather and drought threatening the lives and livelihoods of many groups, particularly marginalised populations including women, children, people living with disabilities, those on low-incomes, those suffering discrimination and Indigenous groups both in the Global North and Global South. "Climate justice is a concept that views climate change and efforts to combat it as having ethical implications and considers how these relate to wider justice concerns" (Robinson & Shine, 2018, p. 564). Distributional inequality of climate change risk and responsibilities has already led to a "double inequality" (Barrett, 2013) affecting the social wellbeing and health of many marginalised groups. While some have highlighted the climate injustice ingrained in North-South relations (Okereke & Coventry, 2016), immigrants, people of colour or low-income groups in the Global North are also under considerable pressure from the increasing costs of climate change-driven impacts such as higher energy bills, costly insurance and infrastructure needed to protect their properties and lives (Heffron, 2022; Porter et al., 2020).

Similarly, the urgency to scale up climate actions in the tourism domain by embracing a low carbon pathway as identified in the Glasgow Declaration (UNWTO, 2021) is in response to the dual realisation of both vulnerability and the contribution of tourism to climate change. However, a critical question to pose is how the transition to a post carbon economy will address social inequalities or tensions that emerge in the process? Or even most importantly, what strategies are in place to avoid harmful actions affecting those who already suffer from climate change impacts? Despite the significant growth of climate change and tourism scholarship over the past 30 years, this work has "failed to prepare the sector for the net-zero transition and climate disruption that will transform tourism over the next three decades" (Scott & Gössling, 2022, p. 1). Scott & Gössling argue this is the preeminent challenge to understand how climate change and the subsequent sustainability transition will transform tourism over the coming decades; this is most critical for tourism-dependent communities in the Global South.

It is difficult to sustain hope for the radical change needed when the same institutions and self-interested political and business players are commonly allowed to set policies and strategies, such as adaptation and mitigation, to address climate change despite being the very creators of the problems. As a result, the response toolkit to climate change prioritises technological solutions and socio-technical processes (some of which are not yet invented or proven such as carbon capture and storage (see Oquist, 2022)) while failing to give due attention to the roots of the crises in our extractive and materialistic socioeconomic systems. The impacts of such narrowly framed transitions are already evident in job losses, population dis/re-locations, rising poverty and lack of funding for local public services such as education, public health and transportation (McCauley & Heffron, 2018; Wang & Lo, 2021) in countries such as Bangladesh (Carrico et al., 2020), China (Lo, 2021) and Australia (Evans & Phelan, 2016).

Additionally, the drive for clean energy technologies, such as solar panels, wind plants and electric cars, brings global social and environmental impacts, with research indicating how such rapid technological change in the Global North has shifted social and environmental problems to the countries of the Global South. Examples can be seen when the increasing demand for minerals and mining in the Global South to provide resources to build these solar panels and electric cars has increased the risk of soil, air and water pollutions, slavery, child labour and sexual abuse (Sovacool, 2021), human rights violations (Marín & Goya, 2021) and even armed conflicts (Church & Crawford, 2020).

Considering the factor of human development in climate actions and transition will be vital. However, there is no country that guarantees climate justice for its people (Furlan & Mariano, 2021). Therefore, the sustainability transition efforts to move to a net zero society must focus more effectively on social and human dimensions (Rastegar & Ruhanen, 2023), particularly when acceptability and adaptation of new policies and practices requires individuals' willingness to change (Martiskainen & Sovacool, 2021). Agyeman et al.'s conceptualisation of "just sustainabilities" is useful guidance in such efforts, which underscores: "The need to ensure a better quality of life for all, now and into the future, in a just and equitable manner, whilst living within the limits of supporting ecosystems" (2003, p. 5).

Identifying climate change as a social and ecological justice issue, a just sustainability transition in tourism research, policy, and practice requires us to "move away from ethnocentrism or the domination of political and economic elite by including the voice of coastal communities, marginalised groups and their traditional knowledge" (Rastegar, 2022, p. 119). Therefore, transitions reaffirming the interests, power and knowledges of privileged elites will not be just or sustainable. Privileged groups are usually more protected and might be less supportive of environmental regulations threatening their advantageous livelihoods. Additionally, such groups can employ new forms of settler- colonial practices by taking over and occupying more protected homes and neighbourhoods while pushing out disadvantaged lower income groups. Viewed from a more macro perspective, the North's response to the climate crisis very much looks like militarisation, as walls, border fences and armed units try to repel those that flee impacted communities in the Global South.

To address these difficult issues, we may need more radical responses, and suggestions have included, degrowth strategies, ending capitalism, building alternative economic models more focused on the public good (see EuroNews, 2023; Hickel, 2021) and more just forms of global order with greater voice and power afforded to the Global South (see Higgins-Desbiolles, 2022). Debates and discussions on degrowth have featured in tourism academia recently and these contest the need for more radical possibilities to better align human needs with planetary boundaries (e.g., Butcher, 2021; Higgins-Desbiolles et al., 2019; Higgins-Desbiolles & Everingham, 2022). Such approaches are under consideration because they may offer greater scope for transformation to address persisting injustice and inequalities.

Arguably this situation requires tourism actors to rethink, redefine and reorient tourism (Higgins-Desbiolles et al., 2019). The transformational opportunities may facilitate the long-heralded responsible, ethical and sustainable transition in tourism. This will not work without tourism actors and decision makers fully and meaningfully adopting justice principles and sustainability values (Rastegar & Ruhanen, 2022) that respect the rights, interests and social needs of vulnerable groups and local communities (Boluk et al., 2019). Therefore, decolonising or de-centring tourism's transition will be necessary by identifying power relations, alternative worldviews, and social realities.

Tourism transformation for a more just and sustainable future

As we have been explaining here, ours is an era of crises. While the COVID-19 global pandemic, for a period in time, upended the world as we knew it, the global crisis caused by climate change gathers even greater attention. We will not be able to look back on climate change as a moment in time as the impacts and effects will be pervasive, constant and altering every sphere of our lives. The climate crisis also portends a multitude of other crises associated with it including conflict, financial recession, resource scarcity, growing authoritarianism and more. Despite these tangibly impactful crises, it is important to also be attentive to cultural perspectives and worldviews on these events and how these may shape how we attend to these many crises and the solutions we may countenance. Val Plumwood (2002) observed, no culture that has set "in motion massive processes of biospheric degradation which it has normalised, and which it cannot respond to or correct can hope to survive for very long" (p. 1). Indigenous Potawatami philosopher and scholar Kyle Whyte noted "[…] the hardships many non-Indigenous people dread most of the climate crisis are ones that Indigenous peoples have endured already due to different forms of colonialism: ecosystem collapse, species loss, economic crash, drastic relocation, and cultural disintegration" (Whyte, 2018, p. 226). Such critiques invite us to step outside the taken-for-granted views and perspectives many of us have held and see the root causes of current crises and injustices. Following this path, however, requires transitional steps to rethink, redefine and reorient tourism futures.

Recognising "worlds and knowledges otherwise" in tourism justice

It is important to recognise that a good deal of the contributions to recent discussion of justice in tourism emerge from the Western tradition. In this era of demands for greater engagement with diversity and inclusion and in the heat of multiple crises, opening up to other ways of understanding is imperative. As Watene (2016) noted, there has been very little engagement with non-Western philosophical traditions, and especially Indigenous ones, in common discussions of justice and ethics. As Arturo Escobar explained, we need recognition of the need for "worlds and knowledges otherwise," that is, for "worlds that are more just and sustainable and, at the same time, worlds that are defined through principles other than those of Eurocentric modernity" (2004, p. 220). For example, scholars argue how understanding gender-leisure nexus in the Global South requires inclusion of women's voices, stories, diverse cultural perspectives, and justice narratives (e.g., Zarezadeh & Rastegar, 2023). This approach of "worlds and knowledges otherwise" invites us to imagine justice in tourism much more widely than has yet occurred. It is an essential agenda that ranges far beyond conventional equality, diversity and inclusion programmes which largely fail to engage with issues of justice. The promise and richness of such diverse worlds of knowledge are many, as seen for example in recent work on "embodied plantation ecologies" (Davis et al., 2019) and ecofeminism (Power, 2016).

Western approaches have dominated for far too long, through a trifecta of forces described by development theorists such as Escobar (2004) as coloniality, development and modernisation. This is a homogenising, universalist agenda that designates Global South communities as "underdeveloped" and purports to set them on a rigid development trajectory whose rules are set by Western institutions, such as the World Bank and the International Monetary Fund. It is operative too in the drive to set tourism as a pathway to development for the developing world and is promoted by institutions such as the UNWTO. Higgins-Desbiolles has critiqued the dependency on tourism that this has fostered that "[...] works against the abilities for the communities of the Global South to achieve just, fair, self-determining and sustainable futures" (2022, p. 7).

The impacts of these dynamics can be described as a form of epistemic injustice as Western approaches are forced on the developing world. Epistemic injustice occurs when a wrong or harm is "done to someone in their capacity as a knower", and justice is secured by counteracting practices of silencing or devaluing alternative forms of knowing and living that do not conform with assumptions about the "authority" of scientific knowledge (Fricker, 2007, p. 1). We can identify cases of epistemic justice in tourism in the projects that ground tourism in Indigenous knowledges and lifeways. Examples can be found in Bhutan's initial tourism programme based in part on Buddhist philosophy found in the "Gross National Happiness" approach (Schroeder, 2018, p. 3) and the "kaupapa tāpoi" framework for Indigenous tourism derived from Māori values and principles offered by Mika and Scheyvens (2022). We can identify epistemic injustices in tourism when tourism undermines the lifeways of local and Indigenous communities, requiring they change to suit tourists and tourism industry interests.

The work in development studies on 'worlds and knowledges otherwise' calls us to be more critical in our considerations of tourism justice issues. It spotlights the ways that 'tourism as modernisation' discourses and practices compel communities to reject their own knowledges and lifeways in the interest of tapping tourism for development. Escobar and others use the term 'pluriversality' as an approach conducive to multiple knowledges and lifeways: "The diversity of mundializacíon is contrasted with the homogeneity of globalisation, aiming at multiple and diverse social orders—in sum, pluriversality" (Escobar, 2004, p. 219). Such approaches to tourism counteract the problems of exploitation, domination, homogeneity and inauthenticity that had long been sources of criticism and opposition to tourism developments. Such concepts and ideas are gaining traction in tourism studies (e.g., Chambers & Buzinde, 2015; Everingham et al., 2021), however further work is needed. For instance, social movements active in and against tourism could be studied through this analytical lens (e.g., Fairbnb, the Global Cruise Activist Network and Network of Southern European Cities against Touristification).

Recognising and protecting "worlds and knowledges otherwise" in tourism offers an important mechanism to ensure diversity, inclusion, justice, and sustainability in tourism as different knowledges are respected and empowered. As we confront multiple crises, epistemic justice approaches help us to access diverse knowledges and innovative approaches that evolve from living in various cultures and places. In contrast, the homogenising and myopic perspectives and approaches derived from Western capital-centric approaches are likely to offer only more of the same.

In this Special Issue, we have a number of contributions that offer insights into tourism from "worlds and knowledges otherwise". Firstly, Benjamin and Laughter use a lens of critical race theory and a "framework of endarkened storywork" to bring to the fore the concerns and experiences of marginalised individuals and communities in tourism, particularly Black, Indigenous and People of Colour (BIPOC). Bigby, Hatley and Jim reflect on Indigenous-led "toxic tours" to the superfund site of Tar Creek in Oklahoma and reveal that an Indigenist lens spotlights action for Indigenous resurgence which differs from Western notions of "justice" in the face of ecological and social crises brought on by settler-colonial injustices. Analysing child labour and orphanage tourism, Canosa and Graham reveal the adult-centric discourses of tourism and advocate for "childist" approaches in tourism, moving work away from patronising and alienating protectionism to respectful engagement with child rights and empowerment.

Solidarity in and through tourism for building collaboration

As we consider justice in tourism, we need to consider the bonds of solidarity and how widely its purview extends because solidarity is essential in addressing the multiple crises now faced, particularly climate change. "While 'justice' refers to rights and duties (Moralität), the concept of solidarity refers to relations of personal commitment and recognition (Sittlichkeit)" (Ter Meulen, 2017, p. 71). The importance of solidarity lies in its relational aspects, particularly its emphasis on cooperation and commonality. Solidarity does not result from a universal human capacity to identify with the needs of others, but instead is a learned "we-ness" that delineates "us" from "others" (Rorty, 1989). However, Schwartz argues that we may be moving towards forms of more global solidarity: "The emerging 'global civil society' of transnational social movements helps raise the visibility of transnational issues" (Schwartz, 2007, p. 143). As the global community confronts crises such as human-induced climate change and global pandemics, the pressures for global solutions through global cooperation and even global order increase.

In such a context, the intercultural and peace-building capacities of tourism may act as a catalyst to greater solidarity. Peace through tourism analyses suggest that tourism may contribute to peace between peoples, communities, and nations. While the intercultural contact theory of tourism has been challenged because tourism is too superficial to induce conflict reduction and peace (e.g., Farmaki, 2019), discussions of solidarity through tourism may advance the possibilities. However, for peace through tourism and solidarity through tourism to be effective and have real meaning, justice underpinnings are essential (Higgins-Desbiolles et al., 2022).

Discussions of solidarity through tourism were revived with the February 2022 Russian invasion of Ukraine. In this context, scholars discuss the need for solidarity tourism action plans (Dolnicar & McCabe, 2022) or understanding place solidarity in supporting Ukraine (Josiassen et al., 2022 & Josiassen et al., 2023). But we are prompted to consider critical questions here at this moment in 2023, as to whether this expression of solidarity is limited to bonds with Ukrainians as newly designated Europeans, while black and brown people have been refused such solidarity in cases where they have sorely needed it (see Lyubchenko, 2022). Here we can argue that "rather than accepting the status quo of global order, this "peace with justice" approach (Rees, 2020) requires that we interrogate tourism's roles and responsibilities in maintaining an unjust global order and act accordingly" (Higgins-Desbiolles, 2022, p. 5). This issue directly links to the Global North and Global South disparities that we have analysed in the previous section concerning how climate change and other crises are experienced and managed.

This is advocacy for a more globalised form of solidarity. Structural injustices derive from unequal power constellations embedded in systemic structures. This perspective presses us to engage with decolonial thinking recognising the colonial, Western roots behind the current global order and economic system, including the tourism industry (Higgins-Desbiolles, 2022). It is also important to understand the complexity and multiple forms of oppression that structural injustices bring. This might be best supported by complex intersectional analyses informed by critical race studies (Crenshaw, 1991).

These discussions on justice, solidarity and what is owed to whom are growing more vital as we are compelled to address complex and global level crises such as the pandemic and climate change. Considerations of solidarity move us from cold and disembodied rules and principles of justice to living and connected relations between people for securing wellbeing and thriving (see Country et al., 2017). As we confront global crises such as pandemics, climate change, financial upheavals and potentially globalised conflicts, we will need cooperation that bridges "civilisational" divides and addresses the structural injustices that have in fact helped create these very crises. Well-considered solidarity through tourism approaches may indeed serve as a helpful pillar in this work.

In this volume of articles, Burrai, Buda and Stevenson take us to the Leeds, UK where refugees and asylum-seekers are being empowered through opportunities as tour guides to not only represent themselves but also co-build more inclusive and diverse city futures. As xenophobia and anti-immigrant politics is stoked, such programmes embody solidarity being built through tourism. Kalisch and Cole suggest the value of combining a feminist ethic of care, social solidarity economy and human rights to transform tourism towards justice. By way of contrast, Seyfi, Hall, Saarinen and Vo-Tranh interviewed tourism actors in Iran to assess the value of international sanctions on tourism in achieving justice, human rights and sustainability goals and found them to be problematic.

The rights of precarious workers in tourism also feature in the analyses in this Special Issue. Wilson and Dashper present an ethnographic study of high-altitude mountaineering tourism in the Himalaya in which they document the precarity and injustices experienced by workers but also their agency through self-organisation and advocacy. Another ethnographic study is offered by Riordan, Robinson and Hoffstaedter focused on the injustices experienced by migrant food delivery workers in Australia who similarly find that these workers are active agents in transforming their circumstances.

Addressing power asymmetries: decolonisation

To build these bonds of solidarity that are essential for confronting the polycrises we face, decolonisation is essential to dismantle structures of exploitation and extraction. Our recent experiences during the global pandemic illustrate this well. In April 2021 at the height of the global pandemic, the World Health Organization reported that 87% of the world's COVID-19 vaccines had gone to people in wealthy countries and just 0.2% had gone to people in low-income countries (UN, 2021). Harman et al. (2021, p. 1) argued: "This reprehensible imbalance is no accident. High-income countries have used neocolonial negotiating power, global policy leverage and capital to procure enough doses to cover 245% of their citizens while leaving few doses for poorer countries".

Such gross injustice is the result of continual failure to dismantle the unequal and unfair global infrastructure that has resulted from the failure to complete decolonisation. Higgins-Desbiolles has called for recognition of the "ongoingness of imperialism" arguing: "Tourism development in such a context of unaddressed and unreformed imperialism, manifests in exploitation, abuse, dispossession, commodification and many other sorts of injustices and inequities" (2022, p. 2). This is evident in the safari tourism in East Africa (see Akama, 2004), white saviourism in volunteer tourism (Bandyopadhyay, 2019) and cultural pollution through tourism seen in Hawai'i in its debasement of Kanaka Maoli (Aikau & Gonzlez, 2019).

Our Special Issue contains a number of articles that address power, injustices, decolonisation and resistance. Buzinde and Caterina-Knorr employ a decolonial lens to interrogate the development policies of Rwanda and Kenya in terms of inclusive development criteria and find the limits imposed by legacies of colonialism and the ongoing impacts of neocolonialism in these countries. Yang analyses the alternative tourism niche growing in China known as "qiongyou", characterised as low-budget travel by some and analysed here as a form of justice tourism in its resistance to being disciplined to commercial tourism regimes. Higgins-Desbiolles, Scheyvens and Bhatia call for "decolonising tourism and development" and offer a case study of the Cambodian Children's Trust to explain how this non-government organisation evolved from paternalistic, white saviourism to decolonising, collaborative approaches corresponding to Freirian praxis which work to build critical consciousness and capacities together. Bigby, Hatley and Jim discuss toxic tours conducted by LEAD agency in terms of Indigenous-led resurgence, offering insights into an Indigenous perspective on crises, the pursuit of justice and recovery from historical and ongoing settler colonialism.

Continuing these important debates, in this volume, Hunt, Barragán-Paladines, Izurieta & Ordóñez offer a historical study of waves of in-migration to the well know ecological destination of the Galapagos Islands, Ecuador. Employing sovereignty theory, they show the complexity of navigating justice issues as these human migrants hold varying values and worldviews on ecology, community, recreation and tourism in this location. Paddison and Hall investigate urban spatial injustices in tourism through a case study of York, UK during the pandemic to assess how transforming policymaking can result in fewer negative tourism impacts. Tomassini and Lamond provide a conceptual analysis rethinking the spatial (in)justices of tourism following the spatial turn in the social sciences. Wu, Wu, Li, Wang and Wang focus on small tourism business entrepreneurs in rural China and investigate their "community citizenship behavior for the environment" arguing this greatly influences the sustainable development of rural tourism communities.

Towards a framework of tourism, crises and justices

The discussions in this introductory article allow for the articulation of a preliminary framework for this topic of tourism, crises and justices (see Figure 1). Firstly, are narratives of truth-telling that must come to the fore in order to address ongoing injustices in tourism. These may, for example, contradict the discourse of tourism as a pathway to development, the benevolence of international aid programmes promoting tourism or the value of charity found in volunteer tourism. These truths will, in many cases, be confronting. For example, Australian scholar Deborah Bird Rose explained: "Settler societies are brought into being through invasion: death and silence pervade and gird the whole project" (Rose, 2004, p. 58). This sheds light on Bigby, Hatley and Jim 's article offered in this Special Issue which narrates the use of toxic tours as a vehicle to reveal that this settler colonial project in still playing out in Oklahoma with deadly consequences. Recent developments in the United States to ban the teaching of critical race studies in schools and colleges reveals how essential and contested this facet is for undoing historic and ongoing injustices (Izaguirre, 2023).

Secondly truth-telling must be followed by transformative actions towards decolonising and finding new pathways forward. Decolonisation is not peripheral to our concerns with tourism, crises and justice. As Perkins explained: "[…] equity, decolonization, and activism are central to building political institutions to reduce carbon emissions and material throughput in human economies, so that humans can again flourish within reciprocal relationships with the rest of life" (Perkins, 2019, p. 184). Tourism analysts and policymakers may want to consider Hickel's advice that "[…] economic sovereignty and self-reliance [a]re essential to real decolonization" (2021, n.p.). To achieve transformative changes in the face of these crises, we may need some radical tools in our tool kit including climate reparations, reparations for colonialism that can replace dependency on debt and approaches to development where recipients are engaged as partners not beneficiaries (see Higgins-Desbiolles, Scheyvens and Bhatia in this volume).

The third and final pillar to our framework is the assertion of diverse and local knowledges, narratives and sovereignties. The theoretical foundations for this pillar we have explained in the discussion offered in Section 3. The practical application in tourism may be found in the arguments on defining and focusing tourism on the rights, benefits and interests of local communities (Higgins-Desbiolles, 2020; Rastegar et al., 2021). This theme arises in a number of analyses found in our volume, including in Buzinde and Caterina-Knorr's call for "locally defined and equitable approaches to development", Canosa and Graham's advocacy for listening to children in tourism dialogues and Paddison and Hall's narrating of York's community reclaiming its voice in tourism, among numerous others.

Figure 1 is intended to offer a flexible tool to guide future research and practice in tourism. The three facets of diverse knowledges, narratives of injustices and transformed relations open up more conducive spaces to build justice, equity and empowerment in communities engaged with tourism. The order in which they are engaged may vary depending on the specific context. For instance, in places where stories of injustices have been long suppressed, their uncovering may need to be prioritised and given greater attention (such as in the United States). In contrast, in other instances diverse and local knowledges, narratives and sovereignties may remain largely strong and therefore offer a sound foundation for proceeding (such as in Indigenous localities). However, no matter which way these three facets are employed in a process of bringing greater justice to communities in tourism, they collectively effectively link past, present and future together so that injustices, oppressions and marginalisations can be addressed. This, we argue, presents more solid foundations for addressing tourism futures likely to be beset by polycrisis.

Conclusion

In this introduction, we have been conversing about crises and even collapse, laying out extraordinary challenges before us. But it is important to not give in to "doomism" which could sap our energy to act for justice. Our discussion is mindful of Haraway's caution on crises discourses which she noted "connotes something approaching apocalypse and its mythologies. Urgencies have other temporalities, and these times are ours. These are the times we must think; these are the times of urgencies that need stories" (Haraway, 2016, p. 37). Such stories are found within this Special Issue; stories of injustices, resistance, conscientisation, resurgence, sovereignty and care, from Tribal Nations in Oklahoma, to Cambodia, to Rwanda, to China and cities in Europe and Australia. Urgencies press us to act with thought and care and the articles in this Special Issue offer insights by introducing new concepts, revealing important case experiences and spotlighting proactive strategies that highlight the ability to act to transform tourism and direct it to building better futures in the face of all of the challenges that we face. We hope that through the papers in this Special Issue, new perspectives, critiques and contemporary debates are sparked as we collectively address this sustainability crises and the role of tourism within it.

In this introduction, we have built on the work of earlier analysts (e.g. Higgins-Desbiolles, 2018; Jamal, 2019; 2021; Rastagar, 2022) but expanded the conversations and debates by specifically considering the voices and interests of the excluded, less powerful and marginalised. We have offered a framework for addressing crises and injustices in tourism towards engaged and transformative outcomes. This is based on listening to narratives of injustice in tourism, engaging with and respecting diverse and local knowledges and sovereignties and then co-designing responses in tourism policy and practices that support just tourism futures. The pandemic has been a portent of a challenging future for both tourism and society. Transformations towards justice are essential in meeting these difficult times and navigating the many crises we collectively face if we are to have any hopes for success.

Disclosure statement

No potential conflict of interest was reported by the authors.

ORCID

Raymond Rastegar http://orcid.org/0000-0002-5794-2096
Freya Higgins-Desbiolles http://orcid.org/0000-0001-6424-0749
Lisa Ruhanen http://orcid.org/0000-0002-8495-8843

References

Agyeman, J., & Evans, T. (2003). Toward just sustainability in urban communities: Building equity rights with sustainable solutions. *The ANNALS of the American Academy of Political and Social Science*, *590*(1), 35–53. https://doi.org/10.1177/0002716203256565

Agyeman, J., Bullard, R. D., & Evans, B. (2002). Exploring the nexus: Bringing together sustainability, environmental justice and equity. *Space and Polity*, *6*(1), 77–90. https://doi.org/10.1080/13562570220137907

Aikau, H. K., & Gonzlez, V. V. (2019). *Detours: A decolonial guide to Hawai'i*. Duke University Press.

Akama, J. (2004). Neocolonialism, dependency and external control of Africa's tourism industry. In C. M. Hall & H. Tucker (Eds.), *Tourism and postcolonialism* (pp. 140–152). Routledge.

Bandyopadhyay, R. (2019). Volunteer tourism and "The White Man's Burden": globalization of suffering, white savior complex, religion and modernity. *Journal of Sustainable Tourism*, *27*(3), 327–343. https://doi.org/10.1080/09669582.2019.1578361

Barrett, S. (2013). Local level climate justice? Adaptation finance and vulnerability reduction. *Global Environmental Change*, *23*(6), 1819–1829. https://doi.org/10.1016/j.gloenvcha.2013.07.015

Becken, S. (2021). Tourism desperately wants a return to the 'old normal' but that would be a disaster. The Conversation. Retrieved 17 October 2022, from https://theconversation.com/tourism-desperately-wants-a-return-to-the-old-normal-but-that-would-be-a-disaster-154182.

Boluk, K., Cavaliere, C. T., & Higgins-Desbiolles, F. (2019). A critical framework for interrogating the United Nations Sustainable Development Goals 2030 Agenda in tourism. *Journal of Sustainable Tourism*, *27*(7), 847–864. https://doi.org/10.1080/09669582.2019.1619748

Butcher, J. (2021). Covid-19, tourism and the advocacy of degrowth. *Tourism Recreation Research*, 1–10. https://doi.org/10.1080/02508281.2021.1953306

Carrico, A. R., Donato, K. M., Best, K. B., & Gilligan, J. (2020). Extreme weather and marriage among girls and women in Bangladesh. *Global Environmental Change*, *65*, 102160. https://doi.org/10.1016/j.gloenvcha.2020.102160

Chambers, D., & Buzinde, C. (2015). Tourism and decolonisation: Locating research and self. *Annals of Tourism Research*, *51*, 1–16. https://doi.org/10.1016/j.annals.2014.12.002

Church, C., & Crawford, A. (2020). Minerals and the metals for the energy transition: Exploring the conflict implications for mineral-rich, fragile states. In M. Hafner & S. Tagliapietra (Eds.), *The geopolitics of the global energy transition* (pp. 279–304). Springer International Publishing. https://doi.org/10.1007/978-3-030-39066-2_12

Country, B., Wright, S., Lloyd, K., Suchet-Pearson, S., Burarrwanga, L., Ganambarr, R., Ganambarr, M., Ganambarr, B., Maymuru, D., & Tofa, M. (2017). Meaningful tourist transformations with Country at Bawaka, North East Arnhem Land, northern Australia. *Tourist Studies*, *17*(4), 443–467. https://doi.org/10.1177/1468797616682134

Crenshaw, K. (1991). Mapping the margins: Intersectionality, identity politics, and violence against women of color. *Stanford Law Review*, *43*(6), 1241–1299. https://doi.org/10.2307/1229039

Dado, N., & Connell, R. (2012). The Global South. *Contexts*, *11*(1), 12–13. https://doi.org/10.1177/1536504212436479

Davis, J., Moulton, A. A., Van Sant, L., & Williams, B. (2019). Anthropocene, Capitalocene, … Plantationocene?: A manifesto for ecological justice in an age of global crises. *Geography Compass*, *13*(5), e12438. https://doi.org/10.1111/gec3.12438

Dolnicar, S., & McCabe, S. (2022). Solidarity tourism how can tourism help the Ukraine and other war-torn countries? *Annals of Tourism Research*, *94*, 103386. https://doi.org/10.1016/j.annals.2022.103386

Escobar, A. (2004). Beyond the Third World: Imperial globality, global coloniality and anti-globalisation social movements. *Third World Quarterly*, *25*(1), 207–230. https://doi.org/10.1080/0143659042000185417

EuroNews. (2023). Seeking solutions beyond growth in 2023. Retrieved 22 May 2023, from https://www.youtube.com/watch?v=CUDQbujZ5xA.

Evans, G., & Phelan, L. (2016). Transition to a post-carbon society: Linking environmental justice and just transition discourses. *Energy Policy*, *99*, 329–339. https://doi.org/10.1016/j.enpol.2016.05.003

Everingham, P., Peters, A., & Higgins-Desbiolles, F. (2021). The (im)possibilities of doing tourism otherwise: The case of settler colonial Australia and the closure of the climb at Uluru. *Annals of Tourism Research*, *88*, 103178. https://doi.org/10.1016/j.annals.2021.103178

Farmaki, A. (2019). Visitor-host encounters in post-conflict destinations. The case of Cyprus. In R. K. Isaac, E. Çakmak, & R. Butler (Eds.), *Tourism and hospitality in conflict-ridden destinations* (pp. 240–252). Routledge.

Fricker, M. (2007). *Epistemic injustice: Power and the ethics of knowing*. Oxford University Press.

Friel, M. (2021). First cruise liner sailing into Venice since the start of the pandemic met by hundreds of protesters demanding 'No Big Ships'. Business Insider. Retrieved 22 May 2023, from https://www.businessinsider.com/video-venice-protesters-oppose-the-first-cruise-ship-since-pandemic-2021-6.

Furlan, M., & Mariano, E. (2021). Guiding the nations through fair low-carbon economy cycles: A climate justice index proposal. *Ecological Indicators*, *125*, 107615. https://doi.org/10.1016/j.ecolind.2021.107615

Haraway, D. J. (2016). *Staying with the trouble: Making kin in the Chthulucene*. Duke University Press.

Harman, S., Erfani, P., Goronga, T., Hickel, J., Morse, M., & Richardson, E. T. (2021). Global vaccine equity demands reparative justice—not charity. *BMJ Global Health*, *6*(6), e006504. https://doi.org/10.1136/bmjgh-2021-006504

Heffron, R. J. (2022). Applying energy justice into the energy transition. *Renewable and Sustainable Energy Reviews*, *156*, 111936. https://doi.org/10.1016/j.rser.2021.111936

Hickel, J. (2021). How to achieve full decolonization. The New Internationalist. Retrieved 17 October 2022, from https://newint.org/features/2021/08/09/money-ultimate-decolonizer-fjf.

Higgins-Desbiolles, F. (2018). The potential for justice through tourism. *Via Tourism Review*, *13*(13). https://doi.org/10.4000/viatourism.2469

Higgins-Desbiolles, F. (2020). Socialising tourism for social and ecological justice after COVID-19. *Tourism Geographies*, *22*(3), 610–623. https://doi.org/10.1080/14616688.2020.1757748

Higgins-Desbiolles, F. (2022). The ongoingness of imperialism: The problem of tourism dependency and the promise of radical equality. *Annals of Tourism Research*, *94*, 103382. https://doi.org/10.1016/j.annals.2022.103382

Higgins-Desbiolles, F. (2022). The question of solidarity in tourism. *Journal of Policy Research in Tourism, Leisure and Events*, 1–10. https://doi.org/10.1080/19407963.2022.2107657

Higgins-Desbiolles, F., & Everingham, P. (2022). Degrowth in tourism: Advocacy for thriving not diminishment. *Tourism Recreation Research*, 1–5. https://doi.org/10.1080/02508281.2022.2079841

Higgins-Desbiolles, F., Blanchard, L.-A., & Urbain, Y. (2022). Peace through tourism: Critical reflections on the intersections between peace, justice, sustainable development and tourism. *Journal of Sustainable Tourism*, *30*(2–3), 335–351. https://doi.org/10.1080/09669582.2021.1952420

Higgins-Desbiolles, F., Carnicelli, S., Krolikowski, C., Wijesinghe, G., & Boluk, K. (2019). Degrowing tourism: Rethinking tourism. *Journal of Sustainable Tourism*, *27*(12), 1926–1944. https://doi.org/10.1080/09669582.2019.1601732

Izaguirre, A. (2023). DeSantis moves to ban critical race theory in state colleges in Florida. Retrieved 22 May 2023, from https://www.nbcmiami.com/news/local/desantis-moves-to-ban-critical-race-theory-in-state-colleges-in-florida/2962569/.

Jamal, T. (2019). *Justice and ethics in tourism*. Routledge.

Josiassen, A., Kock, F., & Assaf, A. G. (2022). In times of war: Place solidarity. *Annals of Tourism Research*, *96*, 103456. https://doi.org/10.1016/j.annals.2022.103456

Josiassen, A., Kock, F., Assaf, A. G., & Berbekova, A. (2023). The role of affinity and animosity on solidarity with Ukraine and hospitality outcomes. *Tourism Management*, *96*, 104712. https://doi.org/10.1016/j.tourman.2022.104712

Levy, B. S., & Patz, J. A. (2015). Climate change, human rights, and social justice. *Annals of Global Health*, *81*(3), 310–322. https://doi.org/10.1016/j.aogh.2015.08.008

Lo, K. (2021). Authoritarian environmentalism, just transition, and the tension between environmental protection and social justice in China's forestry reform. *Forest Policy and Economics*, *131*, 102574. https://doi.org/10.1016/j.forpol.2021.102574

Low, N., & Gleeson, B. (1998). *Justice, society, and nature: An exploration of political ecology*. Routledge.

Lyubchenko, O. (2022). On the frontier of whiteness? Expropriation, war, and social reproduction in Ukraine. LeftEast. Retrieved 6 June 2022, from https://lefteast.org/frontiers-of-whiteness-expropriation-war-social-reproduction-in-ukraine/.

Marín, A., & Goya, D. (2021). Mining—The dark side of the energy transition. *Environmental Innovation and Societal Transitions*, *41*, 86–88. https://doi.org/10.1016/j.eist.2021.09.011

Martiskainen, M., & Sovacool, B. K. (2021). Mixed feelings: A review and research agenda for emotions in sustainability transitions. *Environmental Innovation and Societal Transitions*, *40*, 609–624. https://doi.org/10.1016/j.eist.2021.10.023

McCauley, D., & Heffron, R. (2018). Just transition: Integrating climate, energy and environmental justice. *Energy Policy*, *119*, 1–7. https://doi.org/10.1016/j.enpol.2018.04.014

McGregor, D., Whitaker, S., & Sritharan, M. (2020). Indigenous environmental justice and sustainability. *Current Opinion in Environmental Sustainability*, *43*, 35–40. https://doi.org/10.1016/j.cosust.2020.01.007

Mert, U. (2016). Risk management for sustainable tourism. *European Journal of Tourism, Hospitality and Recreation*, *7*(1), 63–71.

Mika, J. P., & Scheyvens, R. A. (2022). Te Awa Tupua: Peace, justice and sustainability through Indigenous tourism. *Journal of Sustainable Tourism*, *30*(2-3), 637–657. https://doi.org/10.1080/09669582.2021.1912056

Okereke, C., & Coventry, P. (2016). Climate justice and the international regime: Before, during, and after Paris. *WIREs Climate Change*, *7*(6), 834–851. https://doi.org/10.1002/wcc.419

Oquist, B. (2022). The era of the great carbon fraud is upon us. *Australia Institute*. Retrieved 14 January 2023, from https://australiainstitute.org.au/post/the-era-of-the-great-carbon-fraud-is-upon-us/.

Perkins, E. E. (2019). Climate justice, commons, and degrowth. *Ecological Economics*, *160*, 183–190. https://doi.org/10.1016/j.ecolecon.2019.02.005

Plumwood, V. (2002). *Environmental culture: The ecological crisis of reason*. Routledge.

Porter, L., Rickards, L., Verlie, B., Bosomworth, K., Moloney, S., Lay, B., Latham, B., Anguelovski, I., & Pellow, D. (2020). Climate justice in a climate changed world. *Planning Theory & Practice*, *21*(2), 293–321. https://doi.org/10.1080/14649357.2020.1748959

Power, C. (2016). *Multispecies ecofeminism: Ecofeminist flourishing of the twenty-first century* [Thesis University of Victoria]. https://base.socioeco.org/docs/power_chelsea_ma_2020.pdf.

Prayag, G. (2018). Symbiotic relationship or not? Understanding resilience and crisis management in tourism. *Tourism Management Perspectives*, *25*, 133–135. https://doi.org/10.1016/j.tmp.2017.11.012

Rastegar, R. (2022). Towards a just sustainability transition in tourism: A multispecies justice perspective. *Journal of Hospitality and Tourism Management*, *52*, 113–122. https://doi.org/10.1016/j.jhtm.2022.06.008

Rastegar, R., & Ruhanen, L. (2022). The injustices of rapid tourism growth: From recognition to restoration. *Annals of Tourism Research*, *97*, 103504. https://doi.org/10.1016/j.annals.2022.103504

Rastegar, R., & Ruhanen, L. (2023). Climate change and tourism transition: From cosmopolitan to local justice. *Annals of Tourism Research*, *100*, 103565. https://doi.org/10.1016/j.annals.2023.103565

Rastegar, R., Higgins-Desbiolles, F., & Ruhanen, L. (2021). COVID-19 and a justice framework to guide tourism recovery. *Annals of Tourism Research*, *91*, 103161. https://doi.org/10.1016/j.annals.2021.103161

Ritchie, B. W., & Jiang, Y. (2019). A review of research on tourism risk, crisis and disaster management: Launching the annals of tourism research curated collection on tourism risk, crisis and disaster management. *Annals of Tourism Research*, *79*, 102812. https://doi.org/10.1016/j.annals.2019.102812

Robinson, M., & Shine, T. (2018). Achieving a climate justice pathway to 1.5 °C. *Nature Climate Change*, *8*(7), 564–569. https://doi.org/10.1038/s41558-018-0189-7

Rorty, R. (1989). *Contingency, irony, and solidarity*. Cambridge University Press.

Rose, D. B. (2004). *Reports from a wild country: Ethics for decolonisation*. UNSW Press.

Schroeder, K. (2018). *The politics of gross national happiness: Governance and development in Bhutan*. Palgrave Macmillan.

Schwartz, J. M. (2007). From domestic to global solidarity: The dialectic of the particular and universal in the building of social solidarity. *Journal of Social Philosophy*, *38*(1), 131–147. https://doi.org/10.1111/j.1467-9833.2007.00370.x

Scott, D., & Gössling, S. (2022). A review of research into tourism and climate change - Launching the annals of tourism research curated collection on tourism and climate change. *Annals of Tourism Research*, *95*, 103409. https://doi.org/10.1016/j.annals.2022.10340

Sovacool, B. K. (2021). When subterranean slavery supports sustainability transitions? power, patriarchy, and child labor in artisanal Congolese cobalt mining. *The Extractive Industries and Society*, *8*(1), 271–293. https://doi.org/10.1016/j.exis.2020.11.018

Standish, D. (2023). Mass tourism in Venice: Are city officials overreacting? *The Good Tourism Blog*. Retrieved 22 May 2023, from https://www.goodtourismblog.com/2023/04/mass-tourism-in-venice/.

Ter Meulen, R. (2017). *Solidarity and justice in health and social care* (pp. 71–108). Cambridge University Press. https://doi.org/10.1017/9781107707023.004

Tooze, A. (2022). Welcome to the world of the polycrisis. *The Financial Times* (Online). Retrieved January 14, 2023, from https://www.ft.com/content/498398e7-11b1-494b-9cd3-6d669dc3de33.

Treves, A., Santiago-Ávila, F. J., & Lynn, W. S. (2019). Just preservation. *Biological Conservation*, *229*, 134–141. https://doi.org/10.1016/j.biocon.2018.11.018

UN. (2020). *The climate crisis – A race we can win*. Retrieved 29 June 2022, from https://www.un.org/sites/un2.un.org/files/2020/01/un75_climate_crisis.pdf

UN. (2021). Low-income countries have received just 0.2 per cent of all COVID-19 shots given. Retrieved 3 October 2022, from https://news.un.org/en/story/2021/04/1089392.

UNWTO. (2021). *Launching of the Glasgow Declaration: A commitment to a decade of climate action in tourism*. https://www.unwto.org/event/cop-26-launch-of-the-glasgow-declaration-a-commitment-to-a-decade-of-climate-action-in-tourism.

UNWTO. (2022). World Tourism Day 2022 – rethinking tourism. Retrieved 17 October 2022, from https://www.unwto.org/world-tourism-day-2022.

Venn, A. (2019). 24 - Social justice and climate change. In T. M. Letcher (Ed.), *Managing global warming* (pp. 711–728): Academic Press.

Wang, X., & Lo, K. (2021). Just transition: A conceptual review. *Energy Research & Social Science*, *82*, 102291. https://doi.org/10.1016/j.erss.2021.102291

Watene, K. (2016). Indigenous peoples and justice. In K. Watene & J. Drydyk (Eds.), *Theorizing justice: Critical insights and future directions* (pp. 133–151). Rowman & Littlefield International.

Whyte, K. P. (2018). Indigenous science (fiction) for the Anthropocene: Ancestral dystopias and fantasies of climate change crises. *Environment and Planning E: Nature and Space*, *1*(1–2), 224–242. https://doi.org/10.1177/2514848618777621

WWF. (2022). Living planet report 2022 New Zealand. Retrieved 17 October 2022, from https://livingplanet.panda.org/en-NZ/.

Zarezadeh, Z., & Rastegar, R. (2023). Gender-leisure nexus through a social justice lens: The voice of women from Iran. *Journal of Hospitality and Tourism Management*, *54*, 472–480. https://doi.org/10.1016/j.jhtm.2023.02.003

Part I

Activating justice

Toward critical race tourism: valuing counter-narratives and *endarkened storywork*

Stefanie Benjamin and Judson Laughter

ABSTRACT

In this article, we argue for tourism scholars to amplify the stories and lived experiences of traditionally silenced groups through the adaption of critical race theory (CRT) toward a framework of Critical Race Tourism. Critical Race Tourism includes a specific focus on counter-storytelling, *endarkened storywork*, and regular engagement of interest convergence. We believe Critical Race Tourism can contribute to racial reconciliation and healing, particularly during a time of a global pandemic, that exacerbates systems of marginalization and oppression. Using a Critical Race Tourism framework is about realizing the potential of thick, rich data of everyday lived experiences from marginalized communities in order to enrich the theoretical, empiricism, and scholarly insights offered by academic writing. We begin by defining CRT and describe the employment of and application within tourism studies. We introduce the framework of *endarkened storywork*, which responds to the sustainability and justice challenges resulting from past, current, and future inequities for BIPOC communities. We close with implications for tourism contexts and research.

Introduction

Tourism represents more than an industry. Over the past few decades, scholars like Irena Ateljevic, Nigel Morgan, and Annette Pritchard have taken a *critical turn* toward tourism and travel as a social force capable of positive, socially just pursuits. The groundbreaking work of Christine Buzinde, Anna Carr, Donna Chambers, Bryan Grimwood, Freya Higgins-Desbiolles, and Rasul Mowatt demonstrates how tourism scholars "seek to stimulate their audience to transform society and thereby to liberate themselves and others" (Bramwell & Lane, 2014, p. 2). This *critical turn* moves away from materialism, competition, and individualism with a focus on contributions to a social justice agenda. Particularly in a time of global pandemic, which magnifies already-existing systems of oppression, this *critical* research represents a needed and ripe area for growth. However, scholars of color question Eurocentric methodological approaches to *research*, particularly when researching Black, Indigenous, and People of Color (BIPOC; Parker, 2015). As non-BIPOC scholars seeking a dialogue with BIPOC scholars, Denzin et al. (2008) postulated that *we* must "construct stories that are embedded in the landscapes through which we travel" (p. 7). These will be inclusive of narratives and counter-stories where research respects the community norms where the research is conducted) and helps relinquish power and authority over how data are collected and analyzed.

The term *research* is inextricably linked to European imperialism and colonialism (Tuck & Yang, 2014) – qualitative research included – which endorses and privileges Western knowledge systems and their epistemologies (Denzin et al., 2008). However, tourism scholars can disrupt hegemonic Eurocentric research design by adopting critical race theory (CRT) informed methodologies like interest convergence, counter-narratives, and *endarkened storywork*. The potential of counter-narratives as transformative methodology can contribute toward the development of *Critical Race Tourism* - helping to curate scholarship with the potential to change how people understand issues of race and racism, to help to empower racially marginalized communities, and to infuse *joy* into tourism landscapes. Within tourism studies specifically, this focus on healing and justice requires changes in practices and decisions at multiple levels with support from a wide ethical framework (Benjamin et al., 2020).

Although education scholars like Denzin and Lincoln have argued for CRT and critical epistemologies in resisting domination and performing counterhegemonic work for decades, the discipline/field of tourism studies has yet to embrace CRT methods that combine activist scholarship and a participatory approach. While there exists a developed field of critical tourism scholarship, there remains a dearth of scholars using CRT informed methods (Chambers, 2021). Advocating for this type of scholarship cultivates an inclusive tourism landscape that supports co-collaboration, creates agents of change, and influences how educators and stakeholders approach issues faced by vulnerable groups, thereby contributing to just touristic landscapes and experiences.

In this article, we begin by defining CRT and provide examples of the employment of CRT in specific tourism contexts. We then describe the application of CRT in tourism studies with a focus on two fundamental concepts: interest convergence and counter-narratives. While these are common tenets in CRT, a Critical Race Tourism framework represents the opportunity to learn from and provide feedback to the broader CRT community by analyzing and learning from the interactions between guest and host central to tourism experiences. We include findings from tourism research employing CRT and provide implications for moving the field forward as a stronger, more informed social force through the framework of *endarkened storywork* - a framework that has the power to advance critical tourism epistemologies. When scholars learn to center BIPOC lived experiences, they foster Afrofuturistic approaches that help to reclaim and recover injustices from the past. Informed by Toliver's (2021) epistemological design, an *endarkended storywork* framework helps foster new possibilities where science fiction, counter-storytelling, and BIPOC lived experiences can reorient existing inequities to help build a positive current and future global touristic society.

Unpacking CRT

Spearheaded by Derrick Bell, Harvard University's first permanently-appointed Black law professor, CRT originated as a field of legal study in the United States in the 1970s to address shortcomings in understanding how racial discrimination and inequity are perpetuated in the law (Kivel et al., 2009). Focused specifically on the United States, the foundations of CRT posit that race and racism are endemic, permanent, and ingrained in the cultural landscape. Consequently, racism looks and feels ordinary and natural to society (Delgado & Stefancic, 2000) and promotes a culture and history where whiteness is the norm. The term itself first began to gain prominence in the 1990s and early 2000s, as more scholars joined the field.

CRT provides a framework that challenges mainstream neoliberal approaches to racial justice both through theoretical frameworks driving scholarly inquiry and analytical tools for unmasking and unmaking racism. Synthesizing from a collection of CRT articles, Dixson and Rousseau (2005)

Table 1. Tenets of Critical Race Theory (from Laughter & Han, 2019, p. 7).

General Tenet / Author	Racism is systemic	Rebuke of the system and neo-liberalism	Activist theory / interest convergence	Use of voice	Across disciplines
Milner (2007)	Racism endemic, pervasive, widespread, and ingrained		Interest Convergence	Centrality of narrative and counter-narratives	
Delgado & Stefancic (2001)	Racism is ordinary, not aberrational	Race and racism are products of social thought; differential racialization; anti-essentialism	Interest convergence (material determinism) and colorblind concepts of equality	Unique voice of color	
Solórzano & Yosso (2001)	Centrality and intersectionality of race and racism	Challenge to dominant ideology	The commitment to social justice	Centrality of experiential knowledge	Interdisciplinary perspective
Ladson-Billings (1998)	Racism is normal in American society	Critique of liberalism	Interest convergence	Employs storytelling to analyze culture	
Tate (1997)	Racism is endemic	Portrays dominant legal claims of neutrality as camouflage	Reinterprets civil rights law in light of its limitations	Insists on a contextual / historical examination	Racism crosses epistemological boundaries
Ladson-Billings and Tate (1995)	Racism as endemic and deeply ingrained	Challenging claims of neutrality and meritocracy	Reinterpretation of ineffective civil rights law	Naming one's own reality	

identified eight constructs in CRT that scholars use to analyze race and racism: (1) whiteness as Property, (2) Intersectionality, (3) Critique of Liberalism and Colorblindness, (4) Interest Convergence, (5) Racial Realism, (6) Restrictive versus Expansive Notions of Equality, (7) Voice/ Counterstory, and (8) Social Change. Table 1 presents a graphic synthesis of commonly held CRT tenets. But to state clearly, CRT is a *theory* that requires application in context; it does not do work on its own.

When moving into the 21st century, Delgado and Stenfanic (2001) discussed three paths for the continued development of CRT: (1) CRT might become a new civil rights orthodoxy, (2) CRT might be marginalized, or (3) CRT might be analyzed and rejected. In varying contexts and scholarly fields, all three have come to pass. For many, CRT is a primary tool for addressing race and racism (Leonardo, 2013). For others, discussions of race appear divisive and unnecessary (e.g. Tennessee's Divisive Concepts Bill; Paulson, 2022). In many contexts, including our own state of Tennessee, legislative action is forcing CRT to the margins and removing it from K-16 curricula (Kruesi, 2021). As a critical theory, CRT evolves alongside the systemic oppression and institutionalized ideologies it seeks to disrupt. In seeking to learn from and further develop CRT, tourism scholars have new paths of action available, in particular when considering the maxims of counter-narrative and interest convergence.

As explored below, advocates of Critical Race Tourism view historical locations and the interactions between guest and host as a site of struggle between master-narratives that maintain *status quo ante* marginalization and counter-narratives bringing truths of the marginalized back to the center. However, a tourism site must attract money and guests in order to survive, which brings interest convergence to bear on the negotiation between master and counter-narrative. CRT provides a foundation for important scholarship and pedagogy seeking to unmask and unmake systemic racism in the tourism industry, community, and curriculum. The knowledge gleaned from CRT methodologies and the importance of analyzing counter-storytelling and interest convergence helps address the need for socially just tourism development that disrupts the status quo within tourism scholarship.

Disrupting power: interest convergence

The association and power of being white creates a carefully calculated dominant discourse that privileges some and marginalizes others. CRT provides a framework for examining power structures that maintain racial inequities and for developing strategies for action and change (Kolivoski et al., 2014). As such, embedding disruption and refusal of the status quo in research starts with exploring systems and institutions that perpetuate and sustain power (Tuck & Yang, 2014). Within heritage tourism sites, Benjamin and Alderman (2018) concluded that a CRT theoretical lens can influence the reexamination and re-writing of whitewashed narratives to create space for dialogue for digesting difficult and uncomfortable histories.

However, Derrick Bell (1995) argued that for this to actualize, interest convergence must be at play. Interest convergence, as developed by Bell, stipulates that Black people achieve civil rights victories only when white and Black interests converge (Milner IV & McGee) and provides a lens for keener insight into policies that on the surface seem to offer obvious diversity benefits. For instance, one recent tourism example around the commodification of counter-storytelling is the Jack Daniel whiskey distillery.

In 2016, *The New York Times* published a story titled "hidden ingredient - help from a slave," which acknowledged that Nearest Green, an enslaved man, taught Jack Daniel how to make whiskey (Benjamin, 2021). This article inspired author and entrepreneur Fawn Weaver to set out on a quest to reveal Green's full story. After meetings with Green's descendants, Weaver eventually opened the Nearest Green Distillery and pressured the Jack Daniel distillery to make extensive changes to their tour script. Now, Green's legacy as the first known Black master distiller is acknowledged, in addition to an exhibit dedicated to Green and Daniel's relationship (Morgenstern, 2020).

More recently, after the murder of George Floyd in 2020, the Jack Daniel Distillery and the Nearest Green Distillery announced the Nearest & Jack Advancement Initiative to further diversity within the American whiskey industry. As such, the interest of the Jack Daniel Distillery to acknowledge Green and support a diversity initiative was developed and enacted on by outside pressure from Weaver and a desire to appear responsive to the current climate. This example of interest convergence helps to explain how institutions like Jack Daniel Distillery will lose more than prestige if they are perceived as unwelcoming or hostile to BIPOC tourists.

In heritage tourism sites, a constant struggle between interests occurs in the desire to present an accurate historical picture, even if that picture is unpleasant, with the reality that a site must attract tourists (and their money) to survive. Thus, Critical Race Tourism can both learn from and help develop CRT. In Critical Race Tourism, interest convergence is a daily concern, driving the intended interactions between guest and host. CRT researchers should investigate specific examples of this interplay and derive lessons from how to converge interests successfully and authentically. In many respects, the convergence of interest in Critical Race Tourism focuses on the interplay between story and counter-story. Achieving some form of racial reconciliation and social justice requires that tourist destinations and stakeholders represent heritage in non-exclusive ways that demonstrate diversity and expose the inequalities produced through prevailing heritage development. Visitors from around the world now know that an enslaved man named Nearest Green taught Jack Daniel how to make whiskey. The addition of Green's legacy is an example of how tourism sites and destinations can use counter-narratives and interest convergence to help (re)tell difficult histories.

Counter-storytelling through CRT

Expansions of CRT transcend a Black/white racial binary and recognize how racism affects the experiences of various people of color, including Latinx, Indigenous, and Asian/Pacific Islander populations. As a result, different branches, including LatCrit, TribalCrit, and AsianCRT have emerged from CRT (Anzaldúa, 2015). Connecting CRT with Tribal Crit, ensures that research "looks to those who have been marginalized or hurt in the past, as an emancipatory project that forefronts community relationships and interests" (Brayboy et al., 2012, p. 52). Indigenous scholars assert that research is about changing and improving conditions – that they are driven by a "purposeful dream and not a prescription" (Smith, 2005, p. 94) where the desired outcome challenges the worldview of Indigenous people from a Eurocentric perspective. These branches seek to examine specific experiences of oppression by challenging white privilege and exposing deficit-informed research that ignores, and often omits, the scholarship of people of color (Solórzano & Yosso, 2001) yet, this scholarship is sparse within tourism studies (Chambers, 2021).

Critical scholars posit that storytelling can open up an understanding to different aspects of our social world (Fraser, 2004) and provide interpretations from marginalized populations. There are numerous unheard counter-stories; these types of counter-narratives and storytelling have a rich and continuing tradition in Black (Bell, 1987, 1992, 1996; Benjamin & Dillette, 2021; Berkeley Art Center, 1982; Chambers, 2021), Chicana/Chicano (Delgado & & others, 1989; Olivas, 1990), and Indigenous (R. Williams, 1997) communities. Delgado and and others (1989) reminds us that "oppressed groups have known instinctively that stories are an essential tool to their own survival and liberation" (p. 2436). Thus, sharing these experiences through storytelling can help strengthen traditions of social, political, and cultural survival and resistance.

The desire to examine the lived experiences of minoritized people leads CRT scholars to focus on and develop counter-storytelling (Miller et al., 2020), a method asking people of color to name their reality. Naming one's reality is essential, as it can elucidate the constructed nature of social reality, catalyze social change by challenging oppressors' mindsets, and create space for people of color to heal (Toliver, 2020). In this way, counterstories uplift the narratives of minoritized people while also challenging dominant narratives, providing windows into new realities that allow possibilities for a world beyond the one in which we currently live (Delgado & & others, 1989). Miller et al. (2020) described that the theoretical role of CRT in general, and counter-narrative more specifically, "tends to be more evocative than technical" (p. 276), establishing counter-narrative as a theoretical or methodological framing and a research method. For example, Caton (2012) referred to CRT as a "conceptual lens," and counter-storytelling as CRT's "method" (p. 1062).

Counter-storytelling subverts dominant paradigms or epistemologies and presents new solutions or policies that disrupt the status quo (Delgado & & others, 1989). These counter stories expose structures as simultaneously oppressive, marginalizing, and enabling and present stories of possibility – serving to build a much needed community or local capacity. They are narratives of how people see structures, combat them (when necessary), and utilize them (when possible) to create better lives for themselves and others (Brayboy & Chin, 2019). Counter stories are not seen as *apologia*, but as Brayboy and Chin (2019) stated, "they recognize that communities are besieged and people from these communities may participate in these sieges" (pg. 52). Brayboy and Chin (2019) continue:

> When someone listens to a story, they are engaging in the physical action of how one listens. However, hearing stories is quite different, in that it involves listening to what is said, purposeful engagement with the speaker, and taking in the value attached to the story and its meanings. For Tribal Crit and qualitative methods/methodology this means there is value attached to the story; the stories are the collection of history and culture and hold a place in the mindset of group members. The stories represent memories of tradition and reflections on power. The listener has to hear and feel the stories, as well as bear the responsibility of understanding the nuances in the stories and the knowledge embedded in them. (pg. 437)

Brayboy and Chin (2019) argue that TribalCrit honors these stories as legitimate "data" that are not separate from theory: they *are* theory (p. 439). Using counter stories as a method is about realizing the potential of thick, rich data of everyday happenings of marginalized communities to enrich the theoretical, empiricism and scholarly insights offered by academic writing.

LatCrit, grounded in the use of narrative storytelling, has been important in examining how aspects of race, ethnicity, language, and national origin converge and effectively otherize and politically disenfranchise Latinx populations in the United States (Parker, 2015). Critical race feminista scholars like Delgado-Bernal and Humor borrow from multiple methodologies which are grounded in critical race or Chicana/Latina feminist theories such as *testimonios* and *conocimientos* (Bernal et al., 2019). *Testimonio* allows Chicanas/Latinas to "inscribe a social witness account of collective experiences, political injustices, and human struggles that are often erased by dominant discourses and offer a pathway toward healing" (Bernal et al., 2012, p. 115). Chicana/Latina researchers and participants enter themselves – their knowledge, positionalities, and experiences – into the process of theorizing and healing (Delgado et al., 2019).

This co-creation engagement with participants and scholars allows for exploring the intimate relationship "between structure and agency where possibilities of healing and transformation take place, and to bring together the bodymindspirit" (Delgado et al., 2019, p. 117). For instance, Delgado Bernal (2018) shared, "Through *testimonio* we are able to know each other's stories through voices, silences, bodies, and emotions and with the goal of achieving new *conocimientos* – understandings and awareness" (p. 27). Thus through *testimonio* pedagogy, Delgado Bernal (2018) explained, "We can teach each other and learn from each other the *conocimientos* of life ... and it becomes a teaching tool that has the potential to connect people across social positions and build solidarity between both those who are familiar or unfamiliar with the experiences of the *testimonio*" (p. 27).

Counter-storytelling can also be a powerful force when used in political discourse to engage BIPOC citizens in an inclusive process to find solutions to global challenges like climate change (Veland et al., 2018). Consequently, another function of counter stories is that they inspire activism – they are based on collective experiences, can empower people to act against racial oppression (Blaidell, 2021), and highlight the social, (his)torical, and structural dimensions of racism, colonization, patriarchy, and other oppressions in a relatable, allegorical, or insightful narrative (Brayboy & Chin, 2019). Patricia Williams (1991) argued that by telling counter narratives, unrepresented communities have the critical tools to "speak truth to power" (p. 52) as they ground questions of ideologies and social structure in lived experiences.

Such an approach may be particularly appropriate for rural and resource-based communities, particularly when prevailing sustainable tourism development discourse offers a different means to understanding current conditions and future potentials for sustainability (Lowery et al., 2020). Planners can tell stories about their communities that reflect underrepresented citizen concerns about the future and advocate for particular development decisions (Lowery et al., 2020). Furthermore, it is important to stimulate critical reflection on the ultimate goal of tourism equity, and what actions could proceed from the knowledge generated by counter-narratives. In other words, helping participants develop agency as well as their voices, and encouraging them to come up with alternative solutions and further take actions to implement them, will more effectively promote generative transformative action.

CRT composite counter-story methodology

A "story" evolves into a composite counter-story (CCS) when it challenges assumptions of the dominant culture (Solórzano et al., 2002). This collective storytelling method allows reality to be challenged by offering and including the "stock story" from a majoritarian perspective and a "counter story" from a marginalized perspective (Benjamin & Dillette, 2021). The experiences shared in both the stock/majoritarian story and CCS unveil the difficult but extraordinarily important truth that, as a tourism industry, we are still dealing with the remnants of a history steeped in systemic racism. Collective stories are fictionalized narratives drawn from interview transcripts, field notes, memos, lived experiences, and other research data where the authors' use of these data is done creatively in an effort to craft a coherent narrative.

The following points are taken directly from Cook and Dixson (2013) and assist with the interpretation of the collective story:

1. the dialogue, setting, and thoughts of the composite characters come directly from the interviews, field notes, and other data sources but are edited to maintain the flow of the narrative;
2. the analysis of the counter-stories appears later in the paper; and
3. footnotes provide background and factual information for clarity. (p. 1246)

This collective storytelling framework includes the use of composite characters from data collection and personal experience that enable academics to humanize theory and practice (Bernal & Villalpando, 2010). Characters are often representative of several participants rather than a single participant (Delgado & & others, 1989) and "are written into social, historical, and political situations that allow the dialogue to speak to the research findings and creatively challenge racism and other forms of subordination" (Yosso, 2006, p. 11).

This qualitative approach allows scholars to explore, in depth, the underlying themes and nuances within data using "first person, storytelling, narrative, allegory, interdisciplinary treatment of law, and the unapologetic use of creativity" creating a composite counter-story (Bell, 1995, p. 899). In this way, CCS aids in illustrating how marginalized populations share a history with racism, discrimination, and white supremacy and speaks against the majoritarian narrative in U.S. society – a narrative based on the social and cultural history of the dominant race (Yosso, 2006). As Bell (2003) notes, white people hold the power in telling these master narratives that convey "a sense of history as progressive, depict a U.S. society that is basically fair and meritocratic, and assume a trajectory of forward progress in which injustices are eventually recognized and rectified over time" (p. 4). CRT scholars use CCS as a way to confront the master narrative that attempts to erase the struggles and resilience of BIPOC who have challenged the U.S. to live up to its democratic ideals (Cook & Dixson, 2013). However, within sustainable tourism studies, there has been a dearth of scholars who used CRT to inform their research (Chambers, 2021).

CRT methodological framework in sustainable tourism studies

Bright et al. (2021) used CRT to explore the ways in which roadside markers in Tennessee depicted *just representations* of the past. They illustrated that historical roadside marker programs, consciously or unconsciously, ignored Black history, Indigenous history, and white women in history, altering the American collective memory. Dudley et al. (2021) used critical situational analysis to focus on disaster capitalism and tourism development as a strategy in post-Katrina New Orleans. They paired situational analysis with CRT, which informed analyses conducted within the broader study, offering a novel critical approach to tourism research. Lastly, informed by CRT, Dillette et al. (2018) used CRT to unpack the lived experiences of Black travelers using the Twitter hashtag #TravelingWhileBlack. This seminal piece helped inform two additional articles from Dillette and Benjamin where they explored Black Travel Movement leaders' quest for social change (Dillette & Benjamin, 2021) and how racism, white supremacy, and discrimination impact the touristic landscape (Benjamin & Dillette, 2021). However, scholarship inclusive of CRT methods, specifically where data are analyzed within a storytelling and/or counter-storytelling design, are far and few between in tourism studies.

If methodologies were used to silence and marginalize BIPOC, then methodologies can also be used to amplify voices and turn the margins into places of transformative resistance (Solórzano & Delgado Bernal, 2001). For instance, critical tourism scholars, Chambers (2021), along with Benjamin and Dillette (2021), have incorporated the CRT method of fictionalized storytelling to critically assess gender, race, or power dynamics within tourism. Through the lens of CRT, Chambers (2021) provided a consciousness raising (Yosso, 2002) strategy that examined the fictional film *Heading South* from the perspectives of Black women in the film – critiquing and interrogating how Black women are (re)presented in film. She interwove her positionality and interpretation as a Black woman from a postcolonial island in the Caribbean, which decentered white female sex tourism and instead (re)tells a story about "tourism, race, and gender so that it brings Black women to the fore" (p. 6). Centering Blackness and marginalized voices helped advocate for the value of using CRT to "expose intersectional marginalization … with the goals of achieving social justice" (Chambers, 2021, pg. 13). Advocating for gender intersectionality and its importance in sustainable tourism, Chambers used CRT storytelling as the overarching theoretical foundation of her work to amplify the voices of Black female hosts on an island where their voices were silenced within the context of white female sex tourism.

Chambers's interpretation answered the call for "just representation" in the Bright et al. (2021) Special Issue of the *Journal of Sustainable Tourism* on Justice and Tourism – focusing on the context of the (re)presentation of Black women as more than *objects of research*. Chambers (2021) recognized there are "no certainties in research" and argues that storytelling produces "often iterative and fluid ideas which represent a plausible way of interpreting social phenomena" (p. 6). Her work illustrated the power of critiquing film through the CRT method of storytelling where "the focus of this film on female sex tourism has illustrated the power of whiteness" (p. 11) and how tourism practices in the postcolonial world inaccurately (re)present Black women as *vulnerable* or *submissive*. Chambers argued this erasure of Black women as "agentic subjects and the way they are (re)presented as disempowered has implications for the development of tourism that seeks to be sustainable" (p. 11).

Informed by the CRT method of collective storytelling, as a multi-racial research team, Benjamin (first author) and Dillette and Benjamin (2021) helped to uncover the discrimination and racism Black travel leaders faced and continue to experience within the tourism industry. We called for "scholars to get comfortable with being uncomfortable" and to "use privilege, in this case academic and white privilege, to disseminate these difficult conversations around systemic racism, white supremacy, and hegemony within our academic and industry spheres" (p. 9). Consequently, Dillette (identifying as Black, bi-racial) and Benjamin (identifying as white) used collective storytelling to creatively and evocatively design a research study that allowed us to interweave our lived experiences of presenting difficult dialogue to public audiences in addition to centering Black travel leader voices. This research design, not traditionally viewed as *rigorous* in academic settings, provided a blueprint on how narratives and personal stories steeped in racism can be used as evidence of discrimination within tourism landscapes. As such, we interrogated the differences between how Black and white stakeholders interpreted and experienced touristic landscapes through racialized lived experiences.

We incorporated blended voices from interviews, direct quotations from workshop participants, and our own lived experiences to create a two part CRT collective story inclusive of several composite characters discussing racialized aggressions and issues within the tourism industry. Part one unpacked the stock story and was based on data from workshops to predominantly white audiences. This story conveyed how unexamined assumptions guide the design and development of travel and tourism spaces, potentially creating biased and segregated perceptions of the travel industry. Part Two was presented as a composite counter-story (CCS) from the lived experiences of Black Travel Movement (BTM) leaders and workshops presented to Black and audiences of color. The data collected for the stock story were based on observations and experiences at tourism conferences, festivals, and workshops captured through reflexivity journals and analytic memos. We facilitated four workshops on Black travel research at our academic institutions and to public audiences of industry stakeholders in the U.S. during 2019–2020. After each workshop, we documented non-verbal cues, comments, quotations, and questions from participants, helping to inform and design the stock story. Additionally, we included reflexivity and positionality in the manuscript to understand identity, potential biases brought to the field, and the importance of declaring and taking responsibility for positioning as researchers.

For the CCS, we adapted Cook and Dixson (2013) study and conducted semi-structured interviews with leaders of the BTM in 2018. We focused our curiosity around the challenges BTM leaders faced, which helped unpack complexities around race, assumptions, perceptions, and biases. Telling the CCS allowed the exposure and explanation of how marginalized groups continue to feel in predominantly white landscapes. Quotations shared from BTM founders in front of audiences predominantly made up of people of color were well received and created a sense of solidarity in the room amongst attendees. When these stories were shared in a majoritarian room, like in the stock story, the opposite occurred.

In the majoritarian story, many in the room remained silent, aligning with tenets of white fragility (DiAngelo, 2018) that claim white people feel uncomfortable discussing race, acknowledging their part in systemic racism, or continuing to center whiteness within their story (Spingh, 2019). This is indicative of an ideology of whiteness (Mowatt, 2020), where white participants refuse to acknowledge the wealth, power, and privilege passed down generationally or their perpetuation of white supremacy. These dominant Eurocentric worldviews pose a crucial issue as many of them continue to fill rooms around the globe where important decisions about tourism are being made. Though the stock story presented a grim view of Black/white relations in travel and tourism, we hoped that the workshops acted as a platform from which some industry stakeholders left feeling a sense of responsibility to make changes in their various organizations. However, the challenge remains – how can awareness and conversations around race and racism, in a room predominantly full of white people, be heard without white fragility or defensiveness? When will white people hold themselves accountable for white supremacy and racism and actively become anti-racist?

These two stories confirmed the premise within CRT that the master narrative attempts to erase the experiences of the marginalized. The experiences shared in both the stock/majoritarian story and CCS made up the collective story and unveiled the difficult but extraordinarily important truth that, as a tourism industry, we still face the remnants of a history steeped in systemic racism. Being published in one of the top tier tourism academic journals gave this methodological framework legitimacy, validation, and allowed for "empirical space for researchers to recount the stories and experiences of people who are in politically vulnerable positions and working in politically charged environments" (Cook & Dixson, 2013, p. 1253). In return, we were able to disseminate this work within and beyond academia – creating space for honest and difficult dialogue to transpire. We did this virtually through webinars, face-to-face amongst students and communities, and at the dinner table with family and friends. This methodological design and framework of storytelling, inclusive of composite characters, was coherent, easily read, and mirrored social justice issues felt around the world – allowing space for growth to occur. Unfortunately, with the uncovering of difficult dialogue and authentic representation of oppression – conservative politicians are using manipulation and marketing strategies to eliminate the teaching of race, CRT, and discussions of marginalized histories. Ultimately, the controversy over CRT exemplifies a tenet of the theory itself - any racial progress will be met with great resistance.

Backlash against CRT

Like any other approach, CRT can be misunderstood and misapplied. It has been distorted and attacked as it continues to change and evolve. CRT was seldom discussed amongst the general public and mostly taught in law schools – calling for an examination of the U.S. legal system from a racial lens and arguing that the law is not neutral and has been a tool to maintain racial hierarchy (Harris, 2021). U.S. President Bill Clinton, during the early 1990s, nominated the University of Pennsylvania Law School professor Lani Guinier to run the Justice Department's Civil Rights Division – where CRT first entered U.S. public discourse (Harris, 2021). Republicans felt that she was *wrong for the job* and tagged her the *Quota Queen* because she had been "championing a radical school of thought called 'critical race theory" (Harris, 2021, paragraph six). Conservatives further argued that CRT taught Americans to hate America, which influenced GOP-led state legislatures to pass laws to disallow any inclusion of language that has a direct effect on teaching about Indigenous people, colonization, Western expansion, Native sovereignty, and more (Adams, 2021).

In the summer of 2020, after a white police officer was captured on video kneeling on a Black man's neck (George Floyd) for more than nine minutes, the United States became awash in a racial awakening. Centuries of racist policy, both explicit and implicit, perpetuated and sustained a country that "never did a very good job incinerating its traumatic remains ... an education system that fails Black Americans, substandard health care that makes them more vulnerable to death and disease, and an economy that leaves millions without access to a living wage" (Worland, 2020, paragraph 3). But, for many of those who have spent their lives hting for racial justice, this was a moment of reckoning a long time coming. Various anti-racist reading lists, some inclusive of articles that discussed CRT, were popularized in addition to performative actions like posting a Black square on Instagram for solidarity with *Black Out Tuesday* (Heilweil, 2020).

Later that summer, Christopher Rufo, a 36-year-old senior fellow at the Manhattan Institute, a libertarian think tank, received a tip from a municipal employee in Seattle stating that the city was conducting "internalized racial superiority" training sessions for its employees (Harris, 2021). Rufo submitted a Freedom of Information Act request and wrote that, "the city frames the discussion around the idea that black Americans are reducible to the essential quality of 'blackness' and white Americans are reducible to the essential quality of 'whiteness'—that is, the new metaphysics of good and evil" (paragraph 8). Rufo's article — which did not include the phrase *critical race theory* — created a fury of whistleblowers from school districts and federal agencies who voiced their complaints about *forced* attendance in diversity trainings. A month later, Rufo stated, "Critical race theory—the academic discourse centered on the concepts of 'whiteness,' 'white fragility,' and 'white privilege'—is spreading rapidly through the federal government" (Harris, 2021, paragraph 10). *Fox News* host, Tucker Carlson invited Rufo on his show to warn viewers on the evils of CRT and called on then President Donald Trump to ban such training in all federal departments (Adams, 2021); within three weeks, Trump signed an executive order banning the use of CRT by federal departments and contractors in diversity training.

U.S. Republicans and conservatives proposed bills that aimed to stymie educational discussions about race, racism, and systemic oppression in the US — potentially eliminating the conversations altogether (Cienas, 2021). These pervasive and insidious state-sponsored attempts to disempower and disenfranchise marginalized and racialized populations were calculated by right-wing legislative and censorship action (Adams, 2021). Several states passed legislation banning or censoring racial- and gender-based U.S. histories and are working on policies that restrict dialogue and education on racial injustices and related biases, which continues to perpetuate and promote a revisionist whitewashed history (Cienas, 2021). For example, legislators in Alabama filed a bill banning the teaching of any concepts about CRT and argued, "The Alabama State Board of Education believes that the United States of America is not an inherently racist country, and the state of Alabama is not an inherently racist state" (Cienas, 2021, paragraph 11). Similarly, Tennessee Republican Governor Bill Lee signed a measure into law banning teaching certain concepts of race and racism in K-12 public schools, where teachers are at risk of losing state funding if they disobey through withheld funding if teachers connect events to institutional racism (Kruesi, 2021).

Although CRT was developed in a U.S. context, there continues to be global pushback around conversations inclusive of race, nationalism, and difficult histories – especially where conservative policies are advancing (e.g. Australia, Brazil), which continues to cross continents and cultures (e.g. United Kingdom, New Zealand, Columbia) where counter-narratives can be used to illuminate debates on ethnic and racial division, as well as legacies of slavery, colonization, and immigration (e.g. France). This global awakening of research and praxis to guide actions to weave just tourism futures (Jamal & Higham, 2021) is contributing toward the movement of social equity and resistance within sustainable tourism.

For instance, Boukhris (2017) explored The Black Paris Project that deals with the invisibilization of racialized minorities in the French nation and other counter-narratives that have contributed to the production of Paris. Black tourism stakeholders have incorporated components of counter-hegemonic discourse within their tours, which opened "a new space for approaching the challenging questions of racial legacies" (p. 688). However, as Baldwin (1955) said, "I think the past is all that makes the present coherent, and further, that the past will remain horrible for exactly as long as we refuse to access it honestly" (p. 107). As such, the responsibility to acknowledge and own difficult histories, which respect the rights, interests, and social needs of vulnerable groups and local communities (Boluk et al., 2019), helps to expand the public's understanding of how power, race, and nationalism shape and distort public memory. Knowledge created through such counter-narratives can hopefully evoke alternative ways of interpreting history that can contribute toward equitable, inclusive touristic landscapes.

The path toward resistance

Where CRT has been and where it is going regarding tourism research is a path marked by an increasing acknowledgment of the complexity of counter-story and counter-narrative as dynamic explainers of racial conditions and subsequent action. CRT and storytelling as both theory and methodology, by exposing the insidious applications of race, racism, and social equity, contributes toward transformative healing. And, as CRT scholar Ladson-Billings (2013) posits, "the point of storytelling is not to vent or rant or be an exhibitionist regarding one's own racial struggle … but to point out the endemic racism that is extant in our schools, colleges, and other public spaces" (p. 45). Ladson-Billings and Donnor (2005) continue by stating:

> Thus, we argue that the work of critical scholars (from any variety of perspectives) is not merely to try to replicate the work of previous scholars in a cookie-cutter fashion but rather to break new epistemological, methodological, social activists, and moral ground. (p. 291)

Breaking methodological ground by addressing misrepresentation and disrupting dominant discourse is part of CRT and complements Baldwin's dictum toward addressing difficult histories through honest and open dialogue. When will *we* dismantle the hegemonic norms and stories in order to reset our tourism industry? How can these stories serve as a form of resistance that Angela Harris (1994) calls "writing back" – unraveling majoritarian (stock) tales, re-framing historical events, and questioning the very concepts of race and liberation, which remain the center of CRT scholarship (Chapman & DeCuir-Gunby, 2019).

It is through this process of "writing back" that scholars have the *power* to shift discourses to create potential moments for empowerment (DeCuir-Gunby, Chapman, and Schutz, 2019, p. 187), not only within academia but within the tourism landscape. This can start with placing value on re-presenting research where the public can actually access it freely. Research can tell a holistic story where those once at the bottom of the well are lifted up. A story where *we* are no longer afraid and are open to collaborating, co-creating, and making space for difficult dialogue to occur – in order to imagine futuristic touristic landscapes that embrace joy. Toliver (2021) coined this method *endarkened storywork* and views it as a new possibility for qualitative research that hinges on Black storytelling traditions which honors alternative ways of thinking about, doing, and writing scholarship.

Storytelling is an essential component of Black existence, from African nations who had their own storytellers, also known as griots, who acted as teachers, genealogists, historians, advisers, exhorters, witnesses, and praise-singers (Hale, 1998). Griots engage in sacred work by ensuring people's stories are never forgotten and a way for African-American ancestors to retain their stories by, "refusing to let whiteness destroy their connection to the past and to their homelands" (Toliver, 2021, p. xv). Toliver (2021) contended that stories continue to teach, to heal, to bring life; and she shared that storytelling is not a "luxury for Black people, it is our very existence" yet "our storied lives are often shunned in academia" (p. xv). This storytelling extends into other marginalized populations and Dillard (2006) defined it as a *research of responsibility* where scholars need to be accountable to the people and communities in which *they* engage.

Stories allow for "alternative sites of knowledge and being, and honoring the spiritual within the everyday lives of people whose stories have been discounted in traditional academic settings" (Toliver, 2021, p. xvi). *Endarkened storywork* allows community members, who were once ignored in qualitative research, to be invited into the research process and be part of the story. As stories do not always "fit within tight spaces and guiding frameworks," Toliver (2021) advocated for scholars to "consider what is lost, forgotten, and erased … and to consider alternative ways of knowing, those that are often restricted or rejected in academe" (p. xxxiv). She challenged researchers to consider how *we* are moving forward in data collection, data analysis, and data representation and to:

> Focus upon the storied lives of 'our research partners' and engage with them as story listeners. Not only to complete a research endeavor but also to better understand our historic and current world and better situate inquiry for what the future world and future research could look like (pg. xxxvii).

Toliver (2021) refused to use the term *enlightened* when discussing this framework and argued that *endarkened* discards the traditional canon of Eurocentric research and "chooses to liberate the mind by allowing space for different truths about reality- truths that refuse objectivity, require community, and remember responsibility" (p. xvii). This framework chooses to remember the joy, community, connection to land, and traditions of Black existence without requiring the focus on anti-Blackness. The power of storied lives and oral traditions – from those positioned as story listeners – opens a space for co-creation amongst research partners. Toliver also encouraged scholars to locate ourselves within the work and consider the ethical responsibility we have to engage in reciprocity and reverence, "to refuse traditional separation between the researcher and the researched" (p. xx).

Becoming a story listener in her book, *Recovering Black Stories in Qualitative Research*, Toliver (2021) included truths from aspects of her personal life along with the stories of six Black girls. They joined her in an Afrofuturist writing workshop that consisted of eight meetings over the course of two months in the summer of 2019. They read passages from speculative fiction, "focusing on how they can move from the known world to the imagined" (p. xxv) and took insights from the group decisions, mentor texts, and writing exercises to outline story elements for their speculative short stories. Toliver and the six girls wrote and talked together, rejecting any division between the researcher and the researched, which in turn placed the needs of the group over Toliver's personal research agenda. Collectively they wrote texts, where Toliver became the story listener, and returned to the technologies of Black quilt making and worked to find the seam connecting the stories. As such, the girls were able to read and evaluate the story through a shared Google drive folder where Toliver let the story go where it needed to go. She argued that the Black girls deserved research that spoke to them, not just about them and embraced her cultural rituals and traditions.

Interwoven into the framework of *endarkened storywork,* Toliver (2021) challenged scholars to incorporate speculative fiction and *Afrofuturist* practices. Dery (1994) coined this term

Afrofuturism, "speculative fiction that treats African-American themes and addresses African-American concerns in the context of twentieth-century technoculture" (p. 180). Toliver's (2021) interpretation of Afrofuturism showcases Black authors as the creators of speculative texts that, "center Black characters in an effort to reclaim and recover the past, counter negative and elevate positive realities that exist in the present, and create new possibilities for the future" (p. xx). This work encourages stylistic elements prominent in science fiction, fantasy, African cosmology, history, and realism - all modes of communication that allows readers to consider the "intricacies of Black thought and existence across space and time" (p. xxi). Incorporating these technologies of freedom, literacy, spirituality, language, trickery, coding, communal connection, intergenerational links, dance, hope, imagination, and joy can all be classified as Afrofuturist practices.

Bright et al. (2021) argued that "intangible heritage such as cultural rituals and traditions passed down orally is particularly difficult to ensure fairness representation" and as such "from a sustainability perspective, just representation of cultural heritage entails fair consideration and incorporation of intangible heritage along with tangible cultural heritage in heritage tourism development" (p. 430). Thus, by building on Toliver's (2021) *endarkened storywork* framework, we advocate for tourism scholars to weave insights gleaned from counter-narratives and design research that embodies *Afro-touristic* and speculative fiction to reconceptualize the future of touristic landscapes. We seek landscapes that help BIPOC escape physical and psychological bondage - allowing them control of their own actions as a freedom technology. By doing so, *endarkened storywork* can be seen as a freedom technology where "the academic bondage of Eurocentric writing" can make space for "the storied, spiritual work essential to Black life in the past, present, and future" (Toliver, 2021, p. xxiv). Moving beyond the CRT conception of counter-narrative, endarkened storywork specifically seeks to highlight Afrofuturism and Black joy.

As Bright et al. (2021) suggested, Toliver looked to her heritage of Black American quilt makers, who used this method as a form of resistance, as a way to "undermine white power structures in traditional qualitative research" (p. xxxv). Consequently, she uncovered the patterns within the everyday talk and storytelling of Black girls centering Black quilting as the method needed to reconnect and rely on historical methods to better understand the contemporary stories of Black girls. *Endarkend storywork* can make space for alternative representations of data better attuned to the storytelling practices of Black people and people of color, especially within tourism studies. We argue for tourism scholars to embrace *endarkend storywork* and Afrofuturism as an *onto-epistemological tool and ethical premise* that creates new possibilities for the future. We must use the privileges bestowed on us to advocate for equity, fairness, and telling of histories that are authentically dark. With this we ask: How will we – as academics, as management and docents at heritage tourist sites, as stakeholders, as educators – use counter stories, *endarkened storywork*, and new combinations of CRT research methods to address the dearth of marginalized voices and stories to address climate change, to address the sickness exacerbated by tourism, to address our fears, to swing the pendulum toward healing?

Final thoughts

In all the ways that COVID-19 has affected and shut down numerous industries, with hospitality and tourism being hit hardest, a global pandemic only highlights and amplifies the systems of oppression and marginalization at work in tourism sites and tourism research. Adopting a framework informed and infused by CRT allows difficult histories and present challenges to move toward healing during times of global crises. As such, a Critical Race Tourism framwork can potentially influence how educators and stakeholders approach issues faced by vulnerable groups contributing to just touristic landscapes and experiences. Changing the narrative, inviting community into the research design and analysis, and "planning a tourism future not fueled solely by demand but guided by an ethics of care, social and environmental justice, and racial reconciliation" (Benjamin et al., 2020, p. 5) may serve as a dynamic explainer of institutional and structural racism in tourism studies – where our discipline and industry can face and center marginalized voices with the goal of achieving social justice.

In order to move toward equity and justice – tourism scholars and stakeholders can fuse CRT with frameworks of interest convergence, counter storytelling, and *endarkened storywork* to deliberately dismantle whitewashed dominant narratives and instead seek justice within efforts to conduct research that clearly exposes how racism has and continues to shape and sustain racialized harm. Critical race tourism and *endarkened storywork* challenge passive reproduction of established practices, knowledge, and resources to deepen the touristic landscape by understanding the nuances of racism, as well as how research might struggle against racism (Hylton, 2012). We believe that a critical race tourism framework provides a salient testing ground for exploring interest convergence, both as a descriptive methodology and tool for gaining leverage. In short, management at tourism sites have to balance presenting a complete story with the need to get guests in the door. The examination and expansion of this balance toward social justice ends presents an active area for inquiry not always common in other CRT fields. In tourism, interest convergence is not abstract; in tourism interest convergence means actual money.

Likewise, a focus on counter-narrative and *endarkened storywork* may produce valuable research, illuminate ways to end racial discrimination and oppression, and serve as a rallying point for strategic action for building touristic identities that center antiracism and joy through an activist function. Hopefully, tourism scholars and stakeholders will contribute toward the path of liberation, transformation, and equity within the lives of colonized/oppressed people in order to establish our own Theory of Critical Race Tourism. As Derrick Bell (1992) suggested, we need an Afrolantica Awakening that rekindles the soul and reminds us that our collective struggle is in and of itself hope anew. This theoretical and analytical framework connects well with the justice and racial healing lens of narrative change (Toliver, 2021), which uses storytelling to speak the truth about shared histories, hidden hurts, and racial harms helping to design a future celebrating Blackness and the triumphs within Black culture.

These moves need to be made now. Just as discussions of race, racism, and accurate tellings of history are now banned in certain classrooms, the tourism field is likely to experience a similar attack. In fact, even the most powerful tourist sites already feel the backlash of conservative politicians uncomfortable with challenges to the master-narrative, as in the case of Florida Governor Ron DeSantis and Disney (Gore, 2022). Educators and tourism stakeholders might find tourism an attractive area in which to continue working to dismantle dominant Eurocentric and whitewashed narratives at touristic sites and within our classrooms. But this will require significant energy, focus, and organizing to push back against policies trying to silence history, counter-narratives, and *endarkened* storytelling.

Disclosure statement

No potential conflict of interest was reported by the authors.

Funding

No funding or financial grants were used for this research.

ORCID

Stefanie Benjamin http://orcid.org/0000-0002-4653-6754

References

Adams, J. (2021). How the War Over Critical Race Theory Affects Native Americans. *The Nation.* https://www.thenation.com/article/culture/native-americans-crt-debate/

Anzaldúa, G. (2015). Geographies of Selves—Reimagining Identity. In *Light in the dark/Luz en lo oscuro* (pp. 65–94). Duke University Press.

Baldwin, J. (1955). *Notes of a native son*. Beacon Press.

Bell, D. (1987). *And we will not be saved: The elusive quest for racial justice.* Basic Books.

Bell, D. (1992). *Faces at the bottom of the well: The permanence of racism.* Basic Books.

Bell, D. (1996). *Gospel choirs: Psalms of survival for an alien land called home.* Basic Books.

Bell, D. A. (1995). Who's afraid of critical race theory? *University of Illinois Law Review, 4*, 893–910.

Bell, D. (2003). Diversity's distractions. *Columbia Law Review*, 103, 1622.

Benjamin, S. (2021). The story of Nearest Green. *The Conversation.* https://theconversation.com/the-story-of-nearest-green-americas-first-known-black-master-distiller-164311

Benjamin, S., & Alderman, D. (2018). Performing a different narrative: Museum theater and the memory-work of producing and managing slavery heritage at Southern Plantation Museums. *International Journal of Heritage Studies, 24*(3), 270–282. https://doi.org/10.1080/13527258.2017.1378906

Benjamin, S., & Dillette, A. (2021). Black Travel movement: Systemic racism informing tourism. *Annals of Tourism Research, 88*, 103169. https://doi.org/10.1016/j.annals.2021.103169

Benjamin, S., Dillette, A., & Alderman, A. (2020). "We can't return to normal": Committing to tourism equity in the post-pandemic age. *Tourism Geographies, 22*(3), 476– 483. https://doi.org/10.1080/14616688.2020.1759130

Berkeley Art Center. (1982). *Ethnic notions: Black images in the White mind.* Berkeley Art Center.

Bernal, D., Huber, L., Malagón, M., Chapman, T., DeCuir-Gunby, J., & Schutz, P. (2019). Bridging theories to name and claim a critical race feminista methodology. In J. T. DeCuir-Gunby, T. K. Chapman, & P. A. Schutz *(Eds.), Understanding Critical Race Research Methods and Methodologies*, (pp. 109–121). Routledge.

Blaisdell, B. (2021). Counternarrative as strategy: embedding critical race theory to develop an antiracist school identity. *International Journal of Qualitative Studies in Education*, 1–21. https://doi.org/10.1080/09518398.2021.1942299

Boluk, K. A., Cavaliere, C. T., & Higgins-Desbiolles, F. (2019). A critical framework for interrogating the United Nations Sustainable Development Goals 2030 Agenda in tourism. *Journal of Sustainable Tourism, 27*(7), 847–864. https://doi.org/10.1080/09669582.2019.1619748

Boukhris, L. (2017). The Black Paris project: The production and reception of a counter-hegemonic tourism narrative in postcolonial Paris. *Journal of Sustainable Tourism, 25*(5), 684–702. https://doi.org/10.1080/09669582.2017.1291651

Bramwell, B., & Lane, B. (2014). The "critical turn" and its implications for sustainable tourism Research. *Journal of Sustainable Tourism, 22*(1), 1–8. https://doi.org/10.1080/09669582.2013.855223

Brayboy, B. M., Gough, H. R., Leonard, B., Roehl, R. F., & Solyom, J. A. (2012). Reclaiming scholarship: Critical Indigenous research methodologies. In S. D. Lapan, M. T. Quartaroli, & F. J. Riemer (Eds.), *Qualitative research: An introduction to methods and designs* (pp. 423–450). Jossey-Bass/Wiley.

Brayboy, J., & Chin, J. (2019). A match made in heaven: Tribal critical race theory and critical Indigenous research methodologies. In J. T. DeCuir-Gunby, T. K. Chapman, & P. A. Schutz *(Eds.), Understanding Critical Race Research Methods and Methodologies* (pp. 51–63). Routledge.

Bright, C., Foster, K. N., Joyner, A., & Tanny, O. (2021). Heritage tourism, historic roadside markers and "just representation" in Tennessee, USA. *Journal of Sustainable Tourism, 29*(2-3), 428–447. https://doi.org/10.1080/09669582.2020.1768264

Caton, K. (2012). Taking the moral turn in tourism studies. *Annals of Tourism Research*, 39(4), 1906–1928. https://doi.org/10.1016/j.annals.2012.05.021

Chambers, D. (2021). Are we all in this together? Gender intersectionality and sustainable tourism. *Journal of Sustainable Tourism*, *30*, 1586–1601. https://doi.org/10.1080/09669582.2021.1903907

Cienas, F. (2021). What the hysteria over critical race theory is really all about. *Vox News*. https://www.vox.com/22443822/critical-race-theory-controversy

Cook, D. A., & Dixson, A. D. (2013). Writing CRT and method: A composite counterstory on the experiences of Black teachers in New Orleans post-Katrina. *International Journal of Qualitative Studies in Education*, *26*(10), 1238–1258. https://doi.org/10.1080/09518398.2012.731531

DeCuir-Gunby, J., Chapman, T., & Schutz, P. (2019). *Understanding critical race research methods and methodologies*. Routledge.

Delgado Bernal, D. (2018). A testimonio of critical race feminista parenting: Snapshots from my childhood and my parenting. *International Journal of Qualitative Studies in Education*, *31*(1), 25–35. https://doi.org/10.1080/09518398.2017.1379623

Delgado, R. (1989). Storytelling for oppositionists and others: A plea for narrative. *Michigan Law Review*, *87*(8), 2411–2441. https://doi.org/10.2307/1289308

Delgado, R., & Stefancic, J. (2000). *Critical race theory*. Temple University Press.

Delgado, R., & Stefancic, J. (2001). *Critical race theory: An introduction* (Vol. 20). NyU Press.

Denzin, N. K., Lincoln, Y. S., & Smith, L. T. (Eds.). (2008). *Handbook of critical and indigenous methodologies*. SAGE Publications.

Dery, I. (2020). Negotiating positionality, reflexivity and power relations in research on men and masculinities in Ghana. *Gender, Place & Culture*, 27(12), 1766–1784.

Dillard, C. B. (2006). When the music changes, so should the dance: Cultural and spiritual considerations in paradigm 'proliferation'. *International Journal of Qualitative Studies in Education*, 19(1), 59–76. https://doi.org/10.1080/09518390500450185

Dillette, A. K., Benjamin, S., & Carpenter, C. (2018). Tweeting the Black Travel Experience: Social media counternarrative stories as innovative insight on #TravelingWhileBlack. *Journal of Travel Research*, *58*(8), 1357–1372.

Dillette, A., & Benjamin, S. (2021). The Black Travel Movement: A catalyst for social change. *Journal of Travel Research*, *61*, 463–476.

Dixson, A. D., & Rousseau, C. K. (2014). And we are still not saved: Critical race theory in education ten years later. In *Critical race theory in education* (pp. 45–68). Routledge.

Dudley, K., Duffy, L., Terry, W., & Norman, W. (2021). Situational analysis as a critical methodology: mapping the tourism system in post-Katrina New Orleans. *Journal of Sustainable Tourism*, 1–22. https://doi.org/10.1080/09669582.2021.1932930

Fraser, H. (2004). Doing narrative research: Analysing personal stories line by line. *Qualitative Social Work*, 3(2), 179–201. https://doi.org/10.1177/1473325004043383

Gore, D. (2022, 5 May). DeSantis vs. Disney Q&A. *FactCheck.org*. https://www.factcheck.org/2022/05/desantis-vs-disney-qa/

Hale, T. A. (1998). *Griots and Griottes: Masters of words and music*. Indiana University Press.

Harris, A. P. (1994). The jurisprudence of reconstruction. *Columbia Law Review*, 82, 741. https://doi.org/10.2307/3480931

Harris, A. (2021). The G.O.P.'s critical race theory obsession. *The Atlantic*. https://www.theatlantic.com/politics/archive/2021/05/gops-critical-race-theory-fixation-explained/618828/

Heilweil, R. (2020). Why people are posting black squares on Instagram. And why Black Lives Matter organizers want them to stop. *Vox News*. https://www.vox.com/recode/2020/6/2/21278051/instagram-blackout-tuesday-black-lives-matter

Hylton, K. (2012). Talk the talk, walk the walk: Critical race theory in research. *Race Ethnicity and Education*, *15*(1), 23–41. https://doi.org/10.1080/13613324.2012.638862

Jamal, T., & Higham, J. (2021). Justice and ethics: Towards a new platform for tourism and sustainability. *Journal of Sustainable Tourism*, *29*(2–3), 143–157. https://doi.org/10.1080/09669582.2020.1835933

Kivel, B. D., Johnson, C. W., & Scraton, S. (2009). (Re)Theorizing leisure, experience and race. *Journal of Leisure Research*, *41*(4), 473–493. https://doi.org/10.1080/00222216.2009.11950186

Kolivoski, K. M., Weaver, A., & Constance-Huggins, M. (2014). Critical race theory: Opportunities for application in social work practice and policy. *Families in Society: The Journal of Contemporary Social Services*, *95*(4), 269–276. https://doi.org/10.1606/1044-3894.2014.95.36

Kruesi, K. (2021). Tennessee bans teaching critical race theory in schools. *AP news*. https://apnews.com/article/tennessee-racial-injustice-race-and-ethnicity-religion-education-9366bceabf309557811eab645c8dad13

Ladson-Billings, G. (1998). Teaching in dangerous times: Culturally relevant approaches to teacher assessment. *Journal of Negro Education*, 255–267. https://doi.org/10.2307/2668194

Ladson-Billings, G. (2013). Lack of achievement or loss of opportunity. In *Closing the opportunity gap: What America must do to give every child an even chance* (Vol. 11, pp. 11–22).

Ladson-Billings, G., & Donnor, J. K. (2005). Waiting for the call. In *Handbook of critical and indigenous methodologies* (Vol. 61, pp. 279–301).
Ladson-Billings, G., & Tate, W. F. IV. (1995). Toward a critical race theory of education. *Teachers College Record: The Voice of Scholarship in Education, 97*(1), 47–68. https://doi.org/10.1177/016146819509700104
Laughter, J., & Han, K. T. (2019). Critical race theory in teacher education: Coalitions for the future. In K. T. Han & J. Laughter (Eds.), *Critical race theory in teacher education: Informing classroom culture and practice* (p. 1). Teachers College Press.
Leonardo, Z. (2013). *Race frameworks: A multidimensional theory of racism and education*. Teachers College Press.
Lowery, B., Dagevos, J., Chuenpagdee, R., & Vodden, K. (2020). Storytelling for sustainable development in rural communities: An alternative approach. *Sustainable Development, 28*(6), 1813–1826. https://doi.org/10.1002/sd.2124
Milner IV, H. R. (2007). Race, culture, and researcher positionality: Working through dangers seen, unseen, and unforeseen. *Educational Researcher, 36*(7), 388–400. https://doi.org/10.3102/0013189X07309471
Miller, R., Liu, K., & Ball, F. (2020). Critical counter-narrative as transformative methodology for educational equity. *Review of Research in Education, 44*(1), 269–300. https://doi.org/10.3102/0091732X20908501
Morgenstern, E. (2020). The Enslaved Man Who Taught Jack Daniel To Make Whiskey. *The Sporkful*. https://www.sporkful.com/enslaved-man-jack-daniel-whiskey-nearest-green/
Mowatt, M. (2020). Revised notes from a leisure son: Expanding an understanding of White supremacy in leisure. *Annals of Leisure Research, 25*, 291–298.
Olivas, M. (1990). The chronicles, my grandfather's stories, and immigration law: The slave traders chronicle as racial history. *Saint Louis University Law Journal, 34*, 425–441.
Parker, L. (Ed). (2015). CRT and QI: What each has to offer the other now? [Special issue]. *Qualitative Inquiry, 21*(3), 199–205. [Mismatch] https://doi.org/10.1177/1077800414557828
Paulson, K. (2022). Tenn.'s 'divisive concepts' bill: Its impact on campuses. *The Free Speech Center*.https://www.mtsu.edu/first-amendment/post/2793/tenn-s-divisive-concepts-bill-its-impact-on-campuses
Singh, A. A. (2019). *The racial healing handbook: Practical activities to help you challenge privilege, confront systemic racism, and engage in collective healing*. New Harbinger Publications.
Smith, L. T. (2005). On tricky ground. In The Sage handbook of qualitative research (Vol. 3, pp. 85–107).
Solórzano, D. G., & Delgado Bernal, D. (2001). Examining transformational resistance through a critical race and LatCrit theory framework: Chicana and Chicano students in an urban context. *Urban Education, 3*, 308–342.
Solórzano, D., Ceja, M., & Yosso, T. (2002). Critical race theory, racial microaggressions, and campus racial climate: The experiences of African American college students. *Journal of Negro Education, 69*(1/2), 60–73.
Solorzano, D. G., & Yosso, T. J. (2001). Critical race and LatCrit theory and method: Counter-storytelling. *International Journal of Qualitative Studies in Education, 14*(4), 471–495.
Tate IV, W. F. (1997). Chapter 4: Critical race theory and education: History, theory, and implications. *Review of Research in Education, 22*(1), 195–247. https://doi.org/10.3102/0091732X022001195
Toliver, S. R. (2020). Can I get a witness? Speculative fiction as testimony and counterstory. *Journal of Literacy Research, 52*(4), 507–529. https://doi.org/10.1177/1086296X20966362
Toliver, S. R. (2021). *Recovering Black storytelling in qualitative research: Endarkened storywork*. Routledge.
Tuck, E., & Yang, K. W. (2014). Unbecoming claims: Pedagogies of refusal in qualitative research. *Qualitative Inquiry, 20*(6), 811–818. https://doi.org/10.1177/1077800414530265
Veland, S., Scoville-Simonds, M., Gram-Hanssen, I., Schorre, A. K., El Khoury, A., Nordbø, M. J., Lynch, A. H., Hochachka, G., & Bjørkan, M. (2018). Narrative matters for sustainability: The transformative role of storytelling in realizing 1.5C futures. *Current Opinion in Environmental Sustainability, 31*, 41–47. https://doi.org/10.1016/j.cosust.2017.12.005
Williams, P. J. (1991). *The alchemy of race and rights*. Harvard University Press.
Williams, R. (1997). Vampires anonymous and critical race practice. *Michigan Law Review, 95*(4), 741–765. https://doi.org/10.2307/1290045
Worland, J. (2020). America's Long Overdue Awakening to Systemic Racism. *Time*. https://time.com/5851855/systemic-racism-america/
Yosso, T. J. (2002). Toward a critical race curriculum. *Equity & Excellence in Education, 35*(2), 93–107.
Yosso, T. J. (2006). *Critical race counterstories along the chicana/chicano educational pipeline*. Routledge.

The potential of toxic tours: indigenous perspectives on crises, relationships, justice and resurgence in Oklahoma Indian country

Bobbie Chew Bigby, Earl Hatley and Rebecca Jim

ABSTRACT

This article examines Indigenous perspectives on crises, empowerment, justice and resurgence through a case study of Indigenous-led Toxic Tours in the state of Oklahoma, US across both Native and non-Native lands that have been severely polluted. The Tar Creek Superfund Site and Grand Lake watershed comprise an interconnected water and land ecosystem that has been severely contaminated as a result of 20th century lead and zinc mining in the region. This mining legacy has resulted in extreme toxic assaults to water, soil, air, animal and plant life, as well as to human health, cultural lifeways, livelihoods and wellbeing. Through the vehicle of Tar Creek and Grand Lake/Downstream Impacts Toxic Tours led by the community-based organization, LEAD Agency, numerous visitors have been able to witness and engage with these impacted lands, waterscapes, communities, Tribal Nations and non-human living beings over decades. Using an autoethnographic approach to bring forward the perspectives of LEAD co-founders and Toxic Tour guides, Rebecca Jim and Earl Hatley, this research critically explores the ways in which these tours engage with empowerment, justice and resurgence discourses, while centering Indigenous perspectives on the crisis. While the insights and perspectives offered by Rebecca and Earl point to a strong connection between Toxic Tours and their capacity for engaging with empowerment, justice and Indigenous-led resurgence, the analysis of these reflections also point to differences between these discourses and their approach to crises, with resurgence discourses offering unique Indigenous conceptualizations of crisis, priorities and the ways forward.

Introduction

For Native American tribes throughout northeastern Oklahoma, water is sacred and carries a spirit that is powerful. Long before Western, scientific understandings of human physiology explained that our bodies are composed of up to 60% water (USGS, 2019), many Native Nations—along with countless other Indigenous societies around the globe—had understood this deep relationship based on a cultural philosophy that respects water as life-sustaining and deeply interconnected with who we are and how we live our lives. Yet for all of the importance that water holds for sustaining all lifeforms, water sources are under immense and numerous threats from the toxic legacies that humans have left on waterscapes and landscapes across the planet.

From water that has been polluted by heavy metals, unmanaged waste sources, and plastic accumulation to nuclear contamination and the melting of previously frozen water sources, water in all of its manifestations is facing numerous crises and is increasingly becoming unfit for human needs—as well as those of other non-human lifeforms. Furthermore, these myriad water crises are predicted to worsen and multiply with the progression of climate change (Harvey, 2021). Water crises are undoubtedly of great importance to the field of environmental justice that advocates for prevention and better management of environmental harms that arise from economic and development activities. Interwoven into current environmental justice discourses is an emphasis on the social aspects of the environmental context, where humans are centered in considerations of environmental impacts and interventions (Verma, 2018).

In the northeastern corner of Oklahoma, the waterways that cover part of the Cherokee Nation reservation and surrounding nine Tribal Nations, including the Quapaw, Seneca-Cayuga and Wyandotte, have been facing numerous, severe crises for decades. Due to the impacts of lead and zinc mining begun in the early 1900s, the abandoned clean-ups, and the resulting environmental catastrophes left by mining companies, the Tar Creek watershed is saturated with lead and its waters run a rusty red color on the lands of the Quapaw people. But Tar Creek isn't an isolated body of water—its ribbon of rust red contamination, now laced with heavy metals, feeds directly into the Neosho River before draining into the Grand Lake system, one of Oklahoma's largest lakes and tourist attractions—and most importantly, a drinking water source that crosses numerous tribal boundaries. Taken as an interconnected waterway, the toxically assaulted Tar Creek-Grand Lake system is threatened at numerous levels by heavy metal sediment contamination, toxic algae blooms, poultry farm runoff and other pollutants that are spread naturally through the linked watersheds and worsened through recent major flooding in 2019. These hazards extend to all lifeforms and activities connected to the waterways, ranging from tribal harvesting of wild plants for food and fishing to the health of the aquatic life in many areas of the watershed (Garvin et al., 2018).

For several decades now, the community-based organization known as LEAD (Local Environmental Action Demanded) Agency has been at the forefront of advocacy, education and action for these impacted waterways and lands, as well as the communities and Tribal Nations that call them home. This research reflects on the legacy and ongoing efforts of activism for the Tar Creek and Grand Lake watersheds through LEAD Agency's Toxic Tours. Emerging from this foundation of nearly three decades-long Toxic Tourism practice, this paper partners with LEAD Agency with the objective of analyzing the impacts of the Toxic Tours. Utilizing an autoethnographic methodological approach, LEAD co-founders and Toxic Tour leaders, Rebecca Jim (Cherokee Nation) and Earl Hatley (Abenaki Nation), offer their reflections, stories and perspectives on the impacts of these Toxic Tours in understanding crises and conceptualizing crisis response from an Indigenous perspective. Rebecca and Earl reflect on over thirty years of Toxic Tourism experience to address the ways in which these toxic tours intersect with the literature concerning discourses of crisis, empowerment, justice and Indigenous resurgence in relation to tourism. While Rebecca and Earl provide strong links and examples of the ways that Toxic Tours have engaged participants and local community members alike in numerous aspects of relationship building, empowerment, justice and thinking around resurgence, they also highlight points of difference between these processes. Fundamentally, this research points to some of the ways that LEAD Agency's Toxic Tour efforts enable meaningful engagement with crisis, empowerment and justice through tourism, while also highlighting the critical need for Indigenous-led resurgence thinking as an alternative way of conceptualizing and responding to crisis.

Literature review

Toxic tourism as an emerging field

In spite of the fact that across the globe, the number of toxically assaulted places and impacted communities appears to be growing with industrialization, globalization and extractive activities—as do the toxic tours that arise from and inform about this destruction, ranging from tours in the Chernobyl nuclear site to tours of Louisiana's "Cancer Alley"—this specific phenomenon of toxic tourism has not received a great deal of scholarly attention beyond a couple of scholarly voices (Bowers 2013; Di Chiro, 2000; Pezullo 2007; Yankovska & Hannam, 2014). Far less examined are the local perspectives and experiences from the communities most directly impacted by toxic environments, including populations that often identify as Indigenous and people of color. This research thus stands as an attempt to begin filling this notable gap in the scholarly literature. It should be noted that there are some clear parallels between toxic tourism and the field of dark tourism, which is a robust and quickly growing field of literature that similarly analyzes interactions with places of ruin, disaster, destruction and suffering (Lisle, 2007; Sharpley & Stone, 2009; Yankovska & Hannam, 2014). While key overlaps include shared concerns for education, transformative learning, emotional impacts and presenting local perspectives and concerns, dark tourism's prioritized alignment of tourism with concepts of death is problematic for the Indigenous voices at the heart of this paper (Sharpley & Stone, 2009, p. 10). This discomfort stems from the belief that in spite of the massive toxic assaults towards Tar Creek and Grand Lake watersheds, these bodies of water are not dead and cannot be seen as attractions for voyeuristic thrill-seekers; though these waters have been brutally harmed and assaulted, they continue to have agency and spirit and are in need of careful human stewardship. LEAD Agency's toxic tours stand as vehicles that are intended first and foremost to inspire this ethic by educating and sharing Indigenous worldviews informed by goals grounded in culture, resurgence, ethics, justice and healing.

Within the field of toxic tourism, Pezullo (2007)'s writing offers valuable insights to the relevance and power of these tours. Describing "toxic tourism" as "noncommercial expeditions into areas that are polluted by toxins" (2007, p. 5), the author demonstrates that these tours have a history going back to the late 1960s and emphasizes that the sites of toxic tours are generally land and waterscapes that have suffered near irrevocable damage and toxic assault, beyond more commonplace pollution. Through Pezullo's journeys on various toxic tours across the US, she asserts that the importance of these tours is their capacity to generate and communicate a sense of presence, and thus connection, for all actors involved in a tour.

At the most basic level, these toxic tours allow visitors to be present with local people and communities whose homes have become these toxic landscapes. Through local people and knowledgeable toxic tour guides telling stories and experiences, and often sharing personal aspects of the pain, discomfort and changes that the toxic environment has brought to their lives, interpersonal connections begin to be formed. Beyond the interpersonal understanding and sharing, however, is also the potential for connection between visitors and the living lands and waterscapes. As visitors are able to physically experience and be witness to the impacts of the toxic assault—smelling, seeing, and physically engaging with evidence of polluted land and water, as well as plants and animals that may be absent or harmed from contamination—visitors can form connections with the living environment itself, in tandem with the local communities embedded within the polluted landscape. Through Pezullo's eyes, these layered levels of connections that are embedded within toxic tours help those involved to, "challenge feelings of alienation from the land and each other" (2007, p. 10). The view that disconnection between humans and the living world is at the root of industrialization, modern living and toxically harmed landscapes can be seen to be articulated not only by Pezullo, but also shared by Indigenous perspectives such as those of LaDuke and Shade to come in later sections, as well as. Ultimately, Pezzullo argues that toxic tours can be utilized by local communities as tactics of agency, negotiations of power, and as "embodied rhetorics of resistance aimed at mobilizing public sentiment and dissent against material and symbolic toxic patterns" (2007, pp. 3–8).

Toxic tours provide a steppingstone to empowerment and the search for justice

Building off of the link that Pezullo has demonstrated between Toxic Tours as tools for connection, agency and resistance in the midst of crisis, the discourse of tourism and empowerment immediately becomes relevant. In addressing the specific intersection of tourism and empowerment, the theoretical models proposed by Scheyvens have proven relevant and adaptable to changing circumstances. Scheyvens' original 1999 conceptualization of understanding empowerment through tourism focused on four key pillars, including the psychological, social, economic and political dimensions of empowerment for individuals and communities engaged tourism experiences. However, with application of the model to tourism case studies and critiques of her framework over the years, Scheyvens has recently updated this empowerment framework (Scheyvens & van der Watt, 2021). This new model expands the dimensions of empowerment to also include cultural and environmental pillars, leading to a total of six empowerment dimensions. Additionally, the framework centers all six dimensions of empowerment within the contexts of broader enabling conditions, such as governance and customary practices, as well as within the natural environment and diverse ecosystems. This embedding of empowerment dimensions within wider socio-cultural and environmental forces is a significant shift to recognize not only how deeply the natural and cultural contexts inform the tourism-empowerment nexus, but how humans, cultural contexts, environments and tourism are all interlinked. In describing this more holistic approach to understanding empowerment and its link to tourism, the authors write, "empowerment is multi-dimensional, context-specific, both process and outcome-related, and demands redistribution of power to marginalized groups" (2021, p. 2)—a statement that aligns this discourse on empowerment closely with the study of toxic tourism.

In confronting the myriad ecological, political and cultural crises much of the world currently faces and the ways that these challenges and visions of recovery intersect specifically with tourism, consideration of justice discourses is also of extreme importance (Guia, 2021; Jamal, 2019; Jamal & Higham, 2021). As asserted by Rastegar et al., "rethinking tourism necessitates promoting justice and equity at the local level" (2021, p. 2); yet what is equally necessary is critical, robust understandings and approaches to justice in tourism (Jamal & Higham, 2021). Following this vein of thinking, Rastegar et al. identify multiple aspects of justice that fall along a continuum, informing one another. These principles include recognition, procedural, distributive and restorative approaches to justice. Understanding these aspects in relationship to each other, one begins by considering the needs and rights of vulnerable communities and environments, followed by how local rights are included in decision-making processes and in what ways potential benefits and costs are distributed. Based upon these considerations of rights and distribution of costs and benefits, recovery-oriented actions can be put into place; most importantly, these actions must be tailored to the needs of local people, living ecosystems and the relationships that they choose to have with tourism activities.

Seeing beyond crisis: Indigenous perspectives and Indigenous resurgence

In the face of such crises that are the heart of toxic tours, as well as the discourses of empowerment and justice used to guide and conceptualize crisis responses, a turn towards Indigenous viewpoints, traditional knowledges, stories and the notion of resurgence offer critical perspectives for understanding and approaching these crises in new ways. While Indigenous perspectives should never be generalized and oversimplified in their conceptualizations and values, many Indigenous thought systems are fundamentally grounded in a way of seeing the world that deeply prioritizes relationships and responsibilities of stewardship between humans and the non-human living world upon which we depend for all life.

For decades, Winona La Duke (White Earth Anishnaabe) has been an outspoken advocate and frontline activist across numerous issues facing Indigenous peoples and lands. Her book, *All Our Relations: Native Struggles for Land and Life* (2016), is an important entry point in approaching the philosophy shared by many Native Nations that lands, waters, plants and animal lifeforms are considered as relations and teachers to humans. La Duke shares that from her own Anishnaabe worldview, non-human lifeforms are regarded as "our older relatives—the ones who came before [us humans] and taught us how to live" (LaDuke, 2016, p. 18). In acknowledging the agency that these lifeforms possess, along with gratitude for the lessons that these non-human relatives teach us about being in relationship with one another, humans are understood to carry the responsibility of serving as stewards, caretakers and protectors of the environments they inhabit and to which they are related.

Yet in this modern age, LaDuke asserted that these relationships between humans and the living world are the very ones that industrialization and modern development seek to dismantle, whether for purposes of profit, power or both. Clear lines are thus drawn between industrial and mining activities, environmental degradation, pollution and the subsequent loss of both bio and cultural diversity across the globe (2016, p. 22). A significant Cherokee perspective on these relationships comes through from the voice of former Cherokee leader and traditionalist, Hastings Shade (*jigesv*)[1]. In Shade's book entitled *Myths, Legends and Old Sayings* (1994), he addressed what Cherokee Elders in Oklahoma have long been saying about the breakdown in relationships between humans and the living environment. Specifically implicated in his reflection entitled, "What the Elders Say about Nowadays" are the relationships that humans have with machines, metals, mining, diseases and the ways that these relationships have superseded our connections with water and Mother Earth. Shade wrote,

> "We have become dependent on machines—everything we think about doing is metal…
>
> And to make this metal we have stripped Mother Earth—
>
> And dug holes in her.
>
> This is our Mother—
>
> This is why we are paying for the shape we are in.
>
> We are sick with many diseases—
>
> The air is no longer good—
>
> The water is bad—
>
> The trees and animals and plants are dying—
>
> All these things and more are happening because of our disrespect for our Mother Earth and each other…
>
> Sing the songs of our Mother Earth and she will be young again. (1994, pp. 68–69).

Shade's reflection offers an important lens on the roots of the crisis that are not grounded in Western, scientific concepts, but come from a Cherokee way of seeing and being. Shade's thoughts also offer hope that in changing our current ways, these situations can be transformed and relationships and wellbeing can return and resurge.

While concepts such as empowerment, justice, and the rhetoric of rights have been commonly used in the academic discourse and mainstream society to address the crises and uncertainties inherent in the present day, the idea of "resurgence" is one that is increasingly embraced by Indigenous activists, leaders and academics. Resurgence discourses prioritize Indigenous renewal of relationships with land, water, community, cultural and language practices, along with the overarching and interconnected goal of Native Nationhood (Corntassel, 2021, p.2). Emerging from and influenced by the decolonial and Indigenist works of Frantz Fannon, Vine Deloria, Jr. (Standing Rock Sioux), Glen Coulthard (Yellowknives Dene), Gerald Taiaiake Alfred (Kahnawake Mohawk), Jeff Corntassel (Cherokee Nation), Leanne Betasamosake Simpson (Michi Saagiig Nishnaabeg), among others, thinking towards "resurgence" is grounded more fully in Native values, perspectives and worldviews, as opposed to those rooted in Western philosophical and cultural norms. Born from traditional Indigenous viewpoints, languages and concepts, resurgence discourses expand upon, challenge and present alternatives to Western concepts that have usually addressed crises, including notions of rights and justice.

According to Jeff Corntassel, some of the foundational concepts upon which resurgence is built center on strengthening of land and water-based governance, the practice of relational responsibilities and everyday practices of renewal and regeneration, among others (2021, pp. 2–3). Inherent in the notion of resurgence, as opposed to other predominantly Western discourses that emphasize notions of rights and the discrete individual, is the inseparability between humans and the living environment. This viewpoint entails that contaminated water, soils, air and other living beings must be understood as interconnected with and inseparable from humans and communities facing challenges, whether at the level of health or social wellbeing. Elevating the understanding of land, water and non-human beings as relatives with agency and those forces with whom we are in constant relationship, as opposed to an approach that further entrenches dichotomies between humans and the non-human world, is at the root of resurgence discourses. Simpson states, "Indigenous worlds aren't institutions or states, they are relationships, movements, processes—life itself… the theories or stories or philosophies of resurgence inherent in Indigenous thought were the ways my ancestors had always lived" (2016, p. 28). Therefore, imbedded and implied within the notion of Indigenous resurgence is a focus on environmental resurgence, regeneration, renewal and healing that is inextricably tied to the resurgence and healing of humans, as well as animal and plant nations.

Beyond encouraging a nuanced, critical re-thinking and critique of rights-based discourses (Corntassel, 2021, p. 76), resurgence ideas also present alternatives to the notion of justice itself. Simpson explains the need for this re-analysis in the context of critiquing notions of criminal justice by stating that, "I don't think about justice very much. I think about resurgence and movement building… State narratives of justice are processes that are about injustice for us—a system that steals our people and locks them up so the state can access our land for the exploitation of natural resources" (2016, p. 21). By embracing what Simpson terms as, "Indigenous intelligence systems," other possibilities emerge beyond rights and justice-based discourses and practices. In Glen Coulthard's 2014 book, *Red Skin, White Masks: Rejecting the Colonial Politics of Recognition*, he utilizes the term "grounded normativity" to offer a new approach to ethical human relationships and processes with the living worlds around us. Coulthard describes this "grounded normativity" as, "the systems of ethics that are continuously generated by a relationship with a particular place, with land, through the Indigenous processes and knowledges that make up Indigenous life" (2014, p. 60). From this reference point, then, ethical relationships and actions are not centered and adjudicated in the domain of the state, but are grounded in our ways of being and the actions that we take individually and collectively in relating to the living lands and waters around us.

Resurgence can also be seen as closely tied to discourses of resistance, embodied in the legacy of actionable Indigenous resistance that has been maintained over centuries. As argued by Nick Estes (Lower Brule Sioux) (2019), Indigenous peoples in North America have been at the forefront of resistance to the social and environmental harms perpetrated by the settler-colonial state from the very beginning of colonization and have continued undaunted into the present day. Yet resurgence goes beyond mere resistance, offering more and serving as an antidote to movements focused only on resistance as opposition. In its place, resurgence brings forward a more holistic outlook that prioritizes an understanding of human relationships and responsibilities to non-human relatives that make up our living environments. The actions that thus emanate from this resurgence approach are thus focused not only on resistance to assault, pollution and crisis, but also on re-establishing relationships with land and water through remediation and activities centered around renewal, relationships and understanding, including potentially through the vehicle of tourism.

Methodology

The fact that the Tar Creek and Grand Lake watershed systems are located within tribal jurisdictional lands and are intimately connected to the histories of forced removal, allotment and disconnection from land and water that all of the tribes have faced is essential to be acknowledged from the beginning. Furthermore, highlighting that the leadership, involvement and advocacy of LEAD Agency and local community members is composed in large part by members of the ten impacted Tribal Nations is also of deep importance. Thus, these crises facing Tar Creek and Grand Lake, along with the resulting Toxic Tours as tools for engaging crisis and envisioning alternatives, are inextricably interwoven with Native peoples and their connections to land, water and cultural heritages. Based on these deep ties, this research prioritizes Indigenous knowledges, perspectives, histories and approaches as core not only to the research content, but its underlying methodology and approach. By assuming a critical Indigenist methodological approach that "begins with the concerns of Indigenous people" (Denzin et al., 2008, p. 2), as opposed to a focus on damage and ruin (Tuck, 2009), this research aims to imbed and elevate Indigenous values, experiences and visions into all facets of the research journey. The vital importance of these Indigenous perspectives and voices not only to future directions of tourism, but to wider discourses of sustainability and justice, along with their alternatives including resurgence, is made clear in the writing of Boluk et al. (2019). According to these authors, "Indigenous approaches may also offer a positive alternative to the unsustainable trajectories of modernist development models… these [Indigenous] worldviews represent an alternative and a challenge to the myopic, instrumental and exploitative approach of globalizing capitalism" (2019, pp. 7–8).

This research is centered upon decades of work, advocacy and commitment of the LEAD Agency founders, Rebecca and Earl, alongside organization members and academic partners that continues into the present day. This writing thus also takes a collaborative and engaged approach that recognizes, honors and builds on the relationships that LEAD Agency has cultivated over the years, including with local communities, Tribal Nations, government agencies and numerous schools. Further, these relationships encompass connections between people, water, land and other non-human life forms in understanding the harms that have happened and the ways the forward. The primary author, Bobbie, also a Cherokee Nation citizen of Oklahoma, has engaged in relationship building, conversations, interviews and collaborative projects with LEAD co-founders and local community and tribal members over years.

Built upon these long-standing relationships, the study presented is an autoethnographic case study featuring LEAD co-founders and Toxic Tour leaders at its center, Rebecca and Earl. Autoethnography is a research approach that prioritizes personal experiences, as well as the narratives, perspectives and reflections that have been generated as a result from these experiences (Coghlan, 2012). This autoethnographic approach opens up the possibilities for deeply analyzing not only subjectivities and emotions on a personal level, but also personal experiences of social realities that point to broader social and cultural phenomena, including within the tourism context (Ellingson & Ellis, 2008; Ellis et al., 2011; Mair & Frew, 2018). More specifically, this study adopts an Indigenous and collaborative autoethnographic approach (Lapadat, 2017) that is suited to both Rebecca and Earl given their deep connections to Indigenous communities, heritages, stories, tribal administration and advocacy over decades, along with their partnership in work and in life. This Indigenous autoethnographic approach has been adopted and engaged with by an increasing number of Indigenous scholars as a more culturally-grounded method for exploring knowledge, self, relationships and complex phenomena in a way that pushes back on and offers alternatives to strictly Western forms of inquiry (Bishop, 2021; Whitinui, 2014).

Background: Tribal Nations and crisis in the Tar Creek and Grand Lake watersheds

The story of crisis impacting the waterways and land of far Northeastern Oklahoma begin in the Tri-State Mining Region of the US, where the states of Oklahoma, Kansas and Missouri converge. Throughout the 1800s, what is today Oklahoma was land designated as "Indian Territory" by the United States government. Indian Territory served as a place of relocation for tribes from throughout the country who were forcibly removed due to the increased expansion of Anglo-American settlers and subsequent political pressure through legislation such as the Indian Removal Act of 1830 (Clark, 2012). Of the 39 different Tribal Nations that call Oklahoma home today, 10 different tribes maintain a presence in this area of far northeastern Oklahoma. All of the tribes are impacted by the legacy of mining and subsequent environmental harm, but it is the Quapaw Nation that is connected most closely to Tar Creek, while the Cherokee, Seneca-Cayuga and Wyandotte Nations border Grand Lake.

Originally claiming their traditional homelands in what is now the state of Arkansas, the Quapaw, or O-gah-pah, are known as the "Downstream People." In 1834, the Quapaw were forcibly removed to reservation land designated for them in Indian Territory. They were joined by numerous other tribes, including the neighboring Cherokee, Seneca-Cayuga and Wyandotte Nations. Cherokee people called the southeastern US states of Tennessee, the Carolinas and Georgia their original homelands and were forcibly marched to Indian Territory over the Trail of Tears from 1838 to 1839. The Seneca-Cayuga and Wyandotte Nations, both Iroquoian-speaking groups hailing from around the Great Lakes region, were removed and put onto small reservations carved out of the Cherokee Nation's northeastern corner. After several decades of adjustment and rebuilding amidst continuous upheaval, the late 1880s witnessed the passage of the Dawes Act, entailing the forced allotment of tribal lands throughout Indian Territory by assigning parcels of land to individual tribal members as opposed to preserving communal land holdings.

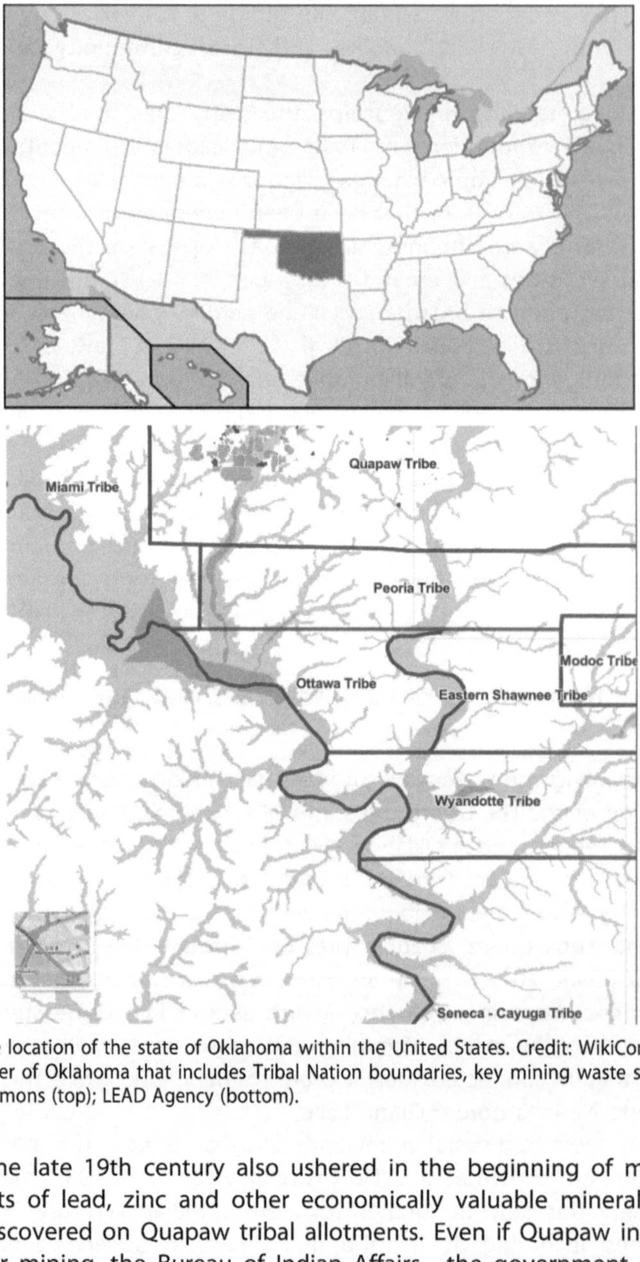

(Top) A map showing the location of the state of Oklahoma within the United States. Credit: WikiCommons. (Below) A map of the far northeastern corner of Oklahoma that includes Tribal Nation boundaries, key mining waste sites, as well as watershed systems. Credit: Wiki commons (top); LEAD Agency (bottom).

This period of the late 19th century also ushered in the beginning of mining operations in the area as deposits of lead, zinc and other economically valuable minerals were found, with much of the ore discovered on Quapaw tribal allotments. Even if Quapaw individuals refused to lease their lands for mining, the Bureau of Indian Affairs—the government entity charged with managing allotments and trust lands for tribal peoples—ruled individual tribal members incompetent in order for mining to proceed (Aber et al., 2010; Johnson, 2009; Myers et al., 2009). The industry began to hum at a rapid pace by 1914 (Adcock, 2018) and quickly became an essential supplier of lead and zinc for the US government's war efforts throughout World Wars I and II. Between 1885 and 1970, this region is credited for having produced nearly 460 million tons of ore, or one-third of the raw materials used during US war campaigns (Manders & Aber, 2014). With all of the ore taken out of the ground, more than 300 miles of underground mine cavities and tunnels were left by the time the mines closed in the 1970s (Adcock, 2018). Once the ore was brought above ground, the materials were highly processed to extract the minerals, leaving behind gigantic piles of rock fragments and mine tailings, known locally as "chat piles." When the mining companies finally left town in the 1970s, no plans or efforts were made for removal or cleanup of the nearly 500 million tons of chat left behind. Thus, over 83 chat piles 63 fine tailing ponds continue to dot the 47 square mile landscape; the problems of chat pile toxicity are compounded by the dangers of severe subsidence as the ground is literally collapsing in on itself and swallowing buildings from being so extensively undermined (Adcock, 2018). This subsidence crisis would be the eventual impetus for the government-sponsored buyout of the area and the relocation of families from Picher and Cardin communities to safer areas of land.

A "chat pile" or tailings pile that is next to Tar Creek. Chat piles are full of toxic metals that pose serious health risks. Credit: Clifton Adcock.

Tar Creek after its convergence with water from the mine shaft and Lytle Creek. Tar Creek flows a rusty orange-red and the banks are lined with chat deposits. Credit: Clifton Adcock.

For years prior to the closure of the mines, communities throughout the Tri-State Mining District were known to have higher rates of sickness compared with neighbouring populations, particularly with tuberculosis and silicosis (Manders & Aber, 2014, pp. 29, 37, 44). But the enormous magnitude of the impacts of soil and air contamination on health was finally revealed in 1991 when blood tests of local children came back with high rates of lead in the blood. High lead exposure for any human can come with health consequences, ranging from elevated cancer rates to anemia, but for children lead exposure can also permanently impact the nervous system and cognitive function (Karkowsi et al., 2014; Neuberger et al., 2009).

The true toll of these extractive industries on the local waterways came to the attention of the wider public in the 1970s when the mining companies moved out and the pumps that removed water from the mine cavities were completely turned off. In 1979, highly contaminated underground water from the mines overflowed and spilled into Tar Creek, turning the water a distinctive, rusty orange-red colour and killing off fish and other lifeforms in the creek (Jim & Scott, 2007). Tar Creek feeds into the Neosho River before eventually emptying into Grand Lake to the south, a key drinking water source and tourism spot in northeastern Oklahoma. This poisoning of Tar Creek is what initially caused the area to gain the attention of the Environmental Protection Agency (EPA) and be listed in 1983 as number one on the Agency's Superfund Site National Priority List according to its Hazard Ranking Model (Manders & Aber, 2014, p. 46; Myers et al., 2009). According to the EPA, a Superfund site is a contaminated site that is eligible for federal funding, administered through the EPA, to clean and de-contaminate the site, particularly when "no viable responsible party" can be located ("What is Superfund?", n.d.). This listing on the National Priority List is a distinction that Tar Creek continues to hold into the present.

In the face of this web of crises impacting water, land and both human and non-human bodies and wellbeing, LEAD Agency has stood at the forefront of telling this story of crisis while continuing the ongoing, protracted fights for remediation. Rebecca and Earl established LEAD in 1997 with the goals of raising awareness and taking action on environmental degradation in Tar Creek Region and the northeastern corner of Oklahoma. For decades now, LEAD has been at the heart of supporting the impacted communities and landscape of the Tar Creek Superfund Site, along with the surrounding and interconnected waterways and communities, with efforts ranging from delivering donations of water filters to families to providing scholarships for local students. LEAD has served as the key advocate for Tar Creek and Grand Lake watersheds in dealing with the government agencies at the levels of local, state, federal and Tribal Nations. In particular, LEAD has been key in continuing to keep pressure on government agencies and representatives to keep the Superfund Site program alive and accessible. LEAD Agency has been responsible for initiating, planning and partnering with experts involved in numerous studies and research projects aimed at deepening understanding of the massive challenges facing this area at the level of environmental remediation, public health, impacts on cultural lifeways and more. On the front of education, LEAD has hosted annual conferences for the past 23 years, while continually engaging community members and visitors alike in understanding the story of Tar Creek and other emerging and interrelated Oklahoma environmental hazards, including the downstream impacts on Grand Lake, recent extreme flooding and toxic drainage, poultry farm runoff, coal ash, asbestos and fracking, among many others.

At the center of LEAD's efforts to educate the wider community about these stories of crisis are their Toxic Tours. Rebecca began these Toxic Tours in 1995 when she first took Earl on a guided drive through the 47 square miles comprising the Tar Creek Superfund Site to show the scale of the devastation. In subsequent years, countless student groups, Sierra Club members, community locals, government officials, Tribal Nations, university researchers and visitors from around the world would pile into school buses or ride bicycles that traced a fairly similar route over this damaged landscape. Rebecca estimates that Tar Creek Toxic Tours have reached nearly 5,000 people. In considering that for the past two years of the COVID-19 pandemic, LEAD Agency has taken their conference and the Tar Creek Toxic Tour to an online format, this has allowed well beyond 5,000 people to experience a toxic tour through a digital space.

Results: an autoethnographic dialogue with LEAD Agency's leaders

Asked to reflect back on the three decades of Toxic Tours that Rebecca and Earl have led across both the Tar Creek and Grand Lake watershed systems, both Rebecca and Earl shared their perspectives, feelings and memories on what these Toxic Tours are capable of bringing about and how they intersect with notions of empowerment and Indigenous resurgence. The lead author,

Bobbie, leads the discussion with Rebecca and Earl and offers questions to prompt some of their reflections and thoughts.

Visitors from the Sierra Club onboard a bus and observing the Tar Creek Superfund Site during a toxic tour in 2018. Credit: Clifton Adcock.

Bobbie: ***In your experience, what is the purpose(s) of the Toxic Tours you have led?***

Rebecca: Each tour depends on how much time people have and what their backgrounds and interests are. So for us, it's been a dance... we try to make each tour relevant to who is on the tour.

From my viewpoint, there are two different moments that are critical: there is the experience on the tour, and then after the tour. On the tour, what I see over and over again is just astonishment that something like this, like what we see at Tar Creek, could happen in America. Then you start breaking the tour up given the group's particular set of interests. So if there are journalists, they see it in a way to be able to tell it to others. If I have teachers, it depends on what they teach. If it's science teachers, they're looking at the science and they look at the tours as what it propels their students to learn.

Earl: I was hired by the Quapaw Tribe and became their first environmental director, tasked with setting up their environmental program from scratch. It was in this role that I really advocated for the Quapaws, demanding that the EPA give the tribe a Superfund program for water, air, and all of that... I did that job while we were getting LEAD Agency up and running. So Rebecca and I led tours to show and educate people about the situation both at the Tar Creek site, as well as downstream. I started the Downstream Tours because when I was the Quapaw Tribe's Environmental Director, I was often taking people around the reservation area and downstream. Working with the other surrounding tribes, I started the Trustee Council for national resources, as well as a Council for dealing with off-site contamination. I got all the tribes, the three states and the Fish and Wildlife Service together to start that council. So you had all the downstream tribes in a council together and I was taking people downstream showing them the wetlands, all the concerns down to the lake. I started those downstream tours so that people could see from the site all the way down through Grand Lake and to the dam, what all the tribal concerns are and the extent of the problems in the area.

Bobbie: ***What do you see as the relationship between Toxic Tours and empowerment? Do you see Toxic Tours as capable of bringing about empowerment in relation to crisis?***

Rebecca: Empowerment is different for every single person and how each individual defines it.

When you think about empowerment, I have seen that experiencing this area on a Toxic Tour often always overtakes people in the need to do action. And action can be educating others. People feel that we have to immediately talk to politicians. We've got to tell EPA. There are many of these action opportunities that people feel motivated towards. And whether they are taken further, I don't know. But in those moments people are empowered in knowing that there is a wrong that needs righting and perhaps they will do their part. Are the actions long-lasting? You'd have to follow each one and see. But at the moment, in my experience these tours take people to a level of engagement that you might not see if you're just doing a tour of Route 66 that intersects the Superfund Site. In fact, the town of Quapaw marks the beginning of Route 66 in Oklahoma.

I was able to give the Picher school kids a tour. That was hard. To explain the dangers they lived in every day. But they wrote reflections afterward that I still have kept.... After the Toxic Tour, those students were many things, but they were empowered and were totally enlightened. Because no other adult had shown them and talked to them about the issues of Tar Creek that were in the newspaper. That tour was before the buyout and before there was clarity about whether there would be a buyout or not. After the tour, we asked the whole classroom of students at that school, as well as students from the other schools in the area, to write about how they felt about the buyout. Then we selected pros and cons, some of each, and invited those students to read their essays and opinions to the Governor, who came here to hear them. So yeah, I think we empowered them.

On the Toxic Tours, we always include the stories of particular local people, their wishes and hopes... On any tour we try to point out anyone who is doing a raised garden bed somewhere, or yards that we know have been remediated. The map that Department of Environmental Quality (DEQ) has, you can see how many pink dots there are where homes or yards have been remediated in the county. And then how many rain barrels you see, indicating who is trying to slow the water down in a town that has toxic floods. I do try to link back to those actions and choices of local empowerment, such as the choices around the kind of fish people are eating and how much lead is in certain fish. Most importantly, I emphasize that you can be in charge of how you're exposed. We distribute fish advisory booklets created by the DEQ to local residents that contain facts on health and fish consumption given the contamination in the area. Putting focus on what you can do and what you're in charge to maintain your own body's health is a form of empowerment.

Earl: Toxic Tours help to bring about empowerment in many different ways. I see the most important part of the Toxic Tour within the Tar Creek Superfund site is that the tour demonstrates just how immense and bad the level of devastation is... words can't describe it, even pictures cannot describe the enormity and the impact of it. That's what I hear from everyone that I take on a tour... What they feel in that moment is the impact of the tour... you have to see it to understand it. Words and pictures can't bring about that feeling you get when you're standing there and seeing just how awful it is and the fact that people have to live there. And then seeing for yourself how it runs through the towns of Commerce, Miami, into the Neosho River, and that it is all running into Grand Lake. There is a fish consumption advisory and we drink from that lake. When you see all that, you realize this isn't right. Following the water down, hearing from the tribes and knowing that every time it floods and all that washes into those floodplains, where folks are trying to harvest the wild carrots, the strawberries, the asparagus and nuts from there, and then the animals that consume those as well, and you realize that we've already sampled all that and it's all so astronomically high that you shouldn't eat any of it. Seeing and hearing and knowing all this, then it's hard for everyone on a tour who is standing in that spot. Whether you're going downstream or you're at the top of the lake and you're still hearing these stories, seeing where tribal people are trying to harvest, learning the stories of generations of tribal members who have been impacted—you realize that this is really bad.

The experience of the tour leads people to do all kinds of things, to get involved in many ways, whether it's donating funds to LEAD or writing letters to agency people or politicians—and some people have written poetry and songs that LEAD has used over the years. They've made paintings and all kinds of art as a result of it, their creativity has come out from it and they've expressed that in all kinds of ways. Reporters from magazines and newspapers have written about it and continue to write about it. Agency people can't deny it when they see it. It's inspiring and empowering in this way—that when people see the sites, they want to do something in their own way.

There's a big list of researchers, collaborators and projects that have come about as a result of participating in a Toxic Tour. Dr. Neil Kirkwood of Harvard University came out several years back, took one Toxic Tour of Tar Creek and hasn't been able to let Tar Creek go since. He's written extensively around phytoremediation and Tar Creek and now devotes one class a year to the topic of remediation and re-designing Tar Creek. There are other researchers from Harvard, Johns Hopkins, as well as Dr. Robert Nairn of the University of Oklahoma who has done a project on ecosystem restoration and wants to plan for another. So many research scientists that we were able to rope in and give a tour to are now regulars at our annual conference. The list goes on and on.

Toxic Tour visitors gazing at the enormous scale of an eroded chat pile. Credit: Clifton Adcock.

Bobbie: ***What about the relationships between Toxic Tours, justice and Indigenous resurgence? Are there any tensions or differences between empowerment, justice and Indigenous resurgence from what you have experienced?***

Rebecca: This Toxic Tour is about justice, wanting it, a place needing justice and a place that has waited long enough to get it.

At LEAD Agency, we'd like to encourage people to get back to our culture, to actively be using the plants, water and soil, but that cultural resurgence, that's been harder. As we learn more on the extent of the areas that are contaminated and not safe, the tours help us see just how many places are risky to engage with. So bringing that back has been delayed because of finding out, well, the plants take up the metals. And then if you're using the plants to make baskets, mats… these remnants of the debris of working with these plants after they've dried, do you let your children play with that? want to face back to our culture, but where is it safe enough to do that? And without guidance on, we know it's bad here, but where is it not? Where can we gather the berries and wild onions? So resurgence could be happening, but can it be done safely? What I want is real guidance on where is far enough away to be safe for cultural activities. How far back does the floodplain go? How does anyone know where the last deposits of metals reach? What else might be sucking it up? So the resurgence has been delayed, but it can come. All of those things can happen.

Earl: The best example of resurgence I can think of that has come about from a Toxic Tour happened a couple of years back during an Indigenous Caucus meeting of the Western Mining Action Network (WMAN), an organization to which LEAD Agency has belonged to for years now. In 2019, I brought the WMAN Indigenous Caucus to Tar Creek and we were hosted at the

Quapaw Casino/Resort. All 25 of those Indigenous leaders from Alaska, Canada and the continental US, they toured Tar Creek and then went all the way downstream, meeting with tribal leaders and communities… and they can't stop talking about that tour until this day. Inspired and moved by what was seen on the Toxic Tour, what we did at the Indigenous Caucus meeting in Quapaw was we wrote an Indigenous Declaration to WMAN over two and a half days. I'm so proud of it… It emphasizes water and culture as fundamental.

Some of the key assertions in the WMAN Declaration include:

Water and culture as fundamental dimensions of our lives
Full exercise of our human and collective rights
Strengthening diverse local economies and territorial stewardship

Indigenous peoples continue to call upon the world to return to dialogue and harmony with Mother Earth, and to adopt a new paradigm of civilization based on Indigenous legal traditions for the greater good of all. In the spirit of humanity and our collective survival, dignity, and well-being, we respectfully offer our cultural worldviews as an important foundation to collectively renew our relationships with each other and Mother Earth and to ensure we proceed with integrity.

Discussion

Toxic Tours as a doorway to connection, relationships and understanding in the midst of crisis

At their most basic level, LEAD's Toxic Tours of Tar Creek and Grand Lake watersheds tell and show stories of complex crises for people, Tribal Nations, health, environments and non-human living beings that are all interrelated with one another. These Toxic Tours are built upon complex layers of connections between people and their environments—whether it is LEAD Agency's long-standing advocacy for the communities and waterways, or partnership with the local people of the area who often accompany the tours to tell stories, point out sink holes and former fishing areas or help to drive the bus of visitors. These local people accompanying Rebecca and Earl on tours are often former residents of the now uninhabited mining towns or current residents of surrounding communities; they often are members of the local Tribal Nations, married to a tribal citizen or are a descendent of a miner.

Beyond the relationships upon which every tour is built and operates, the tours themselves enable and foster experiences of connection and relationship building for visitors to the local people and the living environments that are grappling with major crises. Bearing witness to this toxically-assaulted land and waterscape and the damage it has wrought on the lives, health and wellbeing of people and non-human beings alike can trigger emotions that are difficult, disturbing, meaningful and unforgettable for visitors. As Rebecca states, one of the most common reactions of visitors on the Toxic Tour is utter shock and astonishment that a crisis like this could happen in America. In a similar vein, Earl continually hears from the tours he leads about the importance of being present on a Toxic Tour in grasping the extent of the crisis and all of the implications this environmental destruction has brought about. He shares that so many of the visitors he has led on the Toxic Tours state that photos or writing alone are no substitute for the experience of being present; that it is only through engaging in person with place and people through a tour that a real understanding of the magnitude and depth of devastation begins to form. It is through this presence and engagement that visitors stand to develop a greater awareness, understanding and appreciation of the people and environments that are in the midst of crisis. Thus, these important connections can be seen as characteristic of how Pezullo describes the outcomes of many Toxic Tours that ultimately, "challenge feelings of alienation from the land and each other" (2007, p. 10) for visitors.

Furthermore, the reflections from Rebecca and Earl about the ways in which the tour participants are moved from understanding the situation to being inspired towards action is also in line with Pezullo's characterization of Toxic Tours as "rhetorics of resistance aimed at mobilizing public sentiment and dissent" (2007, pp. 3–8). The capacity of a Toxic Tour for harnessing the power of presence is thus not only towards the ends of generating connections, relationships and understanding among visitors and their surroundings; Toxic Tours, as theorized by Pezullo and echoed by the experiences of Rebecca and Earl, are vehicles for inspiring actionable responses of visitors to the crisis at hand. While Rebecca states openly that not all actions are necessarily sustained or long-lasting, she asserts that each one has to be looked at individually and tracked for the positive impacts it brings about. From this lens of generating understanding and action through Toxic Tours, the analysis easily transitions to an intersection with related discourses on empowerment, justice and tourism.

Toxic Tours in relation to empowerment and justice

Emerging from the connections, understanding and actions that these experiences enable and inspire for tour participants, numerous themes and processes of the Toxic Tours align directly with the intersections of tourism and empowerment that Scheyvens and van der Watt have outlined in their empowerment framework (2021), and importantly, contextualize the tourism-empowerment intersection within the environment of crisis. Moreover, several examples from the Toxic Tours that Rebecca and Earl have led over decades would appear to resonate with all six pillars of the Scheyvens and van der Watt model simultaneously, including a focus on psychological, social, economic, political, cultural and environmental dimensions of empowerment all at once. Given the extent and complexity of the crises at the Tar Creek and Grand Lake watersheds for the human and non-human inhabitants alike and the ways that this Superfund area has impacted all aspects of life in terms of Tribal Nationhood, cultural lifeways, mental and physical health, governance decisions and employment, among other factors, it is clear to see how Toxic Tours can engage simultaneously with diverse issues and forms of empowerment beyond only themes related to the environment.

Yet in any discussion of the potential of Toxic Tours for empowerment, it is essential to first clarify who or what is being empowered through the process of tourism. In one of the stories shared by Rebecca of the tour she led for the students of Picher, the community at the center of the Superfund Site, Rebecca reflects that these local students were enlightened in understanding the toxic assault that had happened to their home community and why it was causing so many problems for everyone living there. This knowledge helped lead to their empowerment and subsequent actions taken. She shares, "That was hard. To explain the dangers they lived in every day… after the Toxic Tour, those students were many things, but they were empowered and totally enlightened." This gift of understanding was very valuable not only because this had been the first time that an adult explained to these school kids what the problems were, but because the knowledge from this tour experience prompted the students to write their opinions and share them with adults in positions of political power, including the Oklahoma Governor who visited them. In sharing their perspectives and understandings as people impacted directly by the crisis engulfing their community, it was the local school students of the Picher community who could experience social and political empowerment that initiated from participation in a Toxic Tour.

This engagement with simultaneous and multifaceted forms of empowerment for local people can come about through other ways during a Toxic Tour that can sometimes be more indirect and through changing the perceptions of visitors toward local people. In one of her reflections, Rebecca emphasized that "including the stories of local people, their wishes and hopes…," during Toxic Tours, including pointing out places that have been remediated, cleaned up and have garden beds, have been a central theme of every single Toxic Tour. For the local community members and Tribal Nations who have been embedded in this painful legacy of toxic contaminants and generations of dispossession, this process of sharing stories with visitors, as well as sites of toxic assault and clean up, allows local people to become more humanized agents, as opposed to passive victims or totally overlooked communities, in the eyes of visitors. This sharing of Indigenous historical and cultural narratives in Toxic Tours resonates strongly with Scheyvens and van der Watt's argument that cultural empowerment is, "central to thinking about empowerment... based on commitment to supporting the rights and wellbeing of Indigenous cultures," (2021, p. 10). Thus, helping to inform and change and perceptions of visitors is one of many steps towards empowering local people, while also empowering the visitors with the understanding of the crisis context and the motivation to do something about it.

Similarly Rebecca emphasizes the importance of sharing with visitors and local people alike during toxic tours information that has been compiled by the Oklahoma DEQ in booklets regarding fish consumption advisories and levels of toxicity. She asserts that, "I do try to link back to those actions and choices of local empowerment... I emphasize that you can be in charge of how you're exposed... Putting focus on what you can do and what you're in charge of to maintain your body's own health is a form of empowerment." These actions that can be encouraged through Toxic Tours and adopted by local people further help to empower local communities in terms of the safe, healthy and appropriate choices to be made to limit exposure to toxins in all aspects of life, bringing about empowerment in the realms of economic and health decisions involved in feeding oneself and one's family.

Earl shares that a direct and repeated outcome of the Toxic Tours he has led has been empowerment of the visitors, first through understanding and knowledge of what they are viewing that is then transformed into an urge towards actions to respond to the crisis at hand. He says, "when you see all that [as a visitor], you realize this isn't right... it's inspiring and empowering in this way—that when people see the sites, they want to do something in their own way." Beyond some of the examples of action on the part of visitors, including writing advocacy letters, creating artwork, writing media pieces and giving donations to LEAD Agency and local communities, Earl highlights that numerous empowering partnerships bringing research, design projects and remediation to Tar Creek have come about as a result of Toxic Tours. Several of these partnerships have been crucial to the physical clean up of the site, monitoring health and wellbeing of local people, plants, animals, soil and water, as well as contributing to overall efforts to continue advocacy for Tar Creek and Grand Lake. These actions have been all the more important given that the Superfund program funding and EPA commitments have ebbed and flowed over the years, ultimately not delivering the full clean-ups that had been promised.

These partnerships with University research institutes and national organizations that directly resulted from participation in Toxic Tours have played a critical piece in understanding the area's ongoing engagement with justice. Over several decades now, government agencies such as the EPA have been tasked with the remediation of the Tar Creek Superfund site, the central nexus of toxic contamination. These remediation efforts unquestionably represent a restorative approach to justice for this complex crisis, with consultation and other actions in the lead up to remediation folding in other elements of justice identified by Rastegar et al., including recognition, procedural and distributive justice. However, as Earl points out in his ongoing dealings with the EPA and other state-based agencies that have the capacity to affect change and keep remediation going, these agency efforts have all fallen short and not delivered the justice that locals have demanded, namely water and land that is safe for human and non-human life. In this gap, several scholarly and community-based efforts have stepped in, including Harvard University's Dr. Kirkwood's phytoremediation and environmental design planning efforts that have involved multiple classes of students (Kirkwood, 2007), as well as University of Oklahoma's Dr. Nairn and his ongoing passive water treatment project, just to name a few. Understanding this broader context of the different actors and challenges involved with the community's quest for justice help to underscore Rebecca's statement that, "This Toxic Tour is about justice, wanting it, a place needing justice and a place that has waited long enough to get it."

Toxic Tours in relation to resurgence and tensions between perspectives

While notions of empowerment and justice have stood as salient themes for the Toxic Tours, and more critically the underlying crises at the heart of these tours, these tour experiences also bring to light perspectives, stories and needs that are grounded in local and Indigenous ways of being and a yearning for Indigenous cultural resurgence—themes which do not necessarily or automatically align with mainstream conceptualizations of empowerment or justice. Working closely with the numerous Tribal Nations whose lands, waters and people have been impacted by decades of toxic assaults, the Toxic Tours that Rebecca and Earl lead are reflective of these tribal needs and stories throughout. Whether it is Earl addressing the story the forced removal of Quapaw people and their connection to land Devil's Promenade crossing close to the tribal powwow grounds, or Rebecca sharing traditional perspectives on seeing Tar Creek as a living entity that has been poisoned, abused and left sick, these Indigenous-informed understandings are a core component of LEAD's Toxic Tours. Furthermore, as Earl had stated in his reasoning behind starting the Grand Lake/Downstream Impacts Toxic Tour, a central motivation had been to demonstrate the concerns of the tribes and the problems impacting the area beyond the Tar Creek Superfund Site alone. Working as the Quapaw Nation's environmental director, he aimed to show that this was not only a Quapaw problem, but one severely impacting all ten Tribal Nations in the area.

These deep relationships that many of the tribal communities have maintained and are attempting to revitalize with land, water animals and plant relatives, go beyond the tribal jurisdiction designations or land owned in trust by the tribe. The stark and painful reality is that for many of the tribes, numerous ceremonial sites, cemeteries, sites of cultural and religious significance and areas for harvesting and hunting are in areas impacted by varying levels of toxic contamination (Garvin et al., 2018). This leads Rebecca to reflect that while these Toxic Tour efforts have been able to promote empowerment and justice more readily, deep and meaningful engagement with Indigenous resurgence and renewal of relationships with land and water has been much harder. Rebecca's questions voicing concern and frustration about not knowing where and what kinds of cultural activities can be done safely speak to the heart of the differences in understanding the context of crisis and Toxic Tours solely in relation to discourses of empowerment and justice versus in relation to Indigenous resurgence. Various forms of empowerment and justice have been and are continuously engaged with in the Toxic Tours, as have been illustrated in the earlier discussion. Yet in initiating resurgence, or Indigenous connections with land and water, cultural lifeways and community that includes human and non-human relations, this objective becomes much more complicated and challenging since it is obstructed by incomplete and undelivered justice-oriented goals such as remediation. Resurgence as a concept is grounded in a holistic return towards Native Nationhood that does not separate out individual objectives of empowerment or justice, such as environmental remediation, cultural revitalization, community political decision-making or emotional-spiritual-social wellbeing. Because of its holistic approach towards conceptualizing and transforming crisis, as well as future beyond crisis, resurgence is thus different from notions of empowerment and justice in its close interweaving of different aspects of human experiences and relations to living world.

While Rebecca's reflections point out the serious challenges and realistic barriers to resurgence that Toxic Tours help to spotlight, Earl's example of the WMAN Declaration helps to paint a more hopeful picture of the way that Toxic Tours can bring about visions and articulations of resurgence for Indigenous people. The 2019 Declaration that was composed as a direct result of Indigenous leaders gathering in the town of Quapaw, Oklahoma and participating in LEAD

Agency's Toxic Tours is an important embodiment of the visions of resurgence that Indigenous people can bring to conceptualizing crisis and attempts to move beyond it. Starting off with the assertion that "water and culture [are] fundamental dimensions of our lives… and [our] world-views [are] based on our sacred relationships to each other and Mother Earth," this Declaration does not exclude the importance of rights discourses or consideration of diverse local economies, but certainly prioritizes a vision that is guided first and foremost in resurgent relationships with land and water. With Indigenous-led resurgence discourses awakening alternate perspectives on priorities, conceptualizations of crisis and crisis responses, and an ethical way of being that is grounded in relationships with a world that is living and is kin—termed as "grounded normativity" by Glenn Coulthard— this Declaration serves as an Indigenous-led vision of the resurgent futures inspired from time on a Toxic Tour amongst wounded land, waterscapes, life-forms and communities.

Conclusion

While the crises facing the Tar Creek-Grand Lake watersheds and the communities that have called them home are severe and widespread, similar crises contaminating and otherwise transforming waterways, land, local communities and non-human relatives are becoming more ubiquitous and widespread across the world as a result of industrialization, resource extraction and climate change-induced disasters. The Toxic Tours provided by LEAD Agency over the past decades have provided an important steppingstone for visitors to better connect with and understand these impacts, while these tours simultaneously allow local communities and Tribal Nations to engage directly and tangibly with empowerment and justice, as well as Indigenous perspectives and resurgence discourses towards crisis. The local communities and Tribal Nations connected to these lands and waters have demonstrated deep resilience and strength in the face of dispossession, removal, loss and impacts of environmental crises that are complex, evolving and still on a path of recovery and healing.

LEAD's Toxic Tours are a vehicle for both conceptualizing and forging pathways towards empowerment and justice for some people connected to the Tar Creek Superfund Site and Grand Lake. On one level, LEAD Toxic Tours open a doorway to forging multiple connections, relationships and understandings between tourists and all elements of engagement on the tour. At another level, these Toxic Tours offer an Indigenous-led platform for addressing crisis, empowerment and justice from locally-grounded perspectives. Finally, by putting the experiences of LEAD Toxic Tours in critical discussion with the literature on crisis, empowerment and justice, some key differences are established in relation to Indigenous perspectives on transforming crisis, navigating justice and seeking resurgence.

Through these Toxic Tours and their emphasis on sharing Indigenous stories, histories and perspectives on living environments that have long faced disaster, new approaches to understanding crisis are opened up and shared. Alongside notions of empowerment and justice, Indigenous-led resurgence discourses thus emerge as a helpful and culturally-informed lens for understanding alternative approaches to crisis, remediation and how all facets of Indigenous lives are interconnected with their living environments. Based in a worldview that prioritizes a renewal of relationships between humans and all non-human forces that comprise the living environments around us, Indigenous ways of seeing and being are thus centered in these Toxic Tours and help to provide a compass for navigating futures beyond crisis and towards resurgence.

Note

1. The Cherokee term *'jigesv'* is an important traditional marker in the language denoting acknowledgement and respect for someone who has passed away.

Acknowledgements

The authors wish to acknowledge all community members and Tribal Nations that have and continue to navigate the complex crises facing the Tar Creek and Grand Lake areas with resilience and strength. Special thanks to Clifton Adcock for sharing his powerful images from the Toxic Tour he took in 2018.

Disclosure statement

No potential conflict of interest was reported by the authors.

ORCID

Bobbie Chew Bigby http://orcid.org/0000-0003-3152-8812

References

"What is Superfund?" (n.d). *United States Environmental Protection Agency (website)*. Retrieved January15, 2021, from https://www.epa.gov/superfund/what-superfund

Aber, J. S., Aber, S. W., Manders, G., & Nairn, R. W. (2010). Route 66: Geology and legacy of mining in the Tri-state district of Missouri, Kansas, and Oklahoma. In K. R. Evans & J. S. Aber (Eds.), *From Precambrian rift volcanoes to the Mississippian shelf margin: Geological field excursions in the Ozark Mountains field guide* (Vol. 17, pp. 1–22). The Geologic Society of America.

Adcock, C. (2018, February 23). "Contaminated, totally": A tour of Tar Creek. *The Frontier*. Retrieved September 15, 2020, from https://www.readfrontier.org/stories/on-a-tour-of-tar-creek/

Bishop, M. (2021). Don't tell me what to do' encountering colonialism in the academy and pushing back with Indigenous autoethnography. *International Journal of Qualitative Studies in Education*, *34*(5), 367–378. https://doi.org/10.1080/09518398.2020.1761475

Boluk, K. A., Cavaliere, C. T., & Higgins-Desbiolles, F. (2019). A critical framework for interrogating the United Nations Sustainable Development Goals 2030 Agenda in tourism. *Journal of Sustainable Tourism*, *27*(7), 847–864. https://doi.org/10.1080/09669582.2019.1619748

Clark, B. (2012). *Indian tribes of Oklahoma: A guide (The Civilization of the American Indian Series Book 261)*. University of Oklahoma Press.

Coghlan, A. (2012). An autoethnographic account of a cycling charity challenge event: Exploring manifest and latent aspects of the experience. *Journal of Sport & Tourism*, *17*(2), 105–124. https://doi.org/10.1080/14775085.2012.729899

Corntassel, J. (2021). Life beyond the state: Regenerating indigenous international relations and everyday challenges to settler colonialism. *The Politics of Indigeneity, Anarchist Praxis, and Decolonization*, *2021*(1), 71–97.

Coulthard, G. (2014). *Red skin, white masks: Rejecting the colonial politics of recognition*. University of Minnesota Press.

Denzin, N. K., Lincoln, Y. S., & Smith, L. T. (2008). *Handbook of critical and indigenous methodologies*. Sage.

Di Chiro, G. (2000). Bearing witness or taking action? Toxic tourism and environmental justice. In R. Hofrichter (Ed.), *Reclaiming the environmental debate: The politics of health in a toxic culture* (pp. 275–300). MIT Press.

Ellingson, L. L., & Ellis, C. (2008). Autoethnography as constructionist project. In J. A. Holstein & J. F. Gubrium (Eds.), *Handbook of constructionist research* (pp. 446–467). The Guildford Press.

Ellis, C., Adams, T.E., & Bochner, A.P. (2011). Autoethnography: An overview. *Historical Social Research*, *36*(4), 273–290. https://doi.org/10.12759/hsr.36.2011.4.273-290

Estes, N. (2019). *Our history is the future: Standing rock versus the Dakota access pipeline, and the long tradition of indigenous resistance*. Verso.

Garvin, E. M., Bridge, C. F., & Garvin, M. S. (2018). Edible wild plants growing in contaminated floodplains: Implications for the issuance of tribal consumption advisories within the Grand Lake watershed of northeastern Oklahoma, USA. *Environmental Geochemistry and Health*, *40*(3), 999–1025. https://doi.org/10.1007/s10653-017-9960-3

Guia, J. (2021). Conceptualizing justice tourism and the promise of posthumanism. *Journal of Sustainable Tourism*, *29*(2–3), 503–519. https://doi.org/10.1080/09669582.2020.1771347

Harper, B. (2008). *Quapaw traditional lifeways scenario*. Retrieved October10, 2020, from http://superfund.oregonstate.edu/sites/default/files/harper_2008_quapaw_scenario_final.pdf

Harvey, F. (2021, August 17) Global water crisis will intensify with climate breakdown, says report. *The Guardian*. Retrieved December15, 2021, from https://www.theguardian.com/environment/2021/aug/17/global-water-crisis-will-intensify-with-climate-breakdown-says-report

Jamal, T. (2019). *Justice and ethics in tourism*. Routledge.

Jamal, T., & Higham, J. (2021). Justice and ethics: Towards a new platform for tourism and sustainability. *Journal of Sustainable Tourism*, *29*(2–3), 143–157. https://doi.org/10.1080/09669582.2020.1835933

Jim, R., & Scott, M. P. (Eds.). (2007). *Making a difference at the tar creek superfund site: Community efforts to reduce risk*. LEAD Agency, Inc.

Johnson, L. G. (2009). *Tar Creek—A history of the Quapaw Indians, the world's largest lead and zinc discovery, and the Tar Creek Superfund Site*. Tate Publishing.

Karkowski, M. P., Just, A. C., Bellinger, D. C., Jim, R., Hatley, E. L., Ettinger, A. S., Hu, H., & Wright, R. O. (2014). Maternal iron metabolism gene variants modify umbilical cord blood lead levels by gene-environment interaction: A birth cohort study. *Environmental Health*, *13*(1), 77–77. https://doi.org/10.1186/1476-069X-13-77

Kirkwood, N. G. (2007). CHAT. In G. P. Macey & J. Z. Cannon (Eds.), *Reclaiming the land* (pp.267-289). Springer. https://doi.org/10.1007/978-0-387-48857-8_10

LaDuke, W. (2016). *All our relations: Native struggles for land and life*. Haymarket Books.

Lapadat, J. C. (2017). Ethics in autoethnography and collaborative autoethnography. *Qualitative Inquiry*, *23*(8), 589–603. https://doi.org/10.1177/1077800417704462

LEAD. (2020, October 16). *WMAN/LEAD agency virtual conference, addressing mining and systemic racism: Staying connected during a pandemic*. Virtual Toxic Tour Promotion Email. In authors' possession.

Lisle, D. (2007). Defending voyeurism: Dark tourism and the problem of global security. In P. Burns & M. Novelli (Eds.), *Tourism and politics: Global frameworks and local realities* (pp. 333–347). Elsevier.

Mair, J., & Frew, E. (2018). Academic conferences: A female duo-ethnography. *Current Issues in Tourism*, *21*(18), 2160–2172. https://doi.org/10.1080/13683500.2016.1248909

Manders, G. C., & Aber, J. S. (2014). Tri-State Mining District legacy in northeastern Oklahoma. *Emporia State Research Studies*, *49*(2), 29–51.

Myers, M. (Producer), Beer, T. (Producer), & Myers, M. (Director). (2009). *Tar Creek* [Film] Bullfrog Films.

Neuberger, J. S., Hu, S. C., Drake, K. D., & Jim, R. (2009). Potential health impacts of heavy-metal exposure at the Tar Creek Superfund site, Ottawa County, Oklahoma. *Environmental Geochemistry and Health*, *31*(1), 47–59. https://doi.org/10.1007/s10653-008-9154-0

Pezzullo, P. C. (2007). *Toxic tourism: Rhetorics of pollution, travel, and environmental justice*. Tuscaloosa: University of Alabama Press.

Rastegar, R., Higgins-Desbiolles, F., & Ruhanen, L. (2021). COVID-19 and a justice framework to guide tourism recovery. *Annals of Tourism Research*, *91*, 103161. https://doi.org/10.1016/j.annals.2021.103161

Scheyvens, R., & van der Watt, H. (2021). Tourism, empowerment and sustainable development: A new framework for analysis. *Sustainability (Basel, Switzerland)*, *13*(22), 12606. https://doi.org/10.3390/su132212606

Shade, H. (1994). *Myths, legends and old sayings*. Cherokee Nation.

Sharpley, R., & Stone, P. (2009). *The darker side of travel: The theory and practice of dark tourism*. Channel View.

Simpson, L. (2016). Indigenous resurgence and co-resistance. *Critical Ethnic Studies*, *2*(2), 19–34.

Tuck, E. (2009). Suspending damage: A letter to communities. *Harvard Educational Review*, *79*(3), 409–427. https://doi.org/10.17763/haer.79.3.n0016675661t3n15

USGS. (2019, May 22). *The water science school. The water in you: Water and the human body*. https://www.usgs.gov/special-topics/water-science-school/science/water-you-water-and-human-body

Verma, M. K. (2018). *Globalisation, environment and social justice: Perspectives, issues and concerns* (1st ed.). Routledge India, an imprint of Taylor and Francis.

Whitinui, P. (2014). Indigenous autoethnography: Exploring, engaging, and experiencing "self" as a native method of inquiry. *Journal of Contemporary Ethnography*, *43*(4), 456–487. https://doi.org/10.1177/0891241613508148

Yankovska, G., & Hannam, K. (2014). Dark and toxic tourism in the Chernobyl exclusion zone. *Current Issues in Tourism*, *17*(10), 929–939. https://doi.org/10.1080/13683500.2013.820260

OPEN ACCESS

Gender justice in global tourism: exploring tourism transformation through the lens of feminist alternative economics

Angela B. Kalisch and Stroma Cole

ABSTRACT

Tourism is one of the sectors most affected by the COVID-19 pandemic. Women make up the majority of tourism and hospitality workers and are bearing the brunt of the impact of COVID-19. Evidence suggests that women's economic and productive lives will be disproportionately and differently affected, with the economic impacts lingering for many years after the end of the pandemic. Feminist political economics critiques the intertwined neo-liberal, patriarchal, capitalist structures, which are deemed to be at the root of the global crises society is facing. Using a critical, feminist perspective, this conceptual article explores alternative economic approaches. We suggest a Feminist Alternative Tourism Economics (FATE) approach, embracing an integration of Feminist Ethic of Care, Social Solidarity Economy and Human Rights Based Economy, critiqued through the lens of decolonisation, could present an alternative pathway to achieving a more gender just society and tourism system. We argue a focus on feminist values of care, solidarity and Human Rights could provide a chance for tackling major global climate and pandemic emergencies by reducing crisis vulnerability and increasing resilience and sustainability.

Introduction

In the face of the current global health and climate crises, society appears on the threshold of a fundamental re-positioning of human and environmental values. The neo-liberal, capitalist system is increasingly being questioned as an appropriate model for creating economic efficiency, equality and justice, and for realising the UN Sustainable Development Goals (Oksala, 2018; Jacobs & Mazzucato, 2016; Bianchi & de Man, 2021).

This conceptual article explores the question to what extent a feminist, alternative vision of a global social and economic order could create structural transformation towards a more just and inclusive system within the tourism sector.

The World Bank estimates the COVID-19 pandemic will push between 71 million and 100 million people into extreme poverty, erasing almost all progress made in the last five years in the fight against extreme poverty, and any achievements on fulfilling the Sustainable Development Goals (UN/DESA, 2020). Gender equality and poverty are intrinsically linked. Levels of poverty can be substantially reduced where women are able to contribute to households with their own independent income, through the restructuring of care responsibilities, the eradication of gender-based violence and input into public policy and decision-making fora (Bessell, 2015; Nieuwenhuis et al., 2018). Like most crises, the coronavirus pandemic has exacerbated existing inequalities, affecting women and other vulnerable groups disproportionately (Moriarty et al., 2020; Nanthini & Nair, 2020). The 2021 World Economic Forum Insight Report on the Global Gender Gap estimated that closing the economic participation gender gap would take 268 years, and that the pandemic may have further slowed progress by 1%–4%.

Women are the majority of tourism and hospitality workers, particularly in the Global South and in the informal sector. Added to existing discrimination, they faced redundancies, business failure, home schooling responsibilities, additional care work, and, worse, increased levels of gender based violence. Globally women earn and save less, hold jobs that are more insecure and lack advancement opportunities (Nanthini & Nair, 2020). Besides the economic impact, the pandemic is likely to affect women's mental well-being, as studies have established an association between insecure employment or unemployment and decrease in psychological wellbeing (Flint et al., 2013). Women of colour, with disabilities, Indigenous and migrant women are even more vulnerable (Grandy et al., 2020).

Issues of gender equality have gained prominence in sustainable tourism and development debates (UNWTO, 2019; UN Women, 2020; World Economic Forum, 2021). Yet, overall, at implementation level in tourism organisations and destinations, genuine commitment to mainstreaming gender equality tends to be lacking (Ferguson, 2018). Furthermore, scant attempts have been made to address the root causes of structural gender injustices, caused by a patriarchal economic system that uses tourism as a tool for neoliberal capital accumulation and development (Britton, 1991; Bianchi, 2011). Bianchi and de Man (2021) critiqued the UNWTO's growth paradigm, unearthing the contradictions between tourism's expansionist aspirations and the achievement of the Sustainable Development Goals. Their argument for a 'radical re-distribution of power and wealth' (p.365) deserves some scrutiny in the context of a feminist political economy perspective.

In this article we examine alternative economic approaches through the lens of a critical feminist analysis, within the framework of Feminist Alternative Economics (FAE), potentially indicating a route towards structural change and gender justice in tourism and hospitality. Our interpretation of critical feminism locates feminist struggles within a critique of capitalist, neo-liberal values, within decolonisation, and transnational alliances, in solidarity with marginalised and Indigenous women in the Global South (Mohanty, 2003a). Tourism, specifically, as a cross-border force of capitalist appropriation and cultural othering requires such alliances. 'Tourism is a key site for the (re)production of difference at the contemporary conjuncture of the transnational and the postcolonial' (Patil, 2011, p. 206).

This article answers Jamal and Higham (2021) call for a richer body of insights to justice, feminism and an ethics of care, and Higgins-Desbiolles (2008) call for alternatives to the corporatised, neo-liberal free market system. The article concludes with a vision for a justice-based future for tourism in the context of a feminist alternative tourism economics (FATE) analysis, and its potential for increasing crisis resilience and sustainability.

Various authors have addressed the need for a more justice-informed future of tourism (Higgins-Desbiolles, 2020, 2008; Jamal & Camargo, 2014; Jamal & Higham, 2021). Camargo et al. (2016) propose an eco-feminist paradigm for ecocultural justice. A recent Special Issue on Gender and Tourism Sustainability casts a challenging light on tourism sustainability from the perspective of feminist epistemologies (Eger et al., 2022). However, in general, gender in tourism is still under-researched; in particular, in the context of a systems-based analysis, which, according to hooks (1994) focuses on the significance of 'the interlocking, interdependent nature of systems of domination' and the recognition of the 'specific ways each system is maintained' (p. 290). Rare, too, is a focus on intersectionality, decolonisation and the consequent understanding of the complexity of women's oppression in the post-colonial Global South (Chambers & Buzinde, 2015; Chambers, 2022), where tourism prevails as a capitalist development tool. Capitalism has burgeoned with the riches from colonial legacies. Structural transformation therefore depends not only on critiquing patriarchy and neo-liberal capitalism, but also on decolonising individual and organisational worldviews and institutional structures. Decolonisation demands a critical anti-racist approach, dismantling the presence of colonial and imperial theory and practice in every societal context, including in feminist movements and economic analysis (Grey, 2004; Chambers & Buzinde, 2015; Saini & Begum, 2020).

Positionality and reflexivity are crucial aspects of decolonising mindsets (Chambers & Buzinde, 2015). We wish to share our positionality as feminist, white Europeans and acknowledge our privilege and the limitations of our lived experiences to informing our analysis, but locate ourselves as allies to women in the Global South (Grey, 2004). We attempt to include critiques and standpoints which emanate from other cultures and from traditionally marginalised women in our analysis, and in the spirit of educating ourselves, we invite critique and comment from majority world voices.

In the following sections we introduce the concept of Feminist Alternative Economics and analyse three models of social and economic organisation: Ethics of Care, Social Solidarity Economy and a Human Rights Based Economy as alternative feminist approaches. We explore their suitability as pathways towards structural transformation of the core tenets of the global tourism system. Our analysis suggests that these three models, albeit requiring further critical scrutiny, offer substantive transformative potential for a critical feminist analysis in tourism.

Feminist alternative economics

Economics is largely organised on the basis of androcentric power relations, alien to the female experience and devoid of the female voice, especially Indigenous and subaltern voices. We argue that economics, in particular tourism economics needs to be re-configured from the perspective of the diversity of women's experiences. Neoliberal approaches to the economy focus exclusively on maximisation of financial returns. Conversely, feminist economic alternatives holistically address relational spheres conventionally considered outside the labour process, such as patriarchy, power dynamics, care, social reproduction, and community as integral constituents of economic analysis (Bakas et al., 2018; Nobel et al., 2020). As discussed by Nobel et al. (2020) 'In contrast to mainstream economics, alternative feminist perspectives strive for holistic understandings of the processes that support life and social provisioning, paying attention to dynamics of power, exclusion and gender and how these play out in economic analysis and decision-making.' (Vol.1:10). This includes exposing the contradictions inherent in capitalism's approach to social reproduction as a necessity for its survival, while simultaneously devaluing it (Fraser, 2016). Feminist political economics critiques the patriarchal structures of exploitative capitalist economics (Burris, 1982), which are also deemed to be at the root of the global crises society is currently facing, such as climate change and the coronavirus pandemic (Wright & Nyberg, 2015; Bump et al., 2021). Within the tourism system, these structures create, amongst others, discrimination along gender and race lines in the labour market, the gender pay gap, commodification and exploitation of sexual and racial relations (Pritchard, 2014), and commodification and appropriation of natural resources, which disadvantage women's rights and livelihoods (Cole, 2014; Neef, 2021). The following three models offer the potential for transforming these exploitative structures.

Ethics of care

The ethics of care arose as a counterpoint to Rawls' theory of justice and Kohlberg's model of the stages of moral development. Emanating from research in the 1980s on the differences between boys' and girls' responses to specific ethical dilemmas, Carol Gilligan concluded that the masculinist bias embedded in the thought processes behind Kohlberg's model prevented and, potentially, negated the acknowledgement of an alternative moral worldview informed by the female experience (Gilligan, 1983).

Gilligan's ethic of care takes a more nuanced, embodied stance. It addresses individual difference, emotional dynamic, and situated complexity and is therefore more fluid and flexible, less rule-bound and abstract. Gilligan conceives the ethic of care as 'an ethic of resistance to the injustices inherent in patriarchy'; an ethic to 'free democracy from patriarchy', transcending the demands of a purely feminist perspective to a concern for all of society (Gilligan, 2011). Care is conceived as a shared human condition, assuring human and planetary survival, an 'inevitable feature of human life' rather than a 'pathology' (Robinson, 2015, p.15), merely related to child- and age care or illness and disability. bell hooks (2001) believes care can only reap societal benefits, if it is conceived through the prism of love; love as ethical living among all sentient beings, 'embracing a global vision wherein we see our lives and our fate as intimately connected to those of everyone else on the planet' (p. 88). Moreover, ethics of care and love are, arguably, inherently interwoven with ethics of justice (Jamal & Camargo, 2014).

Gilligan's anti-patriarchal stance is related to a North American feminism and has evolved over time. It is imperative to reassess it against the lived realities of women of colour and marginalised and Indigenous women, who interpret their gender struggle through a different lens of historical (colonial), class and cultural context. In that context, 'gender is found alongside issues of socio-economic inequality, racism, assimilation, cultural renewal and self-determination' (Grey, 2004, p. 19). The philosophy of human and planetary care as a shared human condition has similarities in other cultures, such as the African notion of '*ubuntu*', and the Maori philosophy of '*whenua*'. Ubuntu holds that 'people can only be human through other people' and the self is interwoven with the community through countless bonds and relationships (Gouws & Van Zyl, 2015). However, Gouws and van Zyl assert that Ubuntu has failed women, as they are still considered subordinate to men. They assert that a feminist ethic of care, integrated with Ubuntu, provides an ethical framework for justice. Similarly *whenua*, a Maori philosophy, places the land, plants, animals and people within an unbreakable, reciprocal bond with the self, and the self within community (Irwin, 2013). Aspects of kinship, respect and the ethics of kaitiakitanga – obligation, sustainability and nurture –are inherent in Maori culture (Barnes et al., 2018). Traditionally, women had an equal role at the centre of this system. However, since colonisation and capitalism have fostered a 'Maori male hegemony' (Hoskins, 1997, p. 40), Maori women are re-interpreting their cultural identity and agency in the context of a post-colonial, feminist analysis, which might integrate a feminist ethic of care within their own cultural context.

Care is a core element of tourism and hospitality. Marketised as customer care and personal service, affective, or emotional, labour, is a crucial component of tourism management. However, care for tourism workers is less of a priority for managers. The majority of women's tourism labour mirrors their taken for granted predisposition for domestic labour and care work, usually outsourced, and, as such, condemned to the lowest ranks by an exploitative, capitalist system. Low pay, rising food prices (inflation often caused by tourism) and budget cuts lead to poverty and disintegration of the family (Federici, 2008), most pertinently in times of crises, such as Covid-19. Women with access to higher career positions face stark choices: outsource domestic and child-care duties to poorer, frequently marginalized migrant women (thus continuing the chain of oppression), while accepting lower wages than their male counterparts (Bakas et al., 2018), or accept the triple burden of a paid career, unpaid care and domestic work, and be branded as a less than 'ideal' worker (Costa et al., 2017). While some domestic tasks can be contracted out, less tangible tasks cannot. Love, affection and discipline tend to fall by the wayside. Ferguson (2010) cautions that women's empowerment through increased income and control over finances is mediated by the stresses of competing demands on household budgets and pressures on family dynamics, caused by both traditional social structures and a capitalist, globalised tourism system.

Yet, to date, the ethic of care has received scant attention in the tourism literature, with varying interpretations. Smith and Duffy (2003) focus on Gilligan's perspective as a guideline for ethical field research in tourism, while Fennell (2006) analyses ethic of care in relation to sex tourism, and Lovelock and Lovelock (2013) devote just one paragraph to it, with a brief reference to disability and tourism provision. Jamal and Camargo (2014) understanding of 'ethic of care' embeds the meaning of care within a range of ethical theories and notions of justice in their discussion on a 'Just Destination'. However, they do not consider ethic of care from a feminist perspective for tourism transformation and structural change.

Jamal, Camargo and Higgins-Desbiolles have produced important analyses of research on ethics and justice in tourism, albeit with scant reference to feminism, a feminist ethic of care or feminist alternative economics, which could be applied in tourism context. Conscious of this gap in sustainable tourism research, Camargo et al. (2016) propose an 'eco-feminist' research paradigm focusing on 'gender awareness and emotions, such as love' for people and nature, which also addresses 'exploitative economic structures and power relations' (p. 70).

However, judging from recent research contributions, feminist tourism paradigms, incorporating justice and ethic of care standpoints, have still not developed substantially to question the socio- economic and political structures that foster inequality in tourism and advance to a decolonised, non-Western, non-patriarchal analysis of gendered power relations in tourism (Jamal & Higham, 2021). Referring to the contributions in the Special Issue on Justice and Tourism, Jamal and Higham emphasise the need for a richer body of multi/post-disciplinary approaches and insights to justice, feminism, and an ethic of care.

As one of the thought leaders in the debate on justice in tourism in the context of neo-liberal globalisation (2008; 2020) and global crises (Rastegar et al., 2021), Higgins-Desbiolles has focused on women as one of several marginalised groups within local communities to be prioritised over tourists and corporates (Higgins-Desbiolles et al.,2019). However, her proposals for a 'radical transformation' of tourism (2008) and a justice framework for post-COVID-19 recovery (2020) are overall generic as alternatives to neo-liberal capitalism and do not include a gendered feminist economic analysis. On the other hand, her most recent collaborative research signals a welcome new analysis of a feminist ethic of care in the context of social entrepreneurship and the purpose economy in the events sector (Higgins-Desbiolles & Monga, 2021). Social Entrepreneurship is part of the Social Solidarity Economy (SSE), which provides an alternative economic model challenging the capitalist, neo-liberal worldview. The following section explores the potential of SSE for a feminist transformation in tourism.

Social solidarity economy

The Social Solidarity Economy (SSE) is an umbrella term for a non-patriarchal alternative to neoliberal capitalism, embracing values of humanism, ethics, democracy, diversity, solidarity, inclusiveness, equality and justice (Réseau Intercontinental de Promotion de L'Economie Sociale Solidaire, RIPESS, 2015; Quiñones, 2008). It is built on the notion of collective, participatory, nonhierarchical organisation and management, and principles of autonomy, self-sufficiency, productive diversification and sustainable regional resource management. Its primary objective is to benefit human and planetary well-being, serving the needs of people and ecological sustainability (RIPESS, 2015; Barkin & Lemus, 2014). As such, SSE consciously invokes the goal of structural, transformational social and economic change. The practical embodiment of this ideology at micro level comprises cooperative structures, social enterprise, social entrepreneurship, legally incorporated social benefit corporations (B-Corps), trade and credit unions (Fonteneau et al.,

2010) and collective mobilisation of resistance struggles (Verschuur et al., 2015; Barkin & Lemus, 2014).

Ideological debates over the nature of social and economic change and structural transformation still indicate some fluidity in interpretations of what genuinely constitutes SSE. For example, while RIPESS are committed to 'transformative, systemic change' (RIPESS, 2015, p.2), Quiñones (2008) more ambiguously emphasises the potential integral relationship of the solidarity economy with traditional market logic 'for as long as its approaches continue to be innovative' (p. 21). Whilst SSE philosophy propounds an equitable ethic, Borowiak et al. (2018) claim that 'racial and class divides suffuse the … .Solidarity Economy movement' (p.582). Historical distrust of white middle-class driven initiatives, combined with constraints of poverty, diminish the positive impact on an alternative socio-economic reality. The UN promote the SSE as a new paradigm for sustainable production and consumption (UNTFSSE. , 2020) and for the realisation of the UN Sustainable Development Goals (UNTFSSE, 2019), without challenging the contradictory nature of sustainable development within a neo-liberal capitalist growth paradigm. Hence, whether SSE is a systemic alternative to hegemonic global capitalism is still open to interpretation (Verschuur et al., 2015).

Research on feminist positions on SSE is extremely limited (Verschuur et al., 2015), although women's participation within SSE is strong, whether at leadership or grassroots levels (UNTFSSE, 2014). The SSE has been associated with women being the driving force, and gender equality being a core value of SSE (Quiñones, 2015). Hillenkamp et al. (2019) propose a 'post-colonial feminist perspective' (p.10) of social enterprise and entrepreneurship, which overcomes the Western feminist informed 'market-household dichotomy'. Their perspective recognises the domestic domain as a seed bed for solidarity and political and economic emancipation, especially for Indigenous and subaltern women. Soler-i-Martí et al. (2021) posit that the very nature of SSE is suitable for challenging gender inequality and patriarchal structures. However, Verschuur et al. (2015) analysis of SSE initiatives in Latin America and India, as one of the rare examples of a feminist perspective on SSE (albeit not tourism-related), critique this very notion. They found that even within SSE, women's (re)productive work was unrecognized and undervalued; work/life balance was blurred, work was poorly paid, and the need to compete with markets in the capitalist economy, which often profit from Social and Solidarity Economy Enterprises and Organisations (SSEEOs), put them at an economic disadvantage. Lack of state support and, at times, collusion between the state and corporations, potentially undermined and threatened their solidarity goals.

Women's motives for entering into entrepreneurship are usually different from men's intentions (Bakas, 2017) and are more aligned with social entrepreneurship, even if not always formalised as such. Deen et al.'s (2016) research on small tourism accommodation establishments suggests they engage less with a profit motive than to make a lifestyle change for a better quality of life, personal growth, and economic benefit combined with social value to society. However, they face multiple obstacles, including societal gender bias, undermining their independence and credibility, restricted access to financial support, and lack of sustainable tourism knowledge and business management training (Kearins & Schaefer, 2017; Deen et al., 2016).

SSE has not featured as such on the radar of tourism researchers. Only Quiñones (2015) mentions the potential of the solidarity economy for creating emancipatory programmes for women in the governance of tourism projects in Lao PDR and Sri Lanka. On the other hand, there is a nascent body of research on the relationship between social capital and social entrepreneurship and tourism, in particular community-based tourism (Dyah-Pramanik et al., 2019; Reindrawati, 2018; Reinke, 2018; Mottiar, 2016; Robinson, 2006). Social capital, incorporates trust, reciprocity, collective, participatory decision-making, solidarity and mutual respect in individual or community economic activity (Ruiz & Adie, 2020).

Mottiar (2016) suggested that tourism social entrepreneurs can create social value and, thereby, exert long-term impact on the structure of the tourism industry. However, this literature does not explore SSE as structurally transformative, nor is gender equality considered as a key component, or even a driver of social capital or social change through tourism. In view of reassessing capitalism, there is a compelling need, for substantially more research in the tourism academy on the SSE from a feminist perspective, taking account of situated realities for women in the Global South. Solidarity is also critical to a Human Rights Based Economy which we now present as our third strand to drive gender justice.

Human rights based economy

Within human rights debates, women's rights are specifically enshrined in the 1979 Convention on the Elimination of all Forms of Discrimination against Women (CEDAW). The 1995 Beijing Declaration and Platform for Action presented an important milestone for 'Human Rights as Women's Rights', which pledges to 'ensure the full implementation of the human rights of women and of the girl child as an inalienable, integral and indivisible part of all human rights and fundamental freedoms' (p.2). Nevertheless, while these are important, binding obligations for UN member states agreed on paper, human rights, and specifically women's rights, have been widely ignored in practice, including in tourism and hospitality settings, namely in respect of unequal pay and labour rights (Baum & Hai, 2019), social, cultural and economic rights, land and water rights (Cole, 2017; 2014; Neef, 2021), and, by implication, international and national trade and socio-economic policies (Toussaint, 2020; Kalisch, 2010).

A Human Rights Based Economy (HRBE), re-interpreted from a critical feminist perspective and based on a widely agreed framework of ethical values and legal obligations provides an alternative to the present capitalist neo-liberal economic system. The human right to safe, empowering, dignified work, for example, will require moving from a fixation on economic growth to a more holistic, ecologically sound and human centred measure of success (Donald, 2020). The neoliberal economy has devalued, exploited and taken for granted the work performed by the majority of women, confined to the lowest ranks, in and outside tourism (Pritchard, 2014; Bakas et al., 2018; Cole, 2018). Taking a Human Rights Based approach would rebalance this inequity and provide economic justice. As Human Rights lie in a comprehensive framework of binding principles and laws, agreed by the majority of governments, a unifying framework for holding companies and governments accountable already exists. Furthermore, international financial institutions are also bound by international obligations and cannot divest themselves of responsibility (Toussaint, 2020). A Human Rights Based Economy would be based on the principles of dignity, equity, solidarity, accountability and justice (Donald, 2020).

In a Human Rights Based Economy dignity would be central to ensuring that workers can have proper working conditions, combined with a decent work/life balance. As Baum and Hai (2019) suggest, all individuals have equal rights to a family life. Presently, tourism can impact family members for example, through child labour within family businesses or the extended absence of parents due to split shifts (Bhat & Rather, 2009; Ferguson, 2010). Tourism and hospitality are renowned for a lack of work place dignity. Workers suffer low pay, emotional exhaustion, and abusive treatment (Winchenbach et al., 2019; Unite the Union Unite Hotel Workers Branch, 2018). Across the globe, this is worsened by low or declining trade union membership in the sector, due to employer hostility and COVID-19 induced unemployment (Wood, 2020). The lack of employee voice in the tourism and hospitality sector is a factor behind the exploitative relations and poor working conditions that many in the sector experience (Curran, 2021). Victories of striking chambermaids, such as Las Kellys in Spain and staff at the Ibis Batignolles in Paris, have shown what the power of collective voice can achieve in improving dignity for female hotel staff. After many months of striking, they managed to achieve improved working conditions and raised wages (Abdelghani, 2020; Hird, 2021).

Businesses (in tourism and every other sector) can intensively combat working poverty by paying a living wage to their suppliers and employees. This includes closing the gender pay gap, which is most pronounced in tourism in higher hierarchical positions dominated by men, but less visible through horizontal segregation in low paid jobs, where women predominate (Bakas et al., 2018). When people are paid adequately for their work, they can experience less dependence on overtime and tips, improved work/life balance, decreasing risks of child labour, increased spending power, overall well-being and a life of dignity. A living wage should, therefore be regarded as a long-term investment, rather than a cost factor.

A fundamental principle running through Human Rights is equality. The inequalities in tourism have been widely discussed in the literature (Cole, 2018; Guimarães & Silva 2016; Bakas et al., 2018; Costa et al., 2011). One prominent area in tourism is the growing precarity in tourism employment (Robinson et al., 2019), and COVID-19 demonstrated this with far greater rapidity than in other sectors. Precarity is driven by the growing practice of outsourcing and flexibilisation, which lowers costs for employers, but is rarely reciprocal (Bolton & Houlihan, 2007), with industry benefitting far more than workers in the long term. Terry (2011) discusses this in relation to cruise workers, and Seifert and Messing (2006) in relation to hotel cleaners in Canada.

The solidarity aspect of a HRBE is not only about workers' rights and collective bargaining, which is critical to ensuring a decent wage and working conditions. It is also about stewardship and fostering collective ownership of the planet. Tourism's record on environmental destruction (Brohman, 1996, Stonich, 1998) and the removal of rights has been debated in tourism in relation to water (Cole, 2014, 2017; Nobel et al., 2012) and the loss of land rights (Neef, 2019). As the Human Rights Council (2021) suggests, our human rights are intertwined with the environment in which we live. Environmental harm interferes with the enjoyment of human rights, and the exercise of human rights helps to protect the environment and to promote sustainable development.

It is the state's duty to uphold Human Rights and for business to protect them (UN's (2011) the Guiding Principles on Business and Human Rights,). The problem is lack of enforcement. Governments and businesses need to be made more accountable. Governments must enforce on businesses to conduct human rights and environmental due diligence, and this must extend to their supply chains and be gender sensitive. Since their publication, the Guiding Principles (GPs) have been promoted to the major tourism businesses to encourage their commitment. For example, to assist hotels, in 2014, the International Tourism Partnership and Green Hotelier published the Know How Guide: Human Rights & the Hotel Industry. It details human rights issues prevalent for a hotel business, including workers' rights, supply chain related issues, community rights, human trafficking risks, customers' rights and governance related issues (International Tourism Partnership & Green Hotelier, 2014). However, thus far action has been limited to single issue causes, e.g. the Modern Slavery Act 2015, with training produced for members of hospitality (Sustainable Hospitality Alliance, 2021) and travel agencies (ABTA, 2018). As Donald (2020) suggests, in order to reflect the indivisibility of rights, the duty of due diligence should instead reflect a broader range of risks to rights.

The final component of a HRBE is justice. Human Rights, if viewed holistically, are profoundly redistributive. Injustice in the tourism system became apparent to everyone during COVID-19. For example, high-profile tourism companies showed little concern for workers' rights, enacting sackings, and leaving cruise ship crews, denied access to ports, trapped on ships at sea for many months, in degrading conditions (Kaji, 2020; Radic et al., 2020; Higgins-Desbiolles, 2020). A vast number of tourism workers in the informal sector (mostly women), were left with little or no social security, lacking any form of protection. Social justice, and working towards a more equal society, lies at the heart of human rights. A HRBE can ensure people are treated with dignity and respect.

As valiant as human rights conventions and treaties are, many feminists in the Global South have critiqued the seemingly universal, androcentric nature of the principles, which they consider as steeped in Western neo-liberal values. As such, they are not considered representative of the diversity of women's lived experiences, particularly in respect of the dichotomy between the private and public spheres, the issues of cultural rights, Islam and feminism, the experience of Indigenous women, and LGBTQ+women (Wichterich, 2021; Women LEAD, 2019; Parisi, 2017; Cheruvallil-Contractor, 2018; Weldon, 2018; Binion, 1995). Women from the Global South demand a focus on Indigenous values systems to inform feminism and justice, and decolonisation of Western values and institutions (Mikaere, 1994; Grey, 2004; Saini & Begum, 2020).

It is beyond the parameters of this paper to delve into a detailed, critical analysis of the many debates on the universality of Human Rights and tourism. We suggest further in-depth future research on this subject. At this stage, we propose a HRB economy, which would offer the scope for a review of HR principles from a feminist perspective that includes the diversity of women's lived experiences globally, and specifically in the context of tourism and hospitality. Moreover, we argue that a decolonised, anti-patriarchal HRBE, would present a space for constructive debates and potential transformation.

Discussion

Transformative implications of feminist alternative economics for the tourism system

In the context of a Feminist Alternative Economics (FAE) paradigm, we have analysed three alternative economy approaches (Feminist Ethic of Care, Social Solidarity Economy, and Human Rights Based Economy). They embrace the core tenets of FAE (see below).

Based on Urban and Pürckhauer (2016); Nobel et al. (2020) (Table 1)

We suggest a Feminist Alternative Tourism Economics (FATE), consisting of an integration of Feminist Ethic of Care (FEoC), Social Solidarity Economy (SSE) and Human Rights Based Economy (HRBE) and critiqued through a transnational feminist, decolonising lens. This could offer an alternative pathway to achieving a more gender just society and, by extension, tourism system (see Figure 1).

At the centre of FATE, a decolonised feminist ethic of care, as an ethic of individual and collective responsibility for both men and women, of resistance and activism for emancipation and structural transformation, provides the ethical foundation that suffuses every aspect of that system. The Social Solidarity Economy offers a practical pathway towards structural transformation, creating equitable and inclusive trade for societal and environmental flourishing. The Human Rights Based Economy provides the overarching international, legally binding framework, which fosters accountability and compliance according to internationally agreed values, guidelines, and commitments.

Tourism is essentially a people-centred industry, focused on service. Therefore, care should be at its core, not just for tourists, but for its workers, for the communities, who are the stewards for the social and natural environment, and all other sentient beings. However, in the drive for profit, privatisation, growth, and increased automation, values of care within the capitalist model of tourism are becoming ever more marginalised. Fraser (2016) argues that the current form of financialised capitalism has spawned a 'crisis of care', resulting from the contradiction that capitalism's success depends on 'free-riding on the life-world' (p.101) while, simultaneously, draining it of the resources it needs to thrive.

A Feminist Ethic of Care in tourism and hospitality re-envisions human and non-human relations from the perspective of women's contextualised and empowered experience. FEoC redefines care and social reproduction, defeminises and destigmatises it. Care becomes a core requirement for social and planetary survival and, as such, is accorded the respect it deserves, allowing women to achieve their full potential, generally and in the tourism environment. Achieving full potential also means recognition and analysis of intersectional dimensions of oppressions, relating, for example, to race, class, (dis)ability, and sexual orientation, 'drawing on

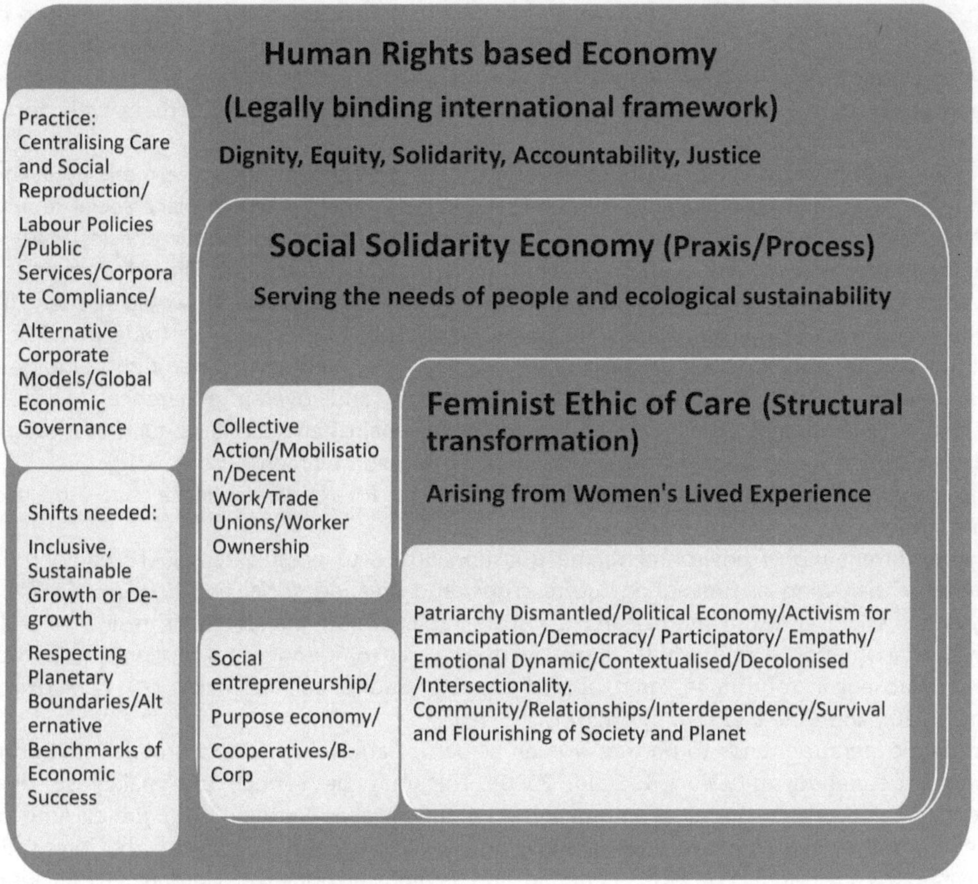

Figure 1. Feminist Alternative Tourism Economics (FATE).

insights from Black feminist literature' (Chambers, 2022, p. 1590), both in tourism consumption and production.

Colonialism has been explored in the tourism literature (Hall & Tucker, 2004; Bandyopadhyay, 2010; Koot, 2016), as has the link between colonialism and gender inequality (Chambers & Buzinde, 2015). Patriarchal relations have been discussed, especially in respect of tourism labour (Sinclair, 1997) and 'the male gaze' (Pritchard & Morgan, 2000). However, the complex relationship between patriarchy, colonialism, capitalism and tourism in the debate on gender justice needs to be more fully explored. We argue that a focus on patriarchal and neo/post-colonial structures in tourism within a critique of capitalism are key priorities for examining the potential for gender equality and justice in tourism.

Overcoming unequal systems of domination requires activism and political engagement, as well as a democratic, participatory environment to allow collective action, solidarity, debate and intellectual diversity, and thus, structural transformation, to flourish (Mohanty, 2003b; Gibson-Graham, 2006). The analysis of SSE demonstrates, the seeds for a reconfiguration of capitalist economic models already exist, where care and social value creation have taken on a central position. Over the past twenty years, the tourism industry has made some progress in respect of sustainable and responsible tourism. As our analysis indicates, a greater awareness of the environment, ethical consumption and justice for destination communities, has spawned an increasing plethora of organisational and market initiatives embracing more just and ethical tourism and hospitality business models. These include social enterprises, community-based tourism, stakeholder ownership, and legally incorporated social benefit companies/corporations (B-Corps)

(Dyah-Pramanik et al., 2019; Zebryte & Jorquera, 2017; Higgins-Desbiolles, 2012). The B-Corp business model in tourism has become particularly prevalent in America and Latin America (Acevedo-Duque et al., 2021). Nevertheless, we contend that, within tourism and hospitality practice and academia they require a more extensive, deeper and critical engagement with decolonised, feminist economic analysis.

A HRBE, with feminist values at its core, would focus on women's empowerment by acknowledging the indivisibility of social and economic rights. Such analysis would place social reproduction and care at the centre of social policies, including tourism and labour policies; care as a moral imperative and a public responsibility, no longer solely the responsibility of women. In the context of global economic governance, a Human Rights Based Economy would reorient international trade rules in tourism, to include human rights obligations and fair trade principles for domestic and foreign investors in respect of employment, land and water rights, Indigenous rights (Kalisch, 2010; Choudhury, 2012; OHCR, 2015), and overall adherence to the UN Sustainable Development Goals. Women would be the main benefactors of such a strategy, as gender equality is integral to all the seventeen SDGs (Moreno-Alarcón & Cole, 2019).

The opportunities of the integrated FATE system for crisis resilience and future crises prevention

In the current global environment, the question arises, to what extent FATE could provide answers for managing or preventing future crises, and creating social and economic resilience. Analysis of gender issues in the research on sustainability, crisis management, resilience, vulnerability and adaptation is seriously lacking (Ravera et al., 2016). Gender and resilience in particular is notably absent in tourism research, apart from rare studies, such as Bakas (2017), Rydzik and Anitha (2020), and Cole and Gympaki (2021).

Academic literature tends to portray women as particularly vulnerable to broader environmental and structural forces (Dean & Stain, 2010). This may be correct, but mainly so due to embedded socio-cultural norms and inequalities within societies (Jordan, 2019). Pre-existing political, social and economic inequalities, poverty and exploitation often come to the fore and are aggravated by crisis-induced socio-economic and ecological upheaval (Jordan, 2019; Lindberg Falk, 2012). Furthermore, as demonstrated by the impact of COVID-19 on the tourism industry, the lack of diversification in tourism-dependent economies exacerbates vulnerability (UNCTAD., 2020).

Resilience implies the ability to recover and 'bounce back' from exogenous shock, 'the capacity of a system to absorb disturbance and reorganize while undergoing change so as to still retain essentially the same function, structure, identity, and feedbacks' (Walker et al., 2004, p.2). More than the ability to withstand and recover from shocks, resilience must be perceived as an opportunity for structural, gendered, and transformative change, which addresses the root causes of people's vulnerabilities (Jordan, 2019), 'the capacity to create a fundamentally new system when ecological, economic, or social (including political) conditions make the existing system untenable' (Walker et al., 2004, p.5).

The proposed, integrated FATE incorporates the opportunity for a transformed tourism system. Greater attention to human rights and equality can reduce vulnerability. A focus on feminist values of care, on planetary boundaries and growth of human and planetary well-being, rather than GDP, will provide a chance for tackling climate change and future pandemics. Lindberg Falk (2012) suggests that social resilience 'implies hardness or resistance in tandem with elasticity and flexibility' (p.179). Women are accustomed to resistance and to maintaining their own and their families' integrity in times of adversity. Research suggests, that, in times of crisis, women don't just negotiate for personal gain, but for the benefit of their families and communities (Bakas, 2017). In business, in leadership positions, or as members of co-operatives, their risk taking and crisis solving approach incorporates a cautious, long-term orientation, tending to avoid debt and high investments, which could bankrupt them during economic hardship (Boffey, 2017). Their ability to be creative and flexible helps with diversifying products and activities. Their focus on

Table 1. Core tenets of Feminist Alternative Economics (FAE).

a.	Recognises the inherent interrelationship between patriarchy, colonialism and capitalism, and critiques these, with the goal of structural transformation and emancipation.
b.	Sees power relations as the central force driving social and economic dynamics, including different forms of inequality on an intersectional basis (race, gender, class).
c.	Considers care as a central element of the economic context.
d.	Understands paid labour, unpaid domestic care, and other unpaid work, as inextricably interlinked.
e.	Includes reproductive labour (raising children, caring for the elderly, feeding the family) and generative labour (bearing children).
f.	Builds on relational and contextual dynamics, acknowledging situated and Indigenous knowledge systems.
g.	Applies qualitative reflexive and embodied, as well as quantitative methodologies, conscious of the subjective, gendered nature of research.

care and social networking skills assists with creating self-help groups, as psychological, economic and knowledge support systems, fostering social cohesion (Lindberg Falk, 2012). The majority of women have to negotiate the triple burden of poverty and/or, employment/entrepreneurship, unpaid care and domestic duties, while resisting patriarchal dominance and socio-economic oppression on a daily basis: sexual harassment, gender-based violence, racial and sexual discrimination, being undervalued and ignored. Such challenges necessitate individual fortitude, as well as solidarity and social networks to create collective power, which generate individual and collective resilience.

Conclusion

Global crises, such as the Covid-19 pandemic and climate change, are exacerbating existing gender inequalities, which are reflected in the tourism and hospitality sector, where women dominate in the lowest ranking positions. The tourism system represents an integral part of the current financialised, neo-liberal, capitalist economy, based on patriarchal and neo-colonial ideology. Our analysis contends that a tourism transformation towards gender justice needs to be envisioned within a critique of that system. Feminist Alternative Economics challenges the existing economic system by incorporating an intersectional, decolonised analysis of power relations, based on the diversity of women's experiences, and social reproduction as inherently linked to production. We have presented an integrated system of Feminist Alternative Tourism Economics (FATE) composed of feminist perspectives on human rights, solidarity and care that could potentially provide the opportunity for social and economic transformation in tourism and, by implication, a more crisis resistant economic and tourism system. FATE addresses the gap in the tourism literature on structural transformation for gender justice and enriches the debate within tourism scholarship on pathways towards sustainable tourism.

Our analysis is not exhaustive, and presents just one alternative vision for a future, where feminist perspectives on economics could lead a path to greater global and gender justice in tourism. By applying an ambitious, macro-economic, multi-disciplinary approach, we have only been able to skim the surface of several important, large fields of research. Each of these, in particular feminist research, would deserve more in-depth, critical scrutiny in relation to tourism and our proposed integrated system. Acknowledging this as a limitation, we suggest a number of future research themes to deepen the analysis of FATE.

We propose the following avenues as particularly pertinent: 1) implementation of FATE in praxis, incorporating decolonised and feminist epistemologies; 2) a critical investigation into claims of universality of feminist principles related to the ethics of care, SSE and HRBE and the present tourism system; 3) a critical feminist analysis of SSEs in tourism and an exploration into how the SSEs support the SDGs; 4) a critical feminist analysis of the UN Human Rights Council's legally binding instruments to regulate the activities of transnational tourism corporations and other business enterprises, including transnational digital platforms.

The challenges of implementing FATE are substantial. For example, Human Rights, although internationally agreed, are interpreted differently, and systematically abused across the globe. At the time of writing, COVID-19 is still prevalent worldwide and climate change is worsening. Rather than planning for recovery, it seems that society will need to adapt to live with crises as a normal state of being. Therefore, a vision for an alternative economic and social system, in which tourism, as a powerful global force, can provide leadership for justice-based transformation, is crucial for society's survival.

Acknowledgements

We are grateful to Maria Gympaki and Antonia Gkiosi for their invaluable research assistance.

Disclosure statement

No potential conflict of interest was reported by the author(s).

ORCID

Stroma Cole http://orcid.org/0000-0002-9135-9339

References

ABTA. (2018). ABTA launches new modern slavery guidance and online training for Members. Available at: https://www.abta.com/news/abta-launches-new-modern-slavery-guidance-and-online-training-members [Accessed 10/09/21].

Abdelghani, S. (2020). Las Kellys demand fair treatment and respect to hotel housekeepers in Spain. In Vizcaino-Suárez, P., Jeffrey, H. & Eger, C. (Eds.), *Tourism and Gender-Based Violence: Challenging Inequalities*, pp. 110–111. CABI.

Acevedo-Duque, Á., Gonzalez-Diaz, R., Vega-Muñoz, A., Fernández Mantilla, M. M., Ovalles-Toledo, L. V., & Cachicatari-Vargas, E. (2021). The role of B-companies in tourism towards recovery from the crisis Covid-19, inculcating social values and responsible entrepreneurship in Latin America. *Sustainability*, *13*(14), 7763. https://doi.org/10.3390/su13147763

Bakas, F. (2017). Community resilience through entrepreneurship: The role of gender. *Journal of Enterprising Communities: People and Places in the Global Economy*, *11*(1), 61–77. https://doi.org/10.1108/JEC-01-2015-0008

Bakas, F. E., Costa, C., Breda, Z., & Durão, M. (2018). A critical approach to the gender wage gap in tourism labor. *Tourism Culture & Communication*, *18*(1), 35–49. https://doi.org/10.3727/109830418X15180180585167

Bandyopadhyay, R. (2010). 'You can do anything in Goa, India.' A visual ethnography of tourism as neo-colonialism. In Burns, P., Lester, J.-A. & Bibbings, L. (Eds.), *Tourism and Visual Culture: Methods and Cases* (Vol. *2*; Ch. 17). CABI International.

Barkin, D., & Lemus, B. (2014). Rethinking the social and solidarity society in light of community practice. *Sustainability*, *6*(9), 6432–6445. https://doi.org/10.3390/su6096432

Barnes, H. M., Eich, E., & Yessilth, S. (2018). Colonization, whenua and capitalism: Experiences from Aotearoa New Zealand. *Continuum*, *32*(6), 685–697. https://doi.org/10.1080/10304312.2018.1525918

Baum, T., & Hai, N. T. T. (2019). Applying sustainable employment principles in the tourism industry: Righting human rights wrongs? *Tourism Recreation Research*, *44*(3), 371–381. https://doi.org/10.1080/02508281.2019.1624407

Bessell, S. (2015). The individual deprivation measure: Measuring poverty as if gender and inequality matter. *Gender & Development*, *23*(2), 223–240. https://doi.org/10.1080/13552074.2015.1053213

Bhat, B. A., & Rather, T. A. (2009). Child labour in the handicrafts home industry in Kashmir: A sociological study. *International NGO Journal*, *4*(9), 391–400. https://doi.org/10.5897/INGOJ.9000020

Bianchi, R. (2011). Tourism, capitalism and Marxist political economy. In Mosedale. J. (Ed.), *Political economy of tourism: A critical perspective*, pp.17–37. Routledge.

Bianchi, R. V., & de Man, F. (2021). Tourism, inclusive growth and decent work: A political economy critique. *Journal of Sustainable Tourism*, *29*(2–3), 353–371. https://doi.org/10.1080/09669582.2020.1730862

Binion, G. (1995). Human rights: A feminist perspective. *Human Rights Quarterly*, *17*(3), 509–526. 10.1353/hrq.1995.0022

Boffey, D. (2017). EU to push for 40% quota for women on company boards. The Guardian. Available at https://www.theguardian.com/world/2017/nov/20/eu-to-push-for-40-quota-for-women-on-company-boards [Accessed 26/11/2021].

Bolton, S., & Houlihan, M. (2007). *Searching for the human in human resource management: Theory, practice and workplace contexts*. Bloomsbury Publishing.

Borowiak, C., Safri, M., Healy, S., & Pavlovskaya, M. (2018). Navigating the fault lines: Race and class in Philadelphia's. *Solidarity Economy. Antipode*, *50*(3), 577–603. https://doi.org/10.1111/anti.12368

Britton, S. (1991). Tourism, capital, and place: Towards a critical geography of tourism. *Environment and Planning D: Society and Space*, *9*(4), 451–478. https://doi.org/10.1068/d090451

Brohman, J. (1996). New directions in tourism for third world development. *Annals of Tourism Research*, *23*(1), 48–70. https://doi.org/10.1016/0160-7383(95)00043-7

Bump, J. B., Baum, F., Sakornsin, M., Yates, R., & Hofman, K. (2021). Political economy of COVID-19: Extractive, regressive, competitive. *BMJ (Clinical Research ed.)*, *372*, n73. https://doi.org/10.1136/bmj.n73.

Burris, V. (1982). The dialectic of women's oppression: Notes on the relation between capitalism and patriarchy. *Berkeley Journal of Sociology*, *27*, 51–74.

Camargo, B. A., Jamal, T., & Wilson, E. (2016). Toward a critical ecofeminist research paradigm for sustainable tourism. In Munar, A.M., & Jamal, T. (Eds.), *Tourism research paradigms: critical and emergent knowledges* (1st Ed.). Emerald Group Publishing Limited.

Chambers, D., & Buzinde, C. (2015). Tourism and decolonisation: Locating research and self. *Annals of Tourism Research*, *51*, 1–16. https://doi.org/10.1016/j.annals.2014.12.002

Chambers, D. (2022). Are we all in this together? Gender intersectionality and sustainable tourism. *Journal of Sustainable Tourism*, *30*(7), 1586–1601. https://doi.org/10.1080/09669582.2021.1903907

Cheruvallil-Contractor, S. (2018). The right to be human: How do Muslim women talk about human rights and religious freedoms in Britain? *Religion & Human Rights*, *13*(1), 49–75. https://doi.org/10.1163/18710328-13011172

Choudhury, B. (2012). *Public services and international trade liberalization: Human rights and gender implications*. Cambridge University Press.

Cole, S. (2014). Tourism and water: From stakeholders to rights holders, and what tourism businesses need to do. *Journal of Sustainable Tourism*, *22*(1), 89–106. https://doi.org/10.1080/09669582.2013.776062

Cole, S. (2017). Water worries: An intersectional feminist political ecology of tourism and water in Labuan Bajo, Indonesia. *Annals of Tourism Research*, *67*, 14–24. https://doi.org/10.1016/j.annals.2017.07.018

Cole, S. (2018). *Gender equality and tourism: Beyond empowerment*. CABI.

Cole, S., & Gympaki, M. (2021). Gender, tourism and resilience: Building forward better on small islands. *Japanese Society for Geographical Sciences*, *76*(2), pp87–105.

Costa, C., Carvalho, I., & Breda, Z. (2011). Gender inequalities in tourism employment: The Portuguese case. *Revista Turismo & Desenvolvimento*, No.15, 37–54. https://doi.org/10.34624/rtd.v0i15.13439

Costa, C., Bakas, F. E., Breda, Z., Durão, M., Carvalho, I., & Caçador, S. (2017). Gender, flexibility and the 'ideal tourism worker'. *Annals of Tourism Research*, *64*, 64–75. https://doi.org/10.1016/j.annals.2017.03.002

Curran, D. (2021). *Inside-out hospitality: A study of working conditions in the hospitality sector in Ireland* (J.E. Cairnes, Ed.). School of Business & Economics.

Dean, J. G., & Stain, H. J. (2010). Mental health impact for adolescents living with prolonged drought. *Australian Journal of Rural Health*, *18*(1), 32–37. https://doi.org/10.1111/j.1440-1584.2009.01107.x

Deen, A., Harilal, V., & Achu, F. N. (2016). Women in Tourism: Experiences and challenges faced by owners of small accommodation establishments. *African Journal of Hospitality, Tourism and Leisure*, Special Edition, *5*(3), 1–15

Donald, K. (2020). A rights-based economy – Putting people and planet first. Christian aid & centre for economic and social rights (CESR). Available at: https://www.cesr.org/rights-based-economy-putting-people-and-planet-first [Accessed 24/06/2021].

Dyah-Pramanik, P., Ingkadijaya, R., & Achmadi, M. (2019). The role of social capital in community based tourism. *Journal of Indonesian Tourism and Development Studies*, *7*(2), 62-73. https://doi.org/10.21776/ub.jitode.2019.007.02.02

Eger, C., Munar, A. M., & Hsu, C. (2022). Gender and tourism sustainability. *Journal of Sustainable Tourism*, *30*(7), 1459–1475. https://doi.org/10.1080/09669582.2021.1963975

Federici, S. (2008). The reproduction of labor power in the global economy and the unfinished feminist revolution. In Federici, S. (Ed.), *Revolution at Point Zero: Housework, Reproduction, and Feminist Struggle* (2nd Ed.). PM Press.

Fennell, D. A. (2006). *Tourism ethics*. Channel View Publications.

Ferguson, L. (2010). Tourism development and the restructuring of social reproduction in Central America. *Review of International Political Economy*, *17*(5), 860–888. https://doi.org/10.1080/09692290903507219

Ferguson, L. (2018). Gender equality and tourism: The global policy context. In Cole (Ed.), *Gender equality and tourism: Beyond empowerment*. CABI.

Flint, E., Bartley, M., Shelton, N., & Sacker, A. (2013). Do labour market status transitions predict changes in psychological well-being? *Journal of Epidemiology and Community Health*, *67*(9), 796–802. https://doi.org/10.1136/jech-2013-202425.

Fonteneau, B., Neamtam, N., Wanyama, F., Pereira Morais, L., & de Poorter, M. (2010). *Social and solidarity economy: Building a common understanding* (1st Ed.). ITCILO. [Accessed 22/09/2021].

Fraser, N. (2016). Contradictions of capital and care. *New Left Review, 100, 99-117.*

Gibson-Graham, J. K. (2006). Imagining and enacting a postcapitalist feminist economic politics. *Women's Studies Quarterly, 34*(1/2), 72–78.

Gilligan, C. (1983). *In a different voice: Psychological theory and women's development.* Harvard University Press.

Gilligan, C. (2011). Interview in ethics of care, 16 July. Available at: https://ethicsofcare.org/carol-gilligan/ [Accessed 10/08/2021].

Gouws, A., & Van Zyl, M. (2015). Towards a feminist ethics of ubuntu: Bridging rights and ubuntu. In *Care ethics and political theory*. Oxford University Press.

Grandy, G., Cukier, W., & Gagnon, S. (2020). (In)visibility in the margins: COVID-19, women entrepreneurs and the need for inclusive recovery. *Gender in Management: An International Journal, 35*(7/8), 667–675. https://doi.org/10.1108/GM-07-2020-0207

Grey, S. (2004). Decolonising feminism: Aboriginal women and the global 'Sisterhood'. Selected Works. Enweyin, Vol. VIII, 2003–2004. Available at: http://works.bepress.com/samgrey/13/.

Guimarães, C. R. F. F., & Silva, J. R. (2016). Pay gap by gender in the tourism industry of Brazil. *Tourism Management, 52*, 440–450. https://doi.org/10.1016/j.tourman.2015.07.003

Hall, M. C., & Tucker, H. (2004). *Tourism and postcolonialism: Contested discourses, identities and representations.* Routledge.

Higgins-Desbiolles, F. (2008). Justice tourism and alternative globalisation. *Journal of Sustainable Tourism, 16*(3), 345–364. https://doi.org/10.1080/09669580802154132

Higgins-Desbiolles, F. (2012). The Hotel Bauen's challenge to cannibalizing capitalism. *Annals of Tourism Research, 39*(2), 620–640. https://doi.org/10.1016/j.annals.2011.08.001

Higgins-Desbiolles, F. (2020). Socialising tourism for social and ecological justice after COVID-19. *Tourism Geographies, 22*(3), 610–623. https://doi.org/10.1080/14616688.2020.1757748

Higgins-Desbiolles, F., & Monga, M. (2021). Transformative change through events business: A feminist ethic of care analysis of building the purpose economy. *Journal of Sustainable Tourism, 29*(11–12), 1989–2007. https://doi.org/10.1080/09669582.2020.1856857

Higgins-Desbiolles, F., Carnicelli, S., Krolikowski, C., Wijesinghe, G., & Boluk, K. (2019). Degrowing tourism: Rethinking tourism. *Journal of Sustainable Tourism, 27*(12), 1926–1944. https://doi.org/10.1080/09669582.2019.1601732

Hillenkamp, I., Lucas dos Santos, L., Eynaud, P., Laville, J. L., Dos Santos, L. L., Banerjee, S., Hulgard, H., & Avelino, F. (2019). The domestic domain within a post-colonial, feminist reading of social enterprise: Towards a substantive, gender-based concept of solidarity enterprise. In *Theory of social enterprise and pluralism: Social Movements, Solidarity Economy, and Global South*, pp. 90–11. Routledge.

Hird, A. (2021). Two-year struggle pays off for sub-contracted hotel workers in Paris. RFI. Available at https://www.rfi.fr/en/france/20210525-22-month-struggle-pays-off-for-sub-contracted-hotel-workers-in-paris-ibis-accor-migrant-labour-strike [Accessed 29/11/2021].

hooks, b (1994). Love as the practice of freedom. In *Outlaw Culture: Resisting Representations* (Ch. 20). Routledge Classics.

hooks, b (2001). *All about love: New visions.* Harper Collins.

Hoskins, C. T. K. (1997). In the interests of Maori women? Discourses of reclamation. *Women's Studies Journal, 13*, 2.

Human Rights Council. (2021). Access to a healthy environment, declared a human right by UN rights council. UN News, Global Perspective Human stories. Available at: https://news.un.org/en/story/2021/10/1102582 [Accessed 10/10/21].

International Tourism Partnership and Green Hotelier. (2014). Know how guide: Human rights & the hotel industry. Available at: https://sustainablehospitalityalliance.org/resource/know-how-guide-human-rights-and-the-hotel-industry/ [Accessed 10/10/2021].

Irwin, R. (2013). Ecological ethics in the context of climate change: Feminist and indigenous critique of modernity. *International Social Science Journal, 64*(211–212), 111–123. https://doi.org/10.1111/issj.12056

Jacobs, M., & Mazzucato, M. (Eds.). (2016). *Rethinking capitalism: Economics and policy for sustainable and inclusive growth*. John Wiley & Sons.

Jamal, T., & Camargo, B. A. (2014). Sustainable tourism, justice and an ethic of care: Toward the just destination. *Journal of Sustainable Tourism, 22*(1), 11–30. https://doi.org/10.1080/09669582.2013.786084

Jamal, T., & Higham, J. (2021). Justice and ethics: Towards a new platform for tourism and sustainability. *Journal of Sustainable Tourism, 29*(2–3), 143–157. https://doi.org/10.1080/09669582.2020.1835933

Jordan, J. C. (2019). Deconstructing resilience: Why gender and power matter in responding to climate stress in Bangladesh. *Climate and Development, 11*(2), 167–179. https://doi.org/10.1080/17565529.2018.1442790

Kaji, M. (2020). 300,000 seafarers still stuck on ships: 'We feel like hostages'. Available at: https://abcnews.go.com/Politics/300000-seafarers-stuck-ships-feel-hostages/story?id=72948111 [Accessed 10/09/2021]

Kalisch, A. (2010). Fair trade in tourism – A marketing tool for social transformation? In Cole, S. & Morgan, N. (Ed.), *Tourism and Inequality: Problems and Prospects* (pp. 85–106). CABI.

Kearins, K., & Schaefer, K. (2017). Women, entrepreneurship and sustainability. In Henry, C., Nelson, T., & Lewis., K. (Eds.), *The Routledge Companion to Global Female Entrepreneurship*. Routledge.

Koot, S. P. (2016). Contradictions of capitalism in the South African Kalahari: Indigenous Bushmen, their brand and baasskap in tourism. *Journal of Sustainable Tourism*, *24*(8–9), 1211–1219. https://doi.org/10.1080/09669582.2016.1158825

Lindberg Falk, M. (2012). Gender, Buddhism and social resilience in the aftermath of the tsunami in Thailand. *South East Asia Research*, *20*(2), 175–190. https://doi.org/10.5367/sear.2012.0099

Lovelock, B., & Lovelock, K. (2013). *The ethics of tourism: Critical and applied perspectives*. Routledge.

Mikaere, A. (1994). Maori women: Caught in the contradictions of a colonised reality. *Waikato Law Review*, *2*, 125–150. HeinOnline.

Mohanty, C. T. (2003a). "Under WesternEyes" revisited:feminist solidarity through anticapitalist struggle. *Signs: Journal of Women in Culture and Society*, *28*(2), 499–535. https://doi.org/10.1086/342914

Mohanty, C. T. (2003b). *Feminism without borders - Decolonizing theory, practicing solidarity*. Duke University Press.

Moreno-Alarcón, D. M., & Cole, S. (2019). No sustainability for tourism without gender equality. *Journal of Sustainable Tourism*, *27*(7), 903–919. https://doi.org/10.1080/09669582.2019.1588283

Moriarty, A., MacRae, M., Singh, N., Nabulivou, N., & Leli, P. (2020). Gendered impacts of Covid-19 on women in Fiji. COVID-19 Response Gender Working Group.

Mottiar, Z. (2016). Exploring the motivations of tourism social entrepreneurs. *International Journal of Contemporary Hospitality Management*, *28*(6), 1137–1154. https://doi.org/10.1108/IJCHM-02-2016-0096

Nanthini, S., & Nair, T. (2020). COVID-19 and the impacts on women NTS insight (IN20-05). RSIS Centre for Non-Traditional Security Studies (NTS Centre). Nanyang Technological University Singapore.

Neef, A. (2021). *Tourism, land grabs and displacement: The darker side of the feel-good industry*. Routledge.

Nieuwenhuis, R., Munzi, T., Neugschwender, J., Omar, H., & Palmisano, F. (2018). Gender equality and poverty are intrinsically linked: A contribution to the continued monitoring of selected sustainable development goals. Discussion Paper, No. 26. UN Women.

Nobel, R., Walker, R., Caballero, L., & Herten, A. (2020). Another world is possible: Advancing feminist economic alternatives to secure rights, justice and autonomy for women and a fair, green, gender equal world. 1, 10. ActionAid UK and ActionAid Netherlands. Available at: https://www.actionaid.org.uk/publications/another-world-possible-advancing-feminist-economic-alternatives [Accessed 12/02/2021].

Nobel, R., Browne, M., Cole, S., Latchford, R., Nang, D., & de Gama, A. (2012). *Water equity in tourism: A human right-A global responsibility*. Tourism Concern, UK.

OHCR. (2015). Land and human rights – Standards and applications. United Nations Human Rights Office of the High Commissioner.

Oksala, J. (2018). Feminism, capitalism, and ecology. *Hypatia*, *33*(2), 216–234. https://doi.org/10.12957/ek.2019.49382

Parisi, L. (2017). Feminist perspectives on human rights. In: *Oxford Research Encyclopedia of International Studies*, 1–30. https://doi.org/10.1093/acrefore/9780190846626.013.48

Patil, V. (2011). Reproducing-resisting race and gender difference: Examining India's online tourism campaign from a transnational feminist perspective. *Signs: Journal of Women in Culture and Society*, *37*(1), 185–210. https://doi.org/10.1086/660181

Pritchard, A., & Morgan, N. (2000). Privileging the male gaze - Gendered tourism landscapes. *Annals of Tourism Research*, *27*(4), 884–905. https://doi.org/10.1016/S0160-7383(99)00113-9

Pritchard, A. (2014). Gender and feminist perspectives in tourism research. In: Lew, A., Hall, M. C. & Williams A. M. (Ed.), *The Wiley Blackwell Companion to Tourism* (1st Ed.). John Wiley & Sons.

Quiñones, B. R. (2008). Facets of solidarity economy. In: Quiñones, B. R., Fretel, A., Poirier, Y. & Kawano, E., Johnson, P. & Delille, T. (Ed.), *A non-patriarchal economy is possible: Looking at solidarity economy from different cultural facets*, pp. 17–86. Alliance for a Responsible, Plural and Solidarity-based Economy (ALOE).

Quiñones, B. R. (2015). *Social and solidarity economy in Asia: A South-South and triangular cooperation perspective*. ILO.

Radic, A., Lück, M., Ariza-Montes, A., & Han, H. (2020). Fear and trembling of cruise ship employees: Psychological effects of the COVID-19 pandemic. *International Journal of Environmental Research and Public Health*, *17*(18), 6741. https://doi.org/10.3390/ijerph17186741

Rastegar, R., Higgins-Desbiolles, F., & Ruhanen, L. (2021). COVID-19 and a justice framework to guide tourism recovery. *Annals of Tourism Research*, *91*, 103161. https://doi.org/10.1016/j.jdmm.2021.100620

Ravera, F., Iniesta-Arandia, I., Martín-López, B., Pascual, U., & Bose, P. (2016). Gender perspectives in resilience, vulnerability and adaptation to global environmental change. *Ambio*, *45*(Suppl. 3), 235–S247. https://doi.org/10.1007/s13280-016-0842-1.

Reindrawati, D. Y. (2018). Social entrepreneurship in tourism: A way to involve locals in tourism development. *KnE Social Sciences*, *3*(11), 173–185. https://doi.org/10.18502/kss.v3i11.2758

Reinke, A. D. (2018). Why tourism social entrepreneurs are good for business and community. Available at: https://www.travindy.com/2018/03/why-tourism-social-entrepreneurs-are-good-for-business-and-community/?fbclid=IwAR0lyq37vUJv-jAO77nZuCfbGBQTrvV-oD-7xhrPULTxCjYkMVaLjAw_t7Y [Accessed 11/09/2021].

RIPESS. (2015). *Global vision for a social solidarity economy: Convergences and differences in concepts, definitions and frameworks*. Réseau Intercontinental de Promotion de l'Économie Sociale Solidaire.

Robinson, F. (2015). Care ethics, political theory, and the future of feminism. In: Engster, D. & Hamington, M. (Eds.), *Care Ethics and Political Theory*, 293–311. Oxford Scholarship on-line. https://doi.org/10.1093/acprof:oso/9780198716341.001.0001

Robinson, J. (2006). Navigating social and institutional barriers to markets: How social entrepreneurs identify and evaluate opportunities. In Mair, J., Robinson, J. & Hockerts, K. (Eds.), *Social entrepreneurship* (pp. 95–120). Palgrave Macmillan. https://doi.org/10.1057/9780230625655_7

Robinson, R., Martins, A., Solnet, D., & Baum, T. (2019). Sustaining precarity: Critically examining tourism and employment. *Journal of Sustainable Tourism*, *27*(7), 1008–1025. https://doi.org/10.1080/09669582.2018.1538230

Ruiz, A., & Adie, B. (2020). Social capital and community-based tourism: The case of San Juan Atzingo, Mexico. In Kumar Walia, S. (Ed.), *The Routledge Handbook of Community-Based Tourism Management: Concepts, Issues & Implications* (1st Ed., pp. 183–193). Routledge. https://doi.org/10.4324/9780429274664

Rydzik, A., & Anitha, S. (2020). Conceptualising the agency of migrant women workers: Resilience, reworking and resistance. *Work, Employment and Society*, *34*(5), 883–899. https://doi.org/10.1177/0950017019881939

Saini, R., & Begum, N. (2020). Demarcation and definition: Explicating the meaning and scope of 'decolonisation' in the social and political sciences. *The Political Quarterly*, *91*(1), 217–221. https://doi.org/10.1111/1467-923X.12797

Seifert, A. M., & Messing, K. (2006). Cleaning up after globalization: An ergonomic analysis of work activity of hotel cleaners. *Antipode*, *38*(3), 557–578. https://doi.org/10.1111/j.0066-4812.2006.00595.x

Smith, M., & Duffy, R. (2003). *The ethics of tourism development*. Routledge.

Soler-i-Martí, R., Trepat, A. C., Oliveras, E., & Andrés, M. S. (2021). The social and solidarity economy: An opportunity for young women's work transitions. *YOUNG*, *29*(5), 529–548. https://doi.org/10.1177/1103308820986057

Stonich, S. C. (1998). Political ecology of tourism. *Annals of Tourism Research*, *25*(1), 25–54. https://doi.org/10.1016/S0160-7383(97)00037-6

Sustainable Hospitality Alliance. (2021). Reporting under the UK Modern Slavery Act: A factsheet for the hospitality industry. Available at: https://sustainablehospitalityalliance.org/resource/reporting-under-the-uk-modern-slavery-act-2/ [Accessed 10/10/2021].

Terry, W. C. (2011). Geographic limits to global labour market flexibility: The human resources paradox of the cruise industry. *Geoforum*, *42*(6), 660–670. https://doi.org/10.1016/j.geoforum.2011.06.006

Toussaint, E. (2020). The World Bank, the IMF and the respect of human rights. Available at: https://www.cadtm.org/The-World-Bank-the-IMF-and-the-respect-of-human-rights [Accessed 05/10/2021].

UNCTAD. (2020). COVID-19 and tourism: Assessing the economic consequences. Available at: https://unctad.org/system/files/official-document/ditcinf2020d3_en.pdf. [Accessed 29/11/2021].

United Nations. (2011). *Guiding principles on business and human rights*. United Nations Human Rights Office of the High Commissioner.

UN/DESA. (2020). Impact of COVID-19 on SDG progress: A statistical perspective. Policy brief No. 81

UNTFSSE. (2020). What role for the social and solidarity economy in the post covid-19 crisis, UNTFSSE statement, June 2020.

UNTFSSE. (2019). *Implementing the Sustainable Development Goals: What Role for Social and Solidarity Economy?*. UNTFSSE International Conference.

UN Women. (2020). COVID-19 and its economic toll on women: The story behind the numbers. Available at: https://www.unwomen.org/en/news/stories/2020/9/feature-covid-19-economic-impacts-on-women [Accessed 12 March 2021].

UNWTO (2019). *Global report on women in tourism* (2nd Edn.). The World Tourism Organisation. Available at: https://www.unwto.org [Accessed 07/10/2021].

Unite the Union Unite Hotel Workers Branch. (2018). Unethical IHG. A decade of commitment to the UN Global Compact. A decade of union avoidance in the UK. Unite the Union &. Available at: https://unitetheunion.org/media/2384/unethical-london-brochure-ihg-crowne-plaza.pdf [Accessed 03/04/2022].

Urban, J., & Pürckhauer, A. (2016). The perspectives of pluralist economics. Feminist Economics. Available at: https://www.exploring-economics.org/en/orientation/feminist-economics/ [Accessed 13/02/2021].

Verschuur, C., Calvão, F., Farah, I., Fournier, M. L., Guérin, I., Haritas, K., … Venkatasubramanian, G. (2015). Feminist analysis of social and solidarity economy practices: Views from Latin America and India. Working paper.

Walker, B., Holling, C. S., Carpenter, S. R., & Kinzig, A. (2004). Resilience, adaptability and transformability in social–ecological systems. *Ecology and Society*, *9*(2), 5. Available at: https://www.ecologyandsociety.org/vol9/iss2/art5 [Accessed 24/11/2021]. https://doi.org/10.5751/ES-00650-090205

Wichterich, C. (2021). Human rights are women's rights are LGBTIQ rights are disabled rights are black rights (People of Colour) are rights of other groups. Human rights and feminist Critique. Available at: https://rosalux-geneva.org/christa-wichterich [Accessed 20/08/2021].

Winchenbach, A., Hanna, P., & Miller, G. (2019). Rethinking decent work: The value of dignity in tourism employment. *Journal of Sustainable Tourism, 27*(7), 1026–1043. https://doi.org/10.1080/09669582.2019.1566346

Women LEAD. (2019). Are women's rights really human rights? Feminist critiques of human rights. September. Available at: http://www.womenleadnetwork.com/?p=154. [Accessed 19/08/2021].

Wood, R. C. (2020). Perspectives on hospitality industry trade unionism in the UK and beyond. *Research in Hospitality Management, 10*(2), 137–147. https://doi.org/10.1080/22243534.2020.1869456

Weldon, S. L. (2018). Gender and global justice: Lu's justice and reconciliation in world politics. *Ethics & Global Politics, 11*(1), 31–41. https://doi.org/10.1080/16544951.2018.1507389

World Economic Forum. (2021). Global Gender gap report. March.

Wright, C., & Nyberg, D. (2015). *Climate change, capitalism, and corporations: Processes of Creative Self-Destruction*. Cambridge University Press.

Zebryte, I., & Jorquera, H. (2017). Chilean tourism sector "B Corporations": Evidence of social entrepreneurship and innovation. *International Journal of Entrepreneurial Behavior & Research, 23*(6), 866–879. https://doi.org/10.1108/IJEBR-07-2017-0218

Justice and community citizenship behavior for the environment: small tourism business entrepreneurs' perspectives

Mao-Ying Wu, Xinfang Wu, Qiu-cheng Li, Jie Wang and Yi Wang

ABSTRACT

Small tourism business entrepreneurs (STBEs) are important stakeholders in rural destinations, so that their community citizenship behavior for the environment (CCBE) greatly influences the sustainable development of rural tourism communities. This paper thus focuses on STBEs' CCBE. Guided by equity theory and social exchange theory, this article examines the impact of perceived justice on CCBE and the mediating roles of place identity and subjective well-being in that process. The survey data was collected from 434 STBEs in rural China. Results indicate that perceived justice positively relates to place identity and subjective well-being. Furthermore, place identity and subjective well-being fully mediate the effect of perceived justice on CCBE. These findings offer a newer understanding of the underlying mechanism between perceived justice and CCBE and offer implications for destination managers to promote sustainable development of rural destinations.

Introduction

The environment is a valuable asset for rural destinations. Environmental issues are, however, complex and difficult to address owing to the common-pool resource attribute of the environment (Briassoulis, 2002; Qiu et al., 2022a). Environmental protection requires the joint efforts of multiple stakeholders of the destination community (Li & Wu, 2020). In recent years, considerable studies have examined the environmentally responsible behavior of tourists (He et al., 2022; Li & Wu, 2020; Qiu et al., 2022b) and hosts (Liu et al., 2022; Song et al., 2019; Wang et al., 2021). However, the discretionary environmental behavior of small tourism business entrepreneurs (STBEs) remains relatively under-researched (Garay & Font, 2013; Smith et al., 2022; Thomas et al., 2011). STBEs describe entrepreneurs and owner-operators of small tourism businesses with no more than 50 employees, which offer lodging and/or catering for tourists (Li et al., 2021). They are critical stakeholders in rural destinations as their businesses dominate the rural tourism and hospitality sector (Getz & Carlsen, 2000). These community-based small businesses can help promote destinations' sustainability by improving the natural environment and landscape, protecting local traditions and heritage, providing authentic and green tourism products and services, bringing new usages to resources, alleviating poverty, and revitalizing rural communities (

Agyeiwaah, 2019; Aquino, 2022; Scheyvens et al., 2021). As such, STBEs act as a powerful engine of sustainable rural tourism and their voluntary environmental behavior plays a crucial role in this process (Font et al., 2016).

STBEs' environmental practices lie within in the context of tourism destinations. This article contends that tourism destinations are not only geospatial entities, but also united organizations that must effectively manage their resources to meet tourist needs and enhance their competitiveness (Wu et al., 2022). Rural tourism destinations are relatively smaller in scale. They usually have identifiable boundaries, continuing bonds, and collective goals. In this sense, they have the major characteristics of organizations (Robbins, 1990). Considering rural destinations as organizations, STBEs' discretionary environmental behavior towards the community is strategically equivalent to employees' organizational citizenship behavior for the environment (OCBE) in conventional organizations (Boiral, 2009). This study thus labels STBEs' discretionary environmental behavior towards the community as the community citizenship behavior for the environment (CCBE). CCBE performed by STBEs can compensate for the deficiencies in formal environmental management systems and improve the cooperative governance of common-pool resources in the destinations; it also helps to form higher quality experience of the tourists and achieve a competitive advantage (Cunha et al., 2020). Therefore, just like OCBE of employees, which helps to improve environmental management and functioning of conventional organizations, CCBE of STBEs greatly contributes to the market competitiveness and the sustainability of rural tourism destinations.

Considering the significance of STBEs' environmental behavior, the literature assesses its potential antecedents, including personal traits and general environmental attitudes (Khan et al., 2022; Kornilaki et al., 2019; Qiu et al., 2022a), social interactional factors (Hallak et al., 2013; Kornilaki & Font, 2019; Wen et al., 2021), and business characteristics (Garay & Font, 2013). Because studies in this area are still inconclusive, more research is warranted to further investigate the systematic mechanisms of STBEs' sustainability practices (Thomas et al., 2011; Wen et al., 2021). To fill these voids, this article explores the critical factors and mechanisms that encourage STBEs to proactively adopt CCBE toward the hosting community. In particular, research shows that perceived justice is a key driver of organizational citizenship behavior of employees (Kloutsiniotis & Mihail, 2020; Teresa et al., 2020) and pro-environmental behavior of residents (He et al., 2018; Liu et al., 2022). Enlightened by this stream of research, this article proposes that perceived justice critically determines STBEs' CCBE, and probes the mechanism of STBEs' engagement in CCBE from a justice perspective. Tourism community research indicates that, if individuals are treated fairly, they develop positive emotions and attachment to the destination and behave in ways that benefit the destination (Ap, 1990). For similar reasons, perceived justice might enable STBEs to generate a sense of subjective well-being (i.e. positive emotional state) and place identity (i.e. attachment to the community), thus motivating them to perform CCBE. As such, this article: (1) investigates the effect of perceived justice on STBEs' CCBE, and (2) examines the mediating effects that subjective well-being and place identity play in the association between perceived justice and CCBE. To this end, this article adopts cognition-emotion-behavior theory to construct an integrated model. This model considers perceived justice as cognitive evaluation, subjective well-being and place identity as the critical emotional factors, and CCBE as the behavioral response.

The exploration of the above issues is especially important in the COVID-19 pandemic. It has exacerbated the challenges faced by many STBEs, especially those in the Global South that depend on international tourists (UNWTO, 2020). The perception of injustice is often aggravated during many crises, which is against the principles of sustainability (Rastegar et al., 2021a). Studies explore how perceived equality/inequality emerges in rural tourism and heritage tourism settings (Rastegar et al., 2021b; Wang et al., 2022). It is equally important to check how perceived justice can bring positive outcomes to both individuals and the destination. This study takes STBEs as the research respondents and explores how perceived justice motivates their CCBE, thus contributing to the field of justice and tourism by explaining how perceived justice can promote sustainability of rural tourism destinations.

Literature review and hypotheses development

STBEs in rural destinations

The rural tourism industry features many small tourism businesses, such as bed and breakfast, home stays, guesthouses, local snack stores, and restaurants. Most small tourism businesses are owned by individuals or families and operate in family houses that the entrepreneurs own or rent (Wang et al., 2019). STBEs are engaged in frontline and managerial practices with their families (Hallak et al., 2012). They shape the directions and decisions of their businesses to the degree that small tourism businesses are an extension of the entrepreneurs' personality and characteristics (Font et al., 2016).

The distinctiveness of STBEs lies in their strong involvement within the local community and the dominant lifestyle motivation of their businesses. Specifically, STBEs depend largely on their social network in the local community and the various resources therein to start and run their businesses (Baù et al., 2019). Also, the local community provides the environment for entrepreneurship, and serves as a meaningful and emotional setting for social interactions (Li et al., 2021). These help STBEs to develop a positive and affective tie with the place (i.e. place identity), which usually leads to their supportive behaviors toward the local community (Hallak et al., 2012).

In addition, many of the STBEs are lifestyle-oriented (Getz & Carlsen, 2000); for them, the pursuit of a certain lifestyle is usually prioritized over maximizing profit (Font et al., 2016). The dominance of lifestyle motivations is mirrored in STBEs' attitudes regarding environment protection and evaluation of business performance. STBEs tend to show concern for the environment and engage in pro-environmental behavior (Font et al., 2016). More importantly, as lifestyle-motivated STBEs aim to sustain and improve their quality of life, they may evaluate business success largely through subjective indicators, such as subjective well-being, personal satisfaction, and health status (Cunha et al., 2020).

These unique characteristics of STBEs suggest that they may engage in varied environmental practices and are motivated by different factors compared to conventional enterprises (Font et al., 2016). Because of their high degree of personalization, local involvement and lower visibility, STBEs tend to engage in more voluntary and informal environmental initiatives rather than adopt merely scaled-down versions of the environmental management systems and tools of large firms (Font et al., 2016). Such discretionary and informal behaviors of STBEs can have a cumulative and significant effect on the environment (Kornilaki & Font, 2019). Research usually conceptualizes STBEs' environmental behaviors under the framework of corporate social responsibility (Garay & Font, 2013) and sustainable behavior (Font et al., 2016), which emphasizes their responsibility for environment, society or intergeneration. These may not comprehensively reflect the discretionary, informal, and tacit nature of such behaviors. As such, we adopt community citizenship behavior for the environment (CCBE) in the following section to describe STBEs' voluntary and positive environmental actions in the community.

STBEs' CCBE

CCBE is derived from OCBE which is deeply rooted in the framework of organizational citizenship behavior (OCB) (Boiral, 2009). OCBE is individual voluntary environmental efforts that are not acknowledged by the official incentive system and, in the aggregate, promote effectively the organizations' environmental performance (Boiral & Paillé, 2012; Robertson & Barling, 2017). OCBE has four main characteristics: firstly, it is performed by individuals by their own volition

(Boiral & Paillé, 2012); secondly, it is neither strictly required by the job description nor rewarded by formal incentives (Boiral, 2009); thirdly, it is an altruistic action that can immediately improve the natural environment, and indirectly benefit the individuals and organizations (Robertson & Barling, 2017); and finally, it is featured by the social attribute of collective action. Notably, addressing environmental issues requires collective efforts owing to their complexity and public nature (Boiral, 2009). Therefore, OCBE contributes to organizations' environmental performance in the aggregate.

Derived from OCBE, CCBE holds under the precondition that tourism destinations can be viewed as united organizations for tourism management and destination competition (Wu et al., 2022). This article focuses on rural tourism destinations not only because STBEs are prevalent in such destinations, but also because rural destinations are more consistent with organizations' characteristics proposed by Robbins (1990). Firstly, rural destination communities are generally small in size, which have geographical and psychological boundaries that can distinguish between members and nonmembers. Secondly, STBEs in rural communities depend more on the local resources, resulting in closer and longer-lasting relationships with local communities (Baù et al., 2019). Also, they usually have stronger familial and cultural ties with the rural community. These characteristics would foster long-term emotional connections between the STBEs and the community, which enable the sustainability of rural tourism (Wu et al., 2022). Finally, STBEs in rural tourism destinations extensively and frequently interact with tourists; these interactions are usually an integral part of the tourists' experience. Therefore, rural destinations require particularly collective efforts of STBEs to achieve their common goals. Thus, the distinctive nature of rural destinations as organizations makes them a significant setting for understanding STBEs' voluntary and altruistic actions towards the community environment.

STBEs' CCBE is vital to enhancing hosts' well-being, tourists' experience and destinations' sustainability. Specifically, unlike conventional organizations that are mainly the workplaces of employees, the rural destination community provides a setting for both work and life for STBEs. As STBEs depend on the resources and environment of the community (Baù et al., 2019), they are more likely to perform CCBE, which is closely related to their immediate and personal interests. STBE's strong emotional attachment to the community where they live and work may also motivate them to adopt more CCBEs (Wu et al., 2022). Engaging in CCBE generates positive emotions, such as enjoyment and a sense of fulfillment, which further contribute to the well-being of STBEs (Robertson & Barling, 2017). Also, CCBE helps to create a better natural environment, improve infrastructure and cultural heritage, and enhance living conditions for both STBEs and locals (Ramkissoon, 2020).

CCBE can also enhance the tourist experience. Because the STBEs' living space and the tourist space overlap spatially, they have extensive interactions with tourists. What STBEs say and do would significantly impact the quality of the tourists' experience. CCBE performed by STBEs could encourage tourists to travel and behave in an environmentally sound manner. This would further contribute to a unique and meaningful experience for tourists, improving their satisfaction, loyalty, and consequently, competitiveness of the destination (Wu et al., 2022).

Given the importance of STBEs' environmental actions, the literature explores various factors that may motivate STBEs to behave sustainably. They comprise personal traits and life-style (Garay & Font, 2013), values (Hassanli & Ashwell, 2020), environmental concern and knowledge (Li & Wu, 2020), and social-interdependent variables such as place attachment and community embeddedness (Wen et al., 2021), and shared norms (Kornilaki & Font, 2019). However, although STBEs' responses can be seriously affected by justice matters pertaining to their businesses and lives (Liu et al., 2022), the dominant attitudinal factor of perceived justice has not received due attention. Specifically, no research examines how STBEs' justice perceptions affect their engagement in CCBE. Therefore, this article addresses this gap and explores STBEs' perceived justice's impacts on their CCBE.

Perceived justice and STBEs' CCBE

Perceived justice is an essential principle to promote the destinations' sustainability and improve the local community members' well-being (Rastegar et al., 2021a; Wang et al., 2022). This article argues that perceived justice also matters for STBEs in rural tourism destinations and acts as an important antecedent of STBEs' engagement in CCBE. Perceived justice is defined as an individual's judgement of how fairly s/he is treated in an organization (Adams, 1965). It consists of three distinctive components: distributive, procedural, and interactional justice (Adams, 1965). Perceived justice predicts OCB in conventional organizations (Podsakoff et al., 2000). Equity theory postulates that individuals will alter their efforts to resolve the stress created by inequity (Adams, 1965). The response to unfair treatment would be a lower level of OCB engagement (Organ et al., 2006). The principles of social exchange theory also suggest that performing OCB can be considered as a kind of reciprocation in exchange relationships (Blau, 1964). When an individual perceives fair treatment that acts as a channel to communicate how the organization values its members, s/he will develop a sense of obligation to reciprocate by performing OCB (Chan & Lai, 2017). As indicated by Kloutsiniotis and Mihail (2020), employees who are treated fairly will develop a sense of responsibility toward the hotel and thereby perform more service-oriented OCBs. Furthermore, perceived justice can develop positive attitudes, such as trust (Podsakoff et al., 2000), organizational commitment (Teresa et al., 2020), and job satisfaction (Nadiri & Tanova, 2010), which in turn positively affect their engagement in OCB.

The rationale between perceived justice and OCB can also be applied to the relationship between perceived justice and CCBE. As an important stakeholder of the destination, STBEs would interact extensively with the destination in several aspects. In this process, they may experience justice or injustice in matters regarding access to information and resources, involvement in policy-making, allocation of costs and benefits of tourism development, and being treated with respect (Rastegar et al., 2021b). Accordingly, STBEs may perceive fairness when: (1) they have access to sufficient information to make informed choices and have their opinions respected, (2) they bear reasonable costs and have an equal opportunity to benefit from tourism development and (3) they are respected and treated with dignity in interactions with destination managers and other stakeholders. Notably, perceived justice is not limited to STBEs' business operations, but is also reflected in their non-economic lives in the community. Such a perception of justice in both business and life domains may profoundly affect STBEs' attitudes and behaviors toward the destination community (Su et al., 2019).

When STBEs are treated fairly in the community, they may perceive that it cares about their needs and values their contributions. This may encourage STBEs to trust their communities (Nunkoo & Ramkissoon, 2012) and generate a high level of community support (Su et al., 2019), which would subsequently lead to their engaging in CCBE reciprocally. Fair treatment can also satisfy STBEs' demands for approbation, affiliation, and esteem, and thus enhance their self-efficacy (Wen et al., 2021). This will further motivate them to protect their communities. Although rarely focused on CCBE from the perspective of STBEs, research identifies the effect of perceived justice on individuals' environmental actions. For instance, Tuan (2019) observes that perceived justice can significantly affect employees' environmental commitment, which is positively related to their OCBE. In tourism destinations, residents who receive fair treatment will form stronger bonds with the community, thus triggering their environmentally responsible behaviors (He et al., 2018). Liu et al. (2022) also confirms the positive effect of perceived justice on residents' environmentally responsible behaviors. Hence, we propose the following hypothesis:

H1: STBEs' perceived justice positively affects their engagement in CCBE.

Effects of perceived justice on CCBE through the mediation of place identity and subjective well-being

Effects of perceived justice on place identity and subjective well-being

Perceived justice is significantly associated with individuals' identity with their organizations. Social identity theory (Tajfel & Turner) suggests that an individual's self-concept strongly correlates with the individual's membership recognition in the social group where one affiliates. Organizations taking fair procedures and resource distribution can earn the respect of their members; this would encourage members to identify themselves more strongly with their organizations (Su et al., 2019). The more fairness an employee receives, the better s/he would identify with the organization (Teresa et al., 2020).

How STBEs identify with a tourism destination is similar to how employees identify with their employing organization. Place identity is a person's emotional tie towards a certain place or community (Zhang & Xu, 2019). As STBEs are involved in the social structure of local place where they live and work, they tend to develop an ingrained attachment to it (Hallak et al., 2013). When STBEs are fairly treated by a destination, they will feel respected, supported, and are proud of being an important part of its community. Such a sense of respect and pride contributes to the development of their place identity (He et al., 2018). In other words, if STBEs feel being treated fairly, they will incorporate the destination into their sense of self and thus view it as part of them (Liu et al., 2022; Wen et al., 2021). He et al. (2018) and Su et al. (2019) also demonstrated a positive impact of residents' perceptions of justice on their community identification. Thus, we suggest:

H2: STBEs' perceived justice positively affects their place identity.

This paper also posits that perceived justice positively links to subjective well-being of STBEs, which describes a broad category of phenomena comprising affective reactions, general life satisfaction, and domain satisfaction (Diener et al., 1999). Self-determination theory (Deci & Ryan, 1980) indicates that the satisfying basic psychological needs contribute to a positive emotional situation, which can improve individuals' subjective well-being. Procedural justice and interpersonal justice can fulfill intrapersonal and interpersonal needs, which increasingly foster individuals' satisfaction (Ilies et al., 2018). Perceived justice can positively influence the satisfaction with job, organizations, and communities (He et al., 2018; Nadiri & Tanova, 2010). In a rural destination, lifestyle-motivated STBEs value the achievement of non-economic goals, such as personal well-being, quality of life, and overall health (Wang et al., 2019). Fair treatment would satisfy the socio-emotional needs of STBEs for self-esteem, self-efficacy, and a sense of belonging, which may contribute to their perceived overall well-being (Liu et al., 2022; Su et al., 2019). On the contrary, perceived injustice may have a negative impact on their subjective well-being as it results in negative emotions (e.g. sadness, anger, anxiety, fear, and depression) (Wang et al., 2022). As such, the following hypothesis is proposed:

H3: STBEs' perceived justice positively affects their subjective well-being.

Effects of place identity and subjective well-being on CCBE

This article proposes that place identity and subjective well-being are influential antecedents of STBEs' CCBE. According to social identity theory, when individuals feel that they are an integral part of an organization, they tend to behave in its collective interests, and choose to conduct OCB or OCBE (Tajfel & Turner, 1979). The tenets of place attachment theory (Bowlby, 1979) also indicate that emotional bonds developed to a specific place can motivate individuals to protect the place. Place identity can foster a strong sense of environmental responsibility, thereby encouraging actions to protect the environment.

Buil et al. (2019) and Zhang et al. (2017) indicated a positive association between organizational identification and OCB. The literature also suggests that employees' organizational identification motivates them more likely to make a "sacrifice" by performing OCBE or pro-environmental behavior (Peng et al., 2020; Su & Swanson, 2019). This is because they will psychologically internalize the benefits of pro-environmental behaviors gained by their organizations when they identify with them (Peng et al., 2020). Similarly, in a destination community, residents who identify with their communities tend to act collectively to improve their community environment. For example, Zhang and Xu (2019) find that place identity positively affects residents' place citizenship behavior in urban destinations. Such emotional ties can also strongly promote residents' environmental citizenship behavior in big cities (Song et al., 2019). Although conducted in different contexts, all these studies confirm place identity as a critical driver of pro-environmental behavior. As for STBEs, they may be more inclined to form emotional ties to the place because of their local position within the social fabric of rural communities (Li et al., 2021); this encourages them to be more responsible in protecting community environment (Wen et al., 2021). Hallak et al. (2012) verify that STBEs' place identity enhances their support for the host community (e.g. protecting local environment). Hence, this paper proposes that a positive association exists between place identity and CCBE. The following hypothesis is postulated:

H4: STBEs' place identity positively affects their engagement in CCBE.

In addition to place identity, subjective well-being is also likely to predict STBEs' CCBE decisions. Social exchange theory reveals that individuals will adjust their behaviors in reaction to a social exchange's negative or positive outcomes (Blau, 1964). The literature indicates that subjective well-being can generate positive behaviors (Su et al., 2018; Su & Swanson, 2019). Specifically, individuals tend to perform pro-environmental behaviors when they are more satisfied with a place (Ramkissoon et al., 2013). This may be because individuals who have satisfactory living conditions incline to feel attached to the community; this results in their greater willingness to put the community's welfare above their immediate private interest (Ramkissoon, 2020).

Research examines the relationship between subjective well-being and individuals' pro-environmental activities. Visitors' satisfaction positively influences their pro-environmental behavioral intentions (Ramkissoon et al., 2013). In a hotel setting, Su and Swanson (2019) report that employees' increased well-being would stimulate their engagement in green workplace behavior. Regarding the destination community, residents' overall community satisfaction positively associates with their support of tourism development (Nunkoo & Ramkissoon, 2011) and environmentally friendly behavior (Su et al., 2018). More relevant to our research, Font et al. (2016) observe that lifestyle-driven STBEs, who value quality of life, exhibit more sustainable actions. Given the foregoing research, subjective well-being could psychologically determine STBEs' CCBE. Thus, we propose that:

H5: STBEs' subjective well-being positively affects their engagement in CCBE.

Mediation of place identity and subjective well-being

While perceived justice is likely to directly influence CCBE, we argue that it can also contribute to CCBE through the mediation of place identity and subjective well-being. Research supports the potential role of place identity and subjective well-being as mediators (Ilies et al., 2018; Teresa et al., 2020). Social identity theory (Tajfel & Turner, 1979) shows that, employees fairly treated by the organization may identify with it, and are thus motivated to behave in the organization's interest. Organ et al. (2006) reveal that perceived justice significantly affects employees' identification and commitment to an organization, which further predicts OCB. Teresa et al. (2020) also examine the relationship among the hotel employees' justice perceptions,

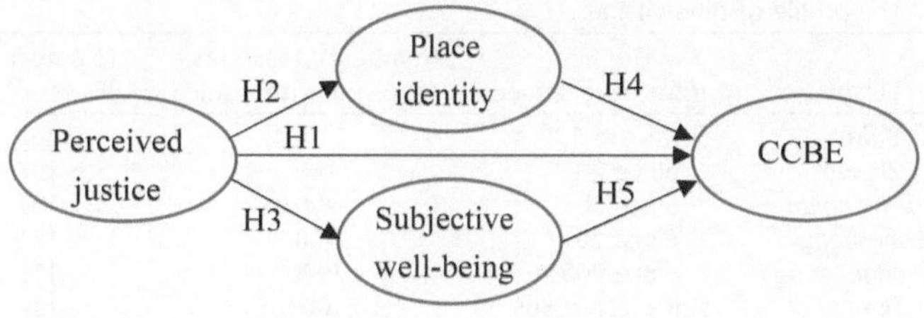

Figure 1. The conceptual model.

organizational commitment (i.e. identifying with and involvement in the organization), and OCB. Although these authors do not investigate the mediation of organizational commitment, they clearly indicate that perceived justice positively impacts organizational commitment, which further significantly affects OCB. Likewise, He et al. (2018) find that residents' perception of justice has a positive effect on their community identification, which in turn predicts environmentally responsible behavior. This finding is reinforced by Liu et al. (2022) study which indicates that residents' psychological ownership mediates the impact of perceived justice on environmentally friendly behavior. Thus, we posit that:

H6: Place identity partially mediates the relationship between perceived justice and CCBE of STBEs.

Subjective well-being can also be a central emotional mediator that links perceived justice and CCBE. When individuals feel fairly treated by the organization, they will develop a high level of subjective well-being, which further encourages them to perform positive behaviors toward the organization in return. Studies discover the mediating role of needs fulfillment and job satisfaction (Ilies et al., 2018) and communication satisfaction (Chan & Lai, 2017) in the perceived justice-OCB relationship. Regarding rural destinations, Wang et al. (2022) confirm that residents' subjective well-being mediates the effect of perceived justice on support for tourism. Although few studies explore subjective well-being as a mediator of the justice-CCBE link, such an exchange-based pathway can be predicted in the light of the above discussions. Therefore, we hypothesize that:

H7: Subjective well-being partially mediates the relationship between perceived justice and CCBE of STBEs.

Figure 1 presents the conceptual model of the study.

Methods

Study sites and sampling

This study was conducted in nine village-level rural tourism destinations in Hangzhou, China. These villages are famous for their stunning mountainous scenery, which are located within one to two hours' driving distance from major tourist origin regions in the Yangtze River Delta of China. These nine villages were selected mainly for two reasons. First, rural tourism plays an important role in these villages with a considerable number of STBEs running their businesses therein. Second, they differ regarding tourism development history, the level of residents' involvement, and the number of small tourism businesses; they can thus contribute to a representative and diversified sample for the study. The profile of the selected village destinations is shown in Table 1.

A questionnaire-based survey was conducted in these nine rural tourism destinations. This article adopted a systematic random sampling approach. Every second small rural tourism businesses were visited to improve representativeness of the sample. Such businesses are jointly operated and managed by the owners, most of whom are married couples (Getz & Carlsen, 2000). The researchers invited a representative of the owners who was in charge and knew best about the business to take part in the research. The paper-and-pencil survey was conducted

Table 1. The profile of nine villages.

No.	Village	Tourism development	Number of residents involved in tourism	Number of STBEs (as of 2020)
V-1	*Baisha*	Since late 1990s	1100	202
V-2	*Zhinan*	Since 2015	700	107
V-3	*Yueliangqiao*	Since 2012	200	102
V-4	*Longshang*	Since 2017	480	131
V-5	*Shangougou*	Since 2003	2000	115
V-6	*Tianmu*	Since late 1990s	1000	122
V-7	*Wushan*	Since 2005	200	71
V-8	*Taigongtang*	Since 2003	210	60
V-9	*Taipingshan*	Since 2005	150	45

Note: Data from local village committee statistics.

from June to September 2020 with the assistance of six university students from local villages. The participants were first informed of the research purposes and the anonymity of the questionnaire. The surveys were mostly self-administrated. For those who were not able to read well owing to limited education or poor eyesight, the research assistants offered help by reading the questions. A total of 478 STBEs (covering more than 50% of the population) were approached and asked to engage in the survey and 454 responses were returned. To avoid biased statistical results, the incomplete surveys were removed from further analysis. In all, 434 valid responses were adopted for further analysis. This satisfies the minimum sample size of 200 and the ratio of 10 cases per variable for accurate use and reliable results of structural equation modeling (Mueller, 1997; Kline, 2011). Table 2 illustrates the participants' profile.

Structure and measurements

The survey comprised three parts to measure CCBE, its antecedents (perceived justice, place identity, subjective well-being), and the demographic factors of the respondents. The items of CCBE and its antecedents were all evaluated on a 5-point Likert scale (1= strongly disagree, 5= strongly agree).

The literature shows that the target-based approach, which classifies OCBE based on the targets to which the action is directed, has advantages in specifying the domains of OCBE (Robertson & Barling, 2017; Williams & Anderson, 1991). This paper follows this approach to evaluate STBEs' CCBE, assessing four dimensions (self-focused CCBE, co-resident focused CCBE, tourists focused CCBE, and community focused CCBE). The approach involved: (1) using three items to measure self-focused CCBE (see Boiral & Paillé, 2012); (2) adapting three items to evaluate co-resident-focused CCBE (see Boiral & Paillé, 2012; Robertson & Barling, 2017); (3) selecting three statements to assess tourist focused CCBE (see Font et al., 2016); and (4) adopting three items to measure community-focused CCBE (see Boiral & Paillé, 2012; Robertson & Barling, 2017).

Additionally, four items were used to measure perceived justice (see Chan & Lai, 2017). Participants were asked to rate how fairly they were treated. To assess STBEs' subjective well-being, three statements were used (see Blore et al., 2011). Respondents evaluated how they perceived the difference between their quality of life at present and (1) their past life, (2) what their relevant others have, and (3) their ideal life. Four items were used to evaluate place identity (see Zhang and Xu (2019). Table 3 presents these items.

Data analysis

This paper used a two-step procedure of analysis. First, a confirmatory factor analysis (CFA) was conducted to assess the measurements' reliability and validity. Structural equation modeling was then adopted to explore the relationship among perceived justice, place identity, subjective well-being and CCBE. This study took Mplus 8.3 software for data analyses.

Table 2. Profile of respondents.

Variable	Category	Distribution, n (%)	Variable	Category	Distribution, n (%)
Gender	Male	213 (49.1)	Registered residence	Local	391 (90.1)
	Female	221 (50.9)		Non-local	43 (9.9)
Age	19–24	2 (0.5)	Household	≤5	93 (21.4)
	25–34	26 (6.0)	annual income	6–10	182 (41.9)
	35–44	73 (16.8)	(ten thousand Chinese Yuan)	11–15	72 (16.6)
	45–54	144 (33.2)		16–20	47 (10.8)
	55–64	138 (31.8)		21–30	19 (4.4)
	≥65	51 (11.8)		≥31	21 (4.8)
Education	Primary School	90 (20.7)	Length of residence	≤10	35 (8.1)
	Junior High	185 (42.6)	years	11–20	38 (8.8)
	Senior High	145 (33.4)		21–30	69 (15.9)
	College or Undergraduate	14 (3.2)		≥31	292 (67.3)
Marital status	Unmarried	13 (3.0)			
	Married	421 (97.0)			

Results

Reliability and validity test

Before assessing the scales' validity and reliability, descriptive statistics of constructs were analyzed. The mean values of CCBE (4.202) (4.122 for self-focused CCBE, 4.065 for co-resident focused CCBE, 4.234 for tourists focused CCBE, and 4.386 for community focused CCBE), perceived justice (3.442), place identity (4.122), subjective well-being (4.122) were all above the midpoint (3.000).

CFA was performed to evaluate the scales' reliability and convergent validity. The measurement model fitted the data well ($\chi^2(209)=415.557$, RMSEA = 0.048, CFI = 0.967, TLI = 0.961). All constructs' standardized loadings were statistically significant ($p < 0.001$) and passed the threshold value (0.5), thus indicating satisfactory convergent validity (Hair et al., 2010) (see Table 3). The constructs' composite reliability (CR) scores ranged from 0.790 to 0.920, all above the recommended value (0.7) (Hair et al., 2010). The average variance extracted (AVE) were all above 0.5 (Hair et al., 2010). The measurements are thus reliable.

Discriminant validity was also tested (see Table 4). Each construct's square root of AVE was higher than the factor's highest correlation with any other constructs, hence demonstrating the discriminant validity (Fornell & Larcker, 1981).

Structural model and testing of hypotheses

Structural equation modeling was undertaken to verify seven proposed research hypotheses. Since this study endeavors to test the relationships between CCBE and the other three concepts, CCBE was regarded as a second-order factor in the model following the internal-consistency approach (Little et al., 2002). Firstly, four parcels were generated using the four dimensions of CCBE as the grouping criteria. Then, each parcel was assigned the average score of all items for a given dimension, and the four parcels were used as manifest indicators of the CCBE construct.

The proposed model well fitted the data (χ^2 (221)=586.711, RMSEA = 0.062, CFI = 0.942, TLI = 0.934). The results of our path analyses show that perceived justice positively affects place identity ($\beta = 0.398$, $p < 0.001$) and subjective well-being ($\beta = 0.375$, $p < 0.001$), hence supporting H2 and H3. Also, place identity is positively linked to CCBE ($\beta = 0.429$, $p < 0.001$), thus supporting H4. H5, which hypothesized that subjective well-being positively correlates with CCBE was also supported by the data ($\beta = 0.341$, $p < 0.001$). However, perceived justice has no statistically significant direct impact on CCBE ($\beta = 0.093$, $p > 0.05$). Thus, H1 was declined. Figure 2 presents hypotheses testing's results (H1–H5).

Table 3. Results of CFA.

Constructs and scale items	Factor loading	CR	AVE
Self-focused CCBE		0.842	0.643
I take the initiative to learn environmental knowledge and skills.	0.823		
I actively participate in training for environmental protection.	0.897		
I rethink the consequence of my behavior before doing things that may impact the environment.	0.668		
Co-resident focused CCBE		0.852	0.658
I persuade other residents to do a good job of garbage classification and sewage treatment.	0.760		
I share my experience and skills in environmental protection with other residents.	0.840		
I take the initiative to communicate with other residents to solve the environment issue.	0.831		
Tourist focused CCBE		0.848	0.652
I share the experience and skills in environmental protection with the tourists.	0.841		
I work with tourists to protect the environment.	0.860		
I actively communicate with the tourists to solve the environment issue.	0.714		
Community focused CCBE		0.790	0.559
I take the initiative to do a good job in garbage classification and sewage treatment to protect the overall environment of our village.	0.786		
I actively protect the village's ancient buildings, trees and folk culture.	0.795		
I spontaneously beautify and clean my courtyard to protect our village's overall environment.	0.653		
Perceived justice		0.920	0.741
The community values my opinions and advice.	0.867		
The management department releases policy information promptly.	0.856		
The government offers fair public affairs related service.	0.859		
The tourism revenue is fairly distributed in the community.	0.862		
Place identity		0.906	0.708
It is more suitable for me to live in this community than other places.	0.769		
This community is very meaningful for me.	0.861		
I feel strongly that I belong to this community.	0.901		
I take pride in this community.	0.829		
Subjective well-being		0.844	0.643
I live a better life than others in the community.	0.840		
I live happier now than in the past.	0.771		
The current life is ideal.	0.793		

Table 4. Test of discriminant validity.

	1	2	3	4	5	6	7
Self-focused CCBE	**0.807**						
Co-resident focused CCBE	0.660	**0.748**					
Tourists focused CCBE	0.625	0.833	**0.802**				
Community focused CCBE	0.611	0.617	0.686	**0.841**			
Perceived justice	0.261	0.387	0.297	0.209	**0.861**		
Place identity	0.433	0.533	0.543	0.524	0.385	**0.841**	
Subjective well-being	0.590	0.523	0.400	0.503	0.359	0.581	**0.802**

Note: Bold numbers in this table represent the squared root of AVE.

The mediating impacts were estimated by bootstrapping analysis with 5000 iterations and a 95% bias-corrected confidence interval (CI) (see Preacher & Hayes, 2008). Table 5 indicates that both place identity ($\beta = 0.171$, $p < 0.001$) and subjective well-being ($\beta = 0.128$, $p < 0.001$) fully mediate the impact of perceived justice on CCBE. Therefore, the data partially supported both H6 and H7.

Discussion

Summary of the research findings

Rural tourism's sustainability greatly relies on the destinations' natural environment, and protection of the environment requires collective efforts of the stakeholders including STBEs. STBEs' community citizenship behavior for the environment (CCBE) is beneficial to shaping the tourists'

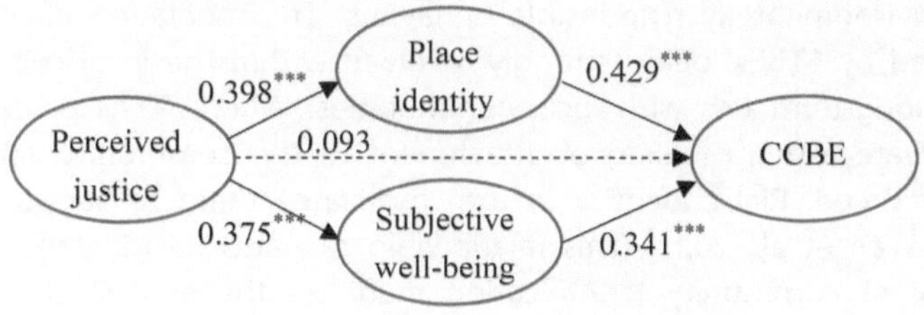

Note. *p<0.1, **p<0.01, ***p<0.001.

Figure 2. Results of structural equation modeling.

Table 5. Bias-corrected bootstrap test on mediating effects.

Paths	Estimate	S.E.	95% CI (Low)	95% CI (High)
PJ→PI→CCBE	0.171***	0.046	0.089	0.271
PJ→SW→CCBE	0.128***	0.035	0.069	0.206

Note. PJ = perceived justice; PI = place identity; SW = subjective well-being.
***$p < 0.001$.

experience, improving the well-being of local community, and enhancing the sustainability of destinations. This paper thus focuses on STBEs' CCBE and its influencing factors. Guided by equity theory and social exchange theory, this paper explores the systematic relationships among perceived justice, place identity, subjective well-being, and STBEs' engagement in CCBE. Seven hypotheses are formulated and verified using a sample of 434 STBEs from nine rural tourism destinations. Most of the proposed hypotheses are proved by the statistical analyses.

The findings reveal that perceived justice positively affects place identity and subjective well-being. These results recall to the literature showing that perceived justice has positive impacts on community identification, community satisfaction, and subjective well-being (He et al., 2018; Wang et al., 2022). This article verifies that perceived justice strengthens the emotional ties of STBEs to destination communities and improves their subjective well-being.

In line with our expectations, place identity and subjective well-being positively affect STBEs' engagement in CCBE. These results coincide with the literature showing that place identity (Liu et al., 2022; Song et al., 2019) and subjective well-being (He et al., 2018; Wang et al., 2022) can motivate individuals to protect the community environment and back up community tourism development. The positive impacts of place identity and subjective well-being on CCBE can also be explained by its collective and altruistic attributes. Environmental protection requires collective participation owing to its public nature. Therefore, STBEs' CCBE can be triggered when they feel that they are integral to the destination community. As indicated by Li and Wu (2020), the positive effect of social-interdependent factors, in this case, place identity might be stronger for CCBE. Also, CCBE is an altruistic action that can do good to the environment, the destination, and other stakeholders (Robertson & Barling, 2017). Altruistic environmental behaviors are more likely to be motivated by a positive emotional state (Wen et al., 2021). This thus implies that subjective well-being could be a crucial driver of CCBE.

Furthermore, perceived justice indirectly impacts STBEs' engagement in CCBE through the mediation of place identity and subjective well-being. This finding conflicts with research that supports the positive and direct influence of perceived justice on OCB (Kloutsiniotis & Mihail, 2020; Nadiri &Tanova, 2010). However, it accords with research indicating that organizational identification and a positive emotional state may mediate the link between perceived justice and OCB (Ilies et al., 2018; Teresa et al., 2020). This article confirms that place identity and subjective well-being act as mediators between perceived justice and CCBE. Specifically, place identity mediates the perceived justice-CCBE relationship. This result is consistent with He et al. (2018) and Liu et al. (2022) who report that perceived justice can enhance community identity, and thereby promote environmentally responsible behaviors. The mediation effect of place identity can also be explained by STBEs' being strongly involved within the local community (Baù et al., 2019). STBEs value long-term ties with local communities. When STBEs perceive fair treatment, they intend to cultivate a deep feeling of attachment to the community, and thus their place identity can be reinforced. Place identity in turn motivates them to support their community (Hallak et al., 2012; Wen et al., 2021). This relationship can also be reflected in Liu et al. (2022) view that the sense of community identification mediates the association between residents' justice perception and their willingness to behave sustainably.

Subjective well-being also functions as a mediator between perceived justice and CCBE. This finding agrees with the literature that reports psychological needs fulfillment and perceived satisfaction mediate the justice perception-OCB link (Chan & Lai, 2017). The mediating role of subjective well-being might be related to STBEs' lifestyle motivation that emphasizes improving and sustaining their quality of life more than maximizing profit (Thomas et al., 2011). They evaluate their success based on subjective well-being rather than financial gains (Wang et al., 2019). Therefore, subjective well-being is important to STBEs' decisions and behaviors. This study shows that STBEs' justice perception can improve their subjective well-being, and thus enable their engagement in CCBE. This finding accords with the research indicating that STBEs who value subjective well-being are more responsible for environment (Wang et al., 2019).

Theoretical and managerial implications

This article extends the literature in four significant ways. First, it focuses on STBEs who are critical yet often neglected stakeholders of rural tourism destinations who pursue sustainable development. The literature mainly emphasizes pro-environmental behaviors of tourists or residents (Qiu et al., 2022a). Very few studies examine the CCBE of STBEs, which is worthy of in-depth discussion, considering that STBEs are prominent in sustainably developing destinations. This paper fills this void by examining CCBE from the perspective of STBEs. Secondly, this paper applies a new lens that views rural tourism destinations as organizations to better understand the dynamics of STBEs' discretionary pro-environmental behaviors toward the community (i.e. CCBE). Thirdly, drawing upon equity theory and social exchange theory, this paper considers perceived justice as an important, but under-studied, antecedent of CCBE. Although the importance of residents' perceived justice in destination management is discussed (Su et al., 2019), few studies explore how perceived justice impacts STBEs' behaviors towards the destination environment. This study finds that perceived justice indirectly affects CCBE, which advances our understanding of the factors that trigger CCBE in rural destinations. Finally, the current paper is among the first to examine the relationship between perceived justice and CCBE by empirically verifying the mediating role of place identity and subjective well-being. While perceived justice is related to individuals' environmental behavior (Liu et al., 2022), no research examines how perceived justice affects CCBE by following different pathways. This study identifies place identity and subjective well-being as dominant mediators, which provides a better understanding of how STBEs' CCBE is formed.

This study also offers advice for rural tourism management bodies and practitioners. The COVID-19 crisis has strengthened the importance for destinations to accord with the 2030 agenda for sustainable development (Ramkissoon, 2020). STBEs' CCBE, which highlights discretionary environmental actions, is an essential means of achieving sustainable development goals. Given the significant influence of perceived justice on CCBE, it is important to address justice-related matters in pursuing sustainable development in rural tourism destinations. Strategies can be developed to ensure procedural, distributive and interactional fairness among STBEs and other stakeholders in rural destinations. Destination managers should establish consistent and unbiased procedures to assure fair benefit and cost sharing, fair access to resources, and substantial participation in decision-making for the STBEs. More specifically, STBEs should be allowed to participate and express their voice about matters regarding their work and lives. A fairly distributed and sharing mechanism should be established to ensure that STBEs share equitable benefits and costs of tourism development. Additionally, to enhance interactional justice, STBEs should be treated with courtesy, decency, and respect. It is also crucial for destination managers to establish transparent communication channels (e.g. meetings or social media) and share information with STBEs in a timely, transparent and honest manner.

This study also shows that place identity can improve STBEs' engagement in CCBE. Destination managers and community leaders can thus devise strategies to strengthen STBEs' identity with the community. On the one hand, destination managers should involve STBEs in community matters as insiders. For instance, local cultural events and festivals can be organized to encourage STBEs' participation and interactions with local residents. On the other hand, various supportive activities can be offered to STBEs especially as has been in the difficult time during the COVID-19 pandemic. Business assistance and revitalization plans should be developed, and policies and incentives to encourage people to travel to the rural destinations should be accepted. Such efforts could possibly improve STBEs' attachment to the hosting destination and lead to more CCBEs.

Finally, considering the positive association between subjective well-being and CCBE, strategies aimed at enhancing STBEs' subjective well-being should also be encouraged. For example, training programs can be developed to improve their professional skills and personal achievements. It is also important to improve STBEs' quality of life regarding the living environment, access to education and healthcare, and recreational and entertainment facilities, given that the destination community provides both their working and living setting.

Limitations and future research

Several limitations worth noting. First, a self-reported survey was used for analysis in this paper. This may have overestimated how respondents evaluated their involvement in CCBE owing to social acceptability. Different sources of information should be considered in future studies to provide a richer perspective. Secondly, this empirical study was carried out in some Chinese rural tourism destinations, where the STBEs suffered moderately from the COVID-19 pandemic. This may result in their positive evaluation of subjective well-being and place identity. However, many STBEs in the Globe South face huge challenges, especially those who depend on international tourists (UNWTO, 2020). Therefore, caution is warranted when generating the findings of this paper to other contexts. Future studies can further verify the findings by using different samples in other tourism destinations. Thirdly, as CCBE is a new concept, more research is needed to examine its other antecedents. Fourth, because this study only explored place identity and subjective well-being as mediators between perceived justice and CCBE, further studies could consider other important attitudes, such as trust and perceived community support. Lastly, this study is cross-sectional by design, and cannot offer any implication of causality. Significant relationships may also exist among the study's constructs in the opposite direction of the proposed hypotheses. For example, a bidirectional causal relationship could possibly exist between subjective well-being and CCBE (Robertson & Barling, 2017). Future studies can benefit from longitudinal and experimental designs to further test the directionality of the causal relationships.

Conclusions

STBEs are a crucial participant in rural destinations and their community citizenship behavior for the environment (CCBE) contributes to sustainable development of the destinations. By integrating equity theory and social exchange theory, this paper proposes and tests an integrated model, explaining how perceived justice influences STBEs' engagement in CCBE through the mediation of perceived justice and subjective well-being. We find that perceived justice positively affects place identity and subjective well-being. Furthermore, place identity and subjective well-being fully mediate the effect of perceived justice on CCBE. Therefore, rural destination managers should better attend to justice matters in pursuing destinations' sustainability. Also, it is crucial to improve STBEs' place identity and subjective well-being as ways to successfully encourage their engagement in CCBE.

Disclosure statement

No potential conflict of interest was reported by the authors.

Funding

The work was supported by the Natural Science Foundation of Zhejiang Province, China (No. LQ20G020001) and the Cultural Project of Zhejiang Province (No. 21WH70080ZD)

ORCID

Jie Wang http://orcid.org/0000-0001-8850-4251

References

Ap, J. (1990). Residents' perceptions research on the social impacts of tourism. *Annals of Tourism Research*, *17*(4), 610–616. https://doi.org/10.1016/0160-7383(90)90032-M

Aquino, R. (2022). Community change through tourism social entrepreneurship. *Annals of Tourism Research*, *95*, 103442. https://doi.org/10.1016/j.annals.2022.103442

Baù, M., Chirico, F., Pittino, D., Backman, M., & Klaesson, J. (2019). Roots to grow: Family firms and local embeddedness in rural and urban contexts. *Entrepreneurship: Theory and Practice*, *43*(2), 360–385.

Blau, P. (1964). *Exchange and power in social life*. John Wiley.

Blore, J. D., Stokes, M. A., Mellor, D., Firth, L., & Cummins, R. A. (2011). Comparing multiple discrepancies theory to affective models of subjective wellbeing. *Social Indicators Research*, *100*(1), 1–16. https://doi.org/10.1007/s11205-010-9599-2

Boiral, O. (2009). Greening the corporation through organizational citizenship behaviors. *Journal of Business Ethics*, *87*(2), 221–236. https://doi.org/10.1007/s10551-008-9881-2

Boiral, O., & Paillé, P. (2012). Organizational citizenship behaviour for the environment: Measurement and validation. *Journal of Business Ethics*, *109*(4), 431–445. https://doi.org/10.1007/s10551-011-1138-9

Bowlby, J. (1979). *The making and breaking of affectional bonds*. Tavistock.

Briassoulis, H. (2002). Sustainable tourism and the question of the commons. *Annals of Tourism Research*, *29*(4), 1065–1085. https://doi.org/10.1016/S0160-7383(02)00021-X

Buil, I., Martínez, E., & Matute, J. (2019). Transformational leadership and employee performance: The role of identification, engagement and proactive personality. *International Journal of Hospitality Management*, *77*, 64–75. https://doi.org/10.1016/j.ijhm.2018.06.014

Chan, S. H. J., & Lai, H. Y. I. (2017). Understanding the link between communication satisfaction, perceived justice and organizational citizenship behavior. *Journal of Business Research*, *70*, 214–223. https://doi.org/10.1016/j.jbusres.2016.08.017

Cunha, C., Kastenholz, E., & Carneiro, M. J. (2020). Entrepreneurs in rural tourism: Do lifestyle motivations contribute to management practices that enhance sustainable entrepreneurial ecosystems? *Journal of Hospitality and Tourism Management*, *44*, 215–226. https://doi.org/10.1016/j.jhtm.2020.06.007

Deci, E. L., & Ryan, R. M. (1980). The empirical exploration of intrinsic motivational processes. *Advances in Experimental Social Psychology*, *13*(8), 39–80.

Diener, E., Suh, E. M., Lucas, R. E., & Smith, H. L. (1999). Subjective well-being: Three decades of progress. *Psychological Bulletin*, *125*(2), 276–302. https://doi.org/10.1037/0033-2909.125.2.276

Font, X., Garay, L., & Jones, S. (2016). Sustainability motivations and practices in small tourism enterprises in European protected areas. *Journal of Cleaner Production*, *137*, 1439–1448. https://doi.org/10.1016/j.jclepro.2014.01.071

Fornell, C., & Larcker, D. (1981). Evaluating structural equation models with unobservable variables and measurement errors. *Journal of Marketing Research*, *18*(1), 39–50. https://doi.org/10.1177/002224378101800104

Garay, L., & Font, X. (2013). Corporate social responsibility in tourism small and medium enterprises: Evidence from Europe and Latin America. *Tourism Management Perspectives*, *7*, 38–46.

Getz, D., & Carlsen, J. (2000). Characteristics and goals of family and owner-operated businesses in the rural tourism and hospitality sectors. *Tourism Management*, *21*(6), 547–560. https://doi.org/10.1016/S0261-5177(00)00004-2

Hair, J. F., Black, W. C., Babin, B. J., & Anderson, R. E. (2010). *Multivariate data analysis* (7th ed.). Pearson.

Hallak, R., Brown, G., & Lindsay, N. J. (2012). The place identity-performance relationship among tourism entrepreneurs. *Tourism Management*, *33*(1), 143–154. https://doi.org/10.1016/j.tourman.2011.02.013

Hallak, R., Brown, G., & Lindsay, N. J. (2013). Examining tourism SME owners' place attachment, support for community and business performance: The role of the Enlightened Self-Interest Model. *Journal of Sustainable Tourism*, *21*(5), 658–678. https://doi.org/10.1080/09669582.2012.709861

Hassanli, N., & Ashwell, J. (2020). The contribution of small accommodations to a sustainable tourism industry. *Current Issues in Tourism*, *23*(3), 261–264. https://doi.org/10.1080/13683500.2018.1530201

He, X., Cheng, J., Swanson, S., Su, L., & Hu, D. (2022). The effect of destination employee service quality on tourist environmentally responsible behavior: A moderated mediation model incorporating environmental commitment, destination social responsibility and motive attributions. *Tourism Management*, *90*, 104470. https://doi.org/10.1016/j.tourman.2021.104470

He, X., Hu, D., & Su, L. (2018). Destination resident's perceived justice, relational quality, and environmentally responsible behavior. *Tourism Tribune*, *33*(9), 117–131. (In Chinese).

Ilies, R., Lanaj, K., Pluut, H., & Goh, Z. (2018). Intrapersonal and interpersonal need fulfillment at work: Differential antecedents and incremental validity in explaining job satisfaction and citizenship behavior. *Journal of Vocational Behavior*, *108*, 151–164. https://doi.org/10.1016/j.jvb.2018.07.005

Khan, O., Bellini, N., Daddi, T., & Iraldo, F. (2022). Effects of behavioral intention and dynamic capabilities on circular economy adoption and performance of tourism SMEs. *Journal of Sustainable Tourism*, 1–20. https://doi.org/10.1080/09669582.2022.2066683

Kline, R. B. (2011). *Principles and practice of structural equation modeling (3rd ed.)*. New York: Guilford Press .

Kloutsiniotis, P. V., & Mihail, D. M. (2020). The effects of high-performance work systems in employees' service-oriented OCB. *International Journal of Hospitality Management*, *90*, 102610.

Kornilaki, M., & Font, X. (2019). Normative influences: How socio-cultural and industrial norms influence the adoption of sustainability practices. *Journal of Environmental Management*, *230*, 183–189.

Kornilaki, M., Thomas, R., & Font, X. (2019). The sustainability behaviour of small firms in tourism: The role of self-efficacy and contextual constraints. *Journal of Sustainable Tourism*, *27*(1), 97–117. https://doi.org/10.1080/09669582.2018.1561706

Li, Q., & Wu, M. (2020). Tourists' pro-environmental behaviour in travel destinations: Benchmarking the power of social interaction and individual attitude. *Journal of Sustainable Tourism*, *28*(9), 1371–1389. https://doi.org/10.1080/09669582.2020.1737091

Li, Q., Zhang, H., Wu, M., Wall, G., & Ying, T. (2021). Family matters: Dual network embeddedness, resource acquisition, and entrepreneurial success of small tourism firms in rural China. *Journal of Travel Research*, 004728752110472. https://doi.org/10.1177/00472875211047275

Little, T., Cunningham, W., Shahar, G., & Widaman, K. (2002). To parcel or not to parcel: Exploring the question, weighing the merits. *Structural Equation Modeling: A Multidisciplinary Journal*, *9*(2), 151–173. https://doi.org/10.1207/S15328007SEM0902_1

Liu, Y., Qu, Z., Meng, Z., & Kou, Y. (2022). Environmentally responsible behavior of residents in tourist destinations: The mediating role of psychological ownership. *Journal of Sustainable Tourism*, *30*(4), 807–823. https://doi.org/10.1080/09669582.2021.1891238

Mueller, R. O. (1997). Structural equation modeling: Back to basics. *Structural Equation Modeling: A Multidisciplinary Journal*, *4*(4), 353–369. https://doi.org/10.1080/10705519709540081

Nadiri, H., & Tanova, C. (2010). An investigation of the role of justice in turnover intentions, job satisfaction, and organizational citizenship behavior in hospitality industry. *International Journal of Hospitality Management*, *29*(1), 33–41. https://doi.org/10.1016/j.ijhm.2009.05.001

Nunkoo, R., & Ramkissoon, H. (2011). Developing a community support model for tourism. *Annals of Tourism Research*, *38*(3), 964–988. https://doi.org/10.1016/j.annals.2011.01.017

Nunkoo, R., & Ramkissoon, H. (2012). Power, trust, social exchange and community support. *Annals of Tourism Research*, *39*(2), 997–1023. https://doi.org/10.1016/j.annals.2011.11.017

Organ, D. W., Podsakoff, P. M., & Mackenzie, S. B. (2006). *Organizational citizenship behavior: Its nature, antecedents, and consequences*. SAGE.

Peng, X., Lee, S., & Lu, Z. (2020). Employees' perceived job performance, organizational identification, and pro-environmental behaviors in the hotel industry. *International Journal of Hospitality Management, 90*, 102632. https://doi.org/10.1016/j.ijhm.2020.102632

Podsakoff, P. M., Mackenzie, S. B., Paine, J. B., & Bachrach, D. G. (2000). Organizational citizenship behaviors: A critical review of the theoretical and empirical literature and suggestions for future research. *Journal of Management, 26*(3), 513–563. https://doi.org/10.1177/014920630002600307

Preacher, K. J., & Hayes, A. F. (2008). Asymptotic and resampling strategies for assessing and comparing indirect effects in multiple mediator models. *Behavior Research Methods, 40*(3), 879–891.

Qiu, H., Wang, X., Wei, W., Morrison, A., & Wu, M.-Y. (2022a). Breaking bad: How anticipated emotions and perceived severity shape tourist civility? *Journal of Sustainable Tourism*, 1–21. https://doi.org/10.1080/09669582.2022.2108039

Qiu, H., Wang, X., Wu, M.-Y., Wei, W., Morrison, A., & Kelly, C. (2022b). The effect of destination source credibility on tourist environmentally responsible behavior. *Journal of Sustainable Tourism*, 1–21. doi: 10.1080/09669582.2022.2067167

Ramkissoon, H. (2020). Perceived social impacts of tourism and quality-of-life: A new conceptual model. *Journal of Sustainable Tourism*, 1–17. https://doi.org/10.1080/09669582.2020.1858091

Ramkissoon, H., Graham Smith, L. D., & Weiler, B. (2013). Testing the dimensionality of place attachment and its relationships with place satisfaction and pro-environmental behaviours. *Tourism Management, 36*(36), 552–566. https://doi.org/10.1016/j.tourman.2012.09.003

Rastegar, R., Higgins-Desbiolles, F., & Ruhanen, L. (2021a). COVID-19 and a justice framework to guide tourism recovery. *Annals of Tourism Research, 91*, 103161. https://doi.org/10.1016/j.annals.2021.103161

Rastegar, R., Zarezadeh, Z., & Gretzel, U. (2021b). World heritage and social justice: Insights from the inscription of Yazd, Iran. *Journal of Sustainable Tourism, 29*(2–3), 521–540. https://doi.org/10.1080/09669582.2020.1782925

Robbins, S. P. (1990). *Organization theory: Structure, design and applications* (3rd ed.). Prentice Hall.

Robertson, J. L., & Barling, J. (2017). Toward a new measure of organizational environmental citizenship behavior. *Journal of Business Research, 75*, 57–66. https://doi.org/10.1016/j.jbusres.2017.02.007

Scheyvens, R., Carr, A., Movono, A., Hughes, E., Higgins-Desbiolles, F., & Mika, J. (2021). Indigenous tourism and the sustainable development goals. *Annals of Tourism Research, 90*, 103260. https://doi.org/10.1016/j.annals.2021.103260

Smith, H., Discetti, R., Bellucci, M., & Acuti, D. (2022). SMEs engagement with the Sustainable Development Goals: A power perspective. *Journal of Business Research, 149*, 112–122. https://doi.org/10.1016/j.jbusres.2022.05.021

Song, Z., Daryanto, A., & Soopramanien, D. (2019). Place attachment, trust and mobility: Three-way interaction effect on urban residents' environmental citizenship behavior. *Journal of Business Research, 105*, 168–177. https://doi.org/10.1016/j.jbusres.2019.08.001

Su, L., Huang, S., & Nejati, M. (2019). Perceived justice, community support, community identity and residents' quality of life. *Journal of Hospitality and Tourism Management, 41*, 1–11. https://doi.org/10.1016/j.jhtm.2019.08.004

Su, L., Huang, S., & Pearce, J. (2018). How does destination social responsibility contribute to environmentally responsible behaviour? A destination resident perspective. *Journal of Business Research, 86*, 179–189. https://doi.org/10.1016/j.jbusres.2018.02.011

Su, L., & Swanson, S. R. (2019). Perceived corporate social responsibility's impact on the well-being and supportive green behaviors of hotel employees: The mediating role of the employee-corporate relationship. *Tourism Management, 72*, 437–450. https://doi.org/10.1016/j.tourman.2019.01.009

Tajfel, H., & Turner, J. (1979). An integrative theory of intergroup conflict. *Social Psychology of Intergroup Relations, 33*, 94–109.

Teresa, A. Q., Yasmina, A. C., & Sangwon, P. (2020). The sequential relationships of hotel employees' perceived justice, commitment, and organizational citizenship behaviour in a high unemployment context. *Tourism Management Perspectives, 35*, 100676.

Thomas, R., Shaw, G., & Page, S. J. (2011). Understanding small firms in tourism: A perspective on research trends and challenges. *Tourism Management, 32*(5), 963–976. https://doi.org/10.1016/j.tourman.2011.02.003

Tuan, L. T. (2019). Catalyzing employee OCBE in tour companies: The role of environmentally specific charismatic leadership and organizational justice for pro-environmental behaviors. *Journal of Hospitality & Tourism Research, 43*(5), 682–711. https://doi.org/10.1177/1096348018817582

UNWTO. (2020). Tourism and rural development. Retrieved September 17, 2021, from https://www.unwto.org/world-tourism-day-2020/tourism-and-rural-development-technical-note

Wang, C., Li, G., & Xu, H. (2019). Impact of lifestyle-oriented motivation on small tourism enterprises' social responsibility and performance. *Journal of Travel Research, 58*(7), 1146–1160. https://doi.org/10.1177/0047287518800389

Wang, J., Wang, S., Wang, H., Zhang, Z., & Ru, X. (2021). Examining when and how perceived sustainability-related climate influences pro-environmental behaviors of tourism destination residents in China. *Journal of Hospitality and Tourism Management, 48*, 357–367. https://doi.org/10.1016/j.jhtm.2021.07.008

Wang, S., Berbekova, A., & Uysal, M. (2022). Pursuing justice and quality of life: Supporting tourism. *Tourism Management, 89*, 104446. https://doi.org/10.1016/j.tourman.2021.104446

Wen, T., Zhang, Q., & Li, Y. (2021). Why small tourism enterprises behave responsibly: Using job embeddedness and place attachment to predict corporate social responsibility activities. *Current Issues in Tourism*, *24*(10), 1435–1450. https://doi.org/10.1080/13683500.2020.1797648

Williams, L. J., & Anderson, S. E. (1991). Job satisfaction and organizational commitment as predictors of organizational citizenship and in-role behaviors. *Journal of Management*, *17*(3), 601–617. https://doi.org/10.1177/014920639101700305

Wu, M.-Y., Wu, X., Li, Q., & Tong, Y. (2022). Community citizenship behavior in rural tourism destinations: Scale development and validation. *Tourism Management*, *89*, 104457. https://doi.org/10.1016/j.tourman.2021.104457

Zhang, H., & Xu, H. (2019). Impact of destination psychological ownership on residents' "place citizenship behavior". *Journal of Destination Marketing & Management*, *14*, 100391. https://doi.org/10.1016/j.jdmm.2019.100391

Zhang, Y., Guo, Y., & Newman, A. (2017). Identity judgements, work engagement and organizational citizenship behavior: The mediating effects based on group engagement model. *Tourism Management*, *61*, 190–197. https://doi.org/10.1016/j.tourman.2017.01.005

Part II

Engaging marginalized tourism stakeholders

Reimagining children's participation: a child rights informed approach to social justice in tourism

Antonia Canosa and Anne Graham

ABSTRACT

Children's historical marginalisation largely grounded in deeply entrenched assumptions about their status as citizens, including the legitimacy of their views on matters central to their wellbeing, is now well argued. As a result, they have often been gate-kept out of research, policy, and practice spheres, including those related to tourism. This limits their potential to address the injustices they often experience, many of which are amplified during global crises. Through a critical discussion of two examples of such injustice for children in tourism-related contexts, child labour and orphanage tourism, we suggest that prevalent "protectionist" stances that circumscribe children's human rights issues in tourism, reflect broader societal and systemic adult-centric views. We propose the need for a child rights informed approach to participation in tourism research, policy and practice that better recognises the contradictions and complexities of childhood experiences in different social and cultural milieus. This offers a starting point for progressing more socially just responses in supply side tourism contexts that are less "adultist" and more "childist" in orientation.

Introduction

The United Nations Sustainable Development Group (2020) and findings from an international consultation with children about their experiences of the pandemic (Global Kids Online, 2020) highlight the unprecedented impacts and resulting social injustice experienced by many children across the globe. This is particularly evident in tourism-dependent locations in the Global South, where economic livelihoods have been significantly impacted by international travel restrictions. Covid-19 has prompted reflections by many critical tourism scholars who have advocated for a rethinking and redirecting of tourism away from neoliberal growth models to a rights-based approach focused on the benefits, interests, and rights of local populations (Canosa et al., 2021; Everingham & Chassagne, 2020; Higgins-Desbiolles, 2020; Lew et al., 2020; Rastegar et al., 2021). Focusing on the rights of often neglected stakeholders such as children and young people living and growing up in host communities is an important endeavour to progress this agenda.

In this article, we extend such debates by focusing on an important dimension of intersectional exclusion and one which is seldom explored – that of being young (Josefsson & Wall, 2020). While Rastegar et al. (2021) rightly argue a "justice turn" is emerging in tourism studies, the theorisation of such justice requires that we look to other disciplines where issues of rights, exclusion, and social justice are being addressed (Jamal & Higham, 2021). Here we draw on the broad interdisciplinary field of childhood studies (Prout & James, 1997) and the ontological lens of "childism" (Wall, 2019) to explore the complexity of childhood experiences as they relate to tourism from a supply-side (i.e. host communities) and in the context of global crises. The notion of "childism" has recently emerged as an analogy to other "ism's" such as feminism, postcolonialism, and environmentalism in an effort to disrupt normative assumptions about childhood and children and promote more child-centred and child-inclusive perspectives on issues impacting children's lives. Wall (2019, p. 1) argues childism offers a needed "critical lens for deconstructing adultism across research and societies and reconstructing more age-inclusive scholarly and social imaginations".

Children have been historically marginalised due to deeply entrenched assumptions that childhood is a time of preparation for adulthood (Prout & James, 1997). Consequently, they have often been excluded from participating in important research about their lives (Canosa, Graham & Wilson, 2019). While significant efforts have been made to date to involve children and young people in tourism research from a demand-side (i.e. children as tourists, see, for example, studies by Cullingford, 1995; Israfilova & Khoo-Lattimore, 2019; Kerr & Price, 2018; Rhoden et al., 2016; Schänzel & Smith, 2011), research exploring children's perspectives and rights from a supply-side of the tourism industry (i.e. living and growing up in host destinations) have often been overlooked (see, for example, Canosa et al., 2017; Seraphin & Green, 2019; Yang et al., 2020). However, ignoring the perspectives of children and limiting their spheres of influence has contributed to global injustices that are particularly evident in the tourism industry (Canosa et al. 2019). Josefsson and Wall (2020, p. 1049) argue that "the lived experiences of global injustice can provide an invaluable resource to imagine new routes towards global empowerment", and this is particularly so during periods of crises. However, children's lived experiences of social injustice in the context of tourism are not widely reported and remain under-theorised. While efforts to invite children and young people (predominantly in the Global North) to contribute to sustainable tourism development are happening through, for example, consultative processes (see the Global Youth Tourism Summit (GYTS) organised by global institutions such as the United Nations World Tourism Organisation (UNWTO), these initiatives are generally adult-driven, and they rarely provide a safe space for young people to express their opinions freely, nor do they involve vulnerable children who experience injustice in the tourism industry.

In this article, we critically examine two examples of such injustice linked to global crises in the context of tourism: child labour in tourism and orphanage tourism connected to child trafficking. We locate this discussion in the context of children's lack of participation more broadly, as well as in the current context of COVID-19, given the impact of the latter on children's already limited participatory rights (Cuevas-Parra, 2021). By critically discussing these examples, we propose a child rights informed theorisation of social justice as a starting point for developing the presence of "childist" standpoints (alongside "adultist" perspectives) as a way forward in promoting more socially just forms of tourism. This is important both in upholding the rights of children as enshrined in the United Nations Convention on the Rights of the Child (United Nations, 1989) and in achieving many of the Sustainable Development Goals (SDGs)[1] adopted by the United Nations Member States in 2015 to achieve a more equitable, just and sustainable world for all (UNICEF, n.d.).

Tourism social injustices in the context of global crises

A "global crisis" is defined as a period of instability and uncertainty which causes negative systemic changes in the human and/or natural environment at a global scale. According to Laydjiev (2014), a global crisis is a systemic failure with worldwide repercussions which are perceived to be dangerous. In other words, the adverse event needs to be symbolically interpreted as such in order for it to be considered a "crisis", in which case global institutions are called into action to reduce uncertainty and return the situation to a new "normal" (Laydjiev, 2014). Similarly, Hall (2010) argues that factors contributing to a crisis are generally related to an "issue-attention cycle" meaning that while the crisis can be ongoing, it is only perceived as such by a particular group of stakeholders when "(a) they are aware of the elements and/or (b) the crisis event occurs in a particular part of the global political–economic system" (Hall, 2010, p. 402). Nevertheless, global crises bring to the surface longstanding and entrenched inequities which often remain veiled (Lew et al., 2020).

In terms of tourism literature, global crises have been explored primarily in relation to the economic and financial repercussions of adverse events such as the global financial crisis or the "9/11" terrorist attacks in the USA in 2001 (see Hall, 2010). Only recently has academic interest shifted to social and environmental concerns, including, for example, the impacts of the current COVID-19 health crisis on human rights issues (Baum & Hai, 2020; Higgins-Desbiolles, 2020; Rastegar et al., 2021).

In the context of tourism, the restrictions to international travel have brought to light the precariousness of economies dependant and over-reliant on tourism, particularly in the Global South. The economic upheaval caused by the pandemic has exacerbated human rights issues already present among, for example, tourism workers, including those with little employment rights in the informal tourism and hospitality sectors (Baum & Hai, 2020) and those working for large international corporations functioning under neo-liberal imperatives. Higgins-Desbiolles (2020, p. 615) argues, for example, that the pandemic has exposed the injustices and exploitations occurring in the cruise tourism sector where poor working conditions and system failures brought to light "the plight of the unfortunate crews" whose rights have often been ignored.

Hence global crises are often a turning point in scrutinising and reassessing the way forward in protecting the human rights of certain stakeholders. In the tourism industry, these stakeholders are represented by the marginalised, vulnerable, and less heard who remain at the margins of public discourse and who are rarely invited to participate in meaningful ways in reimagining and planning for better, more equitable, and just futures. In Australia, the economic impacts of COVID-19, for example, have exacerbated existing inequalities placing financial burdens on families with "deleterious consequences for children" (see, for example, Bessell, 2021, p. 1). In the next section, we critically discuss two examples from tourism activities connected to global crises that highlight how children's perspectives are absent in understanding the pervasive and damaging impact of associated injustice. We propose working collaboratively with young people to develop responses that reflect their views and agency, are more empowering, and potentially more effective in protecting their human rights and addressing injustice.

Child labour in tourism

Child labour has been referred to as a global crisis given the hazardous nature of much work performed by children in poverty-stricken countries. World Vision, for example, lists child labour and trafficking among five global challenges including food insecurity, refugees, climate change and child marriage/gender discrimination (World Vision, 2021). A recent report from the International Labour Organisation (ILO) shows that 160 million children were in "child labour" at the beginning of 2020, with 79 million of these involved in dangerous work (International Labour Organization, 2020). Little is known, however, about the proportion involved in the tourism and hospitality sectors, whether through informal work in family businesses or in more dangerous work such as sexual exploitation, begging and street vending.

Article 32 of the UNCRC stipulates that State Parties must recognise "the right of the child to be protected from economic exploitation and from performing any work that is likely hazardous or to interfere with the child's education, or to be harmful to the child's health of physical, mental, spiritual, moral or social development" (United Nations, 1989). Similarly, Article 2 of the GCET states that "tourism activities should respect the equality of men and women; they should promote human rights and, more particularly, the individual rights of the most vulnerable groups, notably children, the elderly, the handicapped, ethnic minorities and indigenous peoples" (UNWTO, n.d.).

While the UN Sustainable Development Goal 8.7 aims to end child labour in all its forms by 2025, little attention has been paid to children's lived experiences of work in the tourism and hospitality sectors. Child labour in tourism is difficult to quantify, given the often hidden nature of their labour (Canosa & Schänzel, 2021). Apart from limited research into the sexual exploitation of children in the tourism industry, other issues related to children as workers have been neglected (Canosa et al., in press; Yang et al., 2020). In addition, studies exploring tourism and hospitality family entrepreneurship (see Bakas, 2018) and the emotional labour performed by children in this context often fail to profile children's perspectives and voice (see Seymour, 2015 as an exception). The resulting social injustice experienced by children working in the tourism and hospitality sectors remains a neglected topic.

Growing critiques of such injustice point to how human rights treaties and global codes of ethics are often based on universal understandings of childhood and prevalent assumptions of children as "vulnerable" and in need of "saving" (Josefsson & Wall, 2020; Peleg, 2019). Scholarship in the field of childhood studies calls for a "decolonization" of childhood itself, moving away from universal understandings of childhood that fail to reflect the diversity of children's experiences across familial, social, and cultural contexts, including within and across Minority and Majority World contexts (Balagopalan, 2019; Beazley, 2018; Jiménez, 2021). Specifically, Josefsson and Wall (2020, p. 1046) argue that child labour tends to be framed by "vocabularies of victimhood and protection" while many children and young people are lobbying for labour rights in terms of "participation, empowerment, and fairness". These prevailing vulnerability discourses have contributed to the exclusion of children from important research about their lives, including in relation to child labour in the tourism industry. This is particularly problematic in an era when children and young people are increasingly recognised as having important insights that can inform safeguarding policies and practices in other contexts (Moore, 2017; Powell et al., 2021) yet remain excluded from promoting more socially just forms of tourism.

Orphanage tourism and child trafficking

A second persistent form of injustice that is inextricably connected to child labour is the burgeoning practice of orphanage tourism. Orphanage tourism has been defined as "the practice of volunteering and/or visiting children in residential care facilities usually located in less developed countries in the Global South" (Canosa & van Doore, 2022). Orphanage tourism is a subsector of "voluntourism" that has emerged in the last decade and refers to volunteering or "doing good" in the countries and communities visited (Guiney & Mostafanezhad, 2015; Higgins-Desbiolles et al., 2022). While voluntourism appeals to the ethical and moral values of travellers, paradoxically, this practice is contributing to an "orphanage industry" with increased risks of child trafficking and harms to the wellbeing and safety of children institutionalised in these orphanages (Cheney & Ucembe, 2019; Goldman et al., 2020; van Doore, 2016).

Increasingly, links have been made between orphanage tourism and modern slavery (Cheer, 2018, 2019; van Doore, 2019) due to the blurring of boundaries between orphanage tourism and child trafficking. van Doore (2016) argues that often children in these establishments are only orphans on "paper" as they have at least one living parent and/or family member who could care for them with support. In this way, "orphanage trafficking" occurs when "children are recruited from their families into orphanages to pose as 'paper orphans' for the purpose of exploitation and profit" (van Doore, 2019, p. 19). In addition, there is substantial evidence to suggest that the temporary nature of much volunteering with children in orphanages (including activities such as English classes, sports, games) is of a short-term nature and often performed by unskilled volunteers, further compounding the already heightened social-emotional risks to children's wellbeing (Goldman et al., 2020; Richter & Norman, 2010). Further evidence suggests that orphanage tourism is linked to institution-based sexual exploitation as orphanages are the potential "site of or the transit point for exploitation" (Lyneham & Facchini, 2019, pp. 3).

Anti-orphanage tourism campaigns have drawn attention to this global crisis, most evident in in South-East Asia and Africa. Campaigns such as "Children are not Tourist Attractions" by Friends International and the "Love you Give" campaign by the Better Care Network are raising awareness of the potential harms of engaging in orphanage tourism and calling upon travellers to make more responsible and ethical travel choices. In 2019, the United Nations General Assembly Resolution on the Rights of the Child focused on Children without Parental Care and called for State Parties to take appropriate measures to both prevent and address harms related to orphanage volunteering programmes in the context of tourism which could lead to trafficking and exploitation (Canosa & van Doore, 2022). Similarly, Article 2.3 of the GCET seeks to prevent "the exploitation of human beings in any form, particularly sexual, especially when applied to children" (UNWTO, n.d.).

While child trafficking is considered a "global crisis" and direct links to the tourism industry are evident through orphanage tourism, little is known about children's own experiences of life in such institutions and/or of the resulting scale of social injustice. Eurocentric framings of childhood in much of the tourism literature (i.e. the "poor child" icon in volunteer tourism marketing materials, see Benali & Oris, 2019) and resulting vulnerability discourses impede efforts to involve children from these institutions in research around volunteer tourism, and hence there remains a dearth of studies reporting their insights and perspectives (see also Canosa et al., in press; Carpenter, 2019). In addition, as Carpenter (2019, p. 134) suggests, "it is likely that what volunteers do and how they do it, not simply the presence or absence of volunteers, is what has an impact on children and local communities". While "child safe" tourism initiatives exclude children and young people's perspectives, we run the risk such efforts may not be effective in improving their safety and wellbeing.

One of the few studies directly involving children in an orphanage in Cambodia highlights children's resilience in the face of extreme family poverty and lack of education (Miller & Beazley, 2022). While the authors argue that it is important to "evaluate care and development options for impoverished families ... to care for their children", it is an "idealised notion" to think that all families, particularly those migrating for work, will be able to find alternatives to the education and care provided in these residential care institutions (Miller & Beazley, 2022, p. 101). Likewise, Roche's (2020) study in residential care institutions in the Philippines provides evidence directly from 50 young people over the age of 15 years about their experiences of adapting to life in these orphanages and the agentic ways in which they make sense of their lives in the face of considerable adversity and hardship. Thus, while child trafficking for orphanage tourism must be curbed by closing exploitative orphanages (van Doore, 2019), for those that remain open, it is essential to implement safeguarding policies and practices developed in partnership with the children and young people they seek to protect.

Children's lack of participation in tourism policy and planning

In the above sections, we have explored how child labour and child trafficking crises are now receiving closer public attention in an effort to highlight associated injustices and curb human rights violations. A less evident form of social injustice highlighted by these examples is the lack of participation by children and young people in tourism policy and planning. This reflects broader societal norms that perpetuate children's prevalent exclusion from what is seen to be the political spheres of adult decision-making. Underlining the need for such injustice to be challenged and changed, Riddell and Tisdall (2021, p. 5) argue that the pandemic represents "a crucial test to explore whether children and young people's participation rights are protected and fulfilled as inalienable human rights."

While the United Nations Convention on the Rights of the Child (UNCRC) (United Nations, 1989) is now one of the most ratified international human rights treaties, recent research shows that neither the spirit nor the principles of the Convention are reflected in research focused on children and young people in the tourism industry (Canosa et al., in press). The UNCRC comprises 54 Articles that recognise children's evolving capacities and the need for care, protection, and participation. Commonly referred to as the 3Ps, the 54 Articles in the Convention are often broadly grouped into provision rights (e.g. Article 28 – the right to education), protection rights (e.g. Article 36 – the right to protection from exploitation), and participation rights (e.g. Article 12 – children's right to voice in all matters that affect them). Although the articles of the Convention are indivisible, interdependent, and interrelated (Cuevas-Parra, 2021; Peleg, 2019), the participation rights of children are those most often eclipsed by concerns around their vulnerability (Lundy, 2019).

There is little doubt the COVID-19 crisis has amplified this marginalised status of children. Recent research on the impacts of the pandemic on children's human rights shows that the response measures implemented across more than 200 countries and territories have actually been "retrogressive for children and young people's rights to participation" (Cuevas-Parra, 2021, p. 81). Cuevas-Parra (2021, p. 81) argues that the 'common denominator' across these countries has been the "reduced spaces for children and young people to influence decision-making processes and policy responses associated with COVID-19". While crises bring to the fore existing social inequalities and injustices, they also trigger responses geared towards protecting the most basic of human rights (e.g. health) at the expense of other important human rights (e.g. freedom of expression). While school closures and lockdowns were aimed at protecting children's health and wellbeing, they also provoked immeasurable negative consequences for children undermining their "rights to participate on the premise that their protection is more relevant in crisis situations" (Cuevas-Parra, 2021, p. 81). This not only underlines the tensions and complexities in navigating children's protection *and* participation but also supports the view that children's participatory rights are the most contentious and most challenging to enact for adults, adult institutions, and adult systems (McMellon & Tisdall, 2020) . Such tensions are evident in a tourism context, where children and young people remain largely neglected as stakeholders (Canosa et al., 2019), and predominantly "adultist" stances prevail in policies and practices.

While the pandemic has surfaced existing lack of regard for human rights and social justice in tourism related contexts such as the two discussed above, it has also provoked a rethinking of tourism towards de-growth models and more socially and environmentally just approaches – albeit perhaps only in aspirational terms and within academic circles (Higgins-Desbiolles, 2020; Lew et al., 2020; Rastegar et al., 2021). In the next section, we propose a child rights informed theorisation of social justice to help inform how children and young people's participation might be meaningfully advanced in the tourism industry going forward. This theorisation places a core emphasis on addressing human rights issues from the perspective of local host communities in more relational, child-inclusive, and ethical ways.

Towards a child rights informed theorisation of social justice in tourism

As flagged above, the social injustice experienced by children in the tourism industry remains a neglected topic. In the discussion below, we build on childhood studies principles to argue for a theorisation of social justice in tourism which is less "adultist" and more "childist" or child-inclusive, responsive to the rights, wellbeing, and safety of children and young people (Josefsson &

Wall, 2020). Whereas feminist and postcolonial approaches in tourism have engaged with issues of gender, class, and ethnicity (Cole, 2017; Everingham et al., 2021; Stinson et al., 2021), age is seldom addressed (Canosa et al., 2019; Sedgley et al., 2011). The issue of intergenerational equity is an important one, given youth are wearing these crises disproportionately and with little democratic input (see, for example, Jourdan & Wertin, 2020).

The question is, are children too vulnerable to participate in research, policy, and planning to inform sustainable tourism solutions? If not, are there other reasons why their participation has not been seriously enacted? We have argued that "vulnerability" should not eclipse children's agency and capacity to influence decisions that affect them (Lundy, 2019). Evidence shows that even in the most complex situations, children can be supported to participate in research and influence decision-making in positive ways. Recent research from the Royal Commission into Institutional Child Sexual Abuse in Australia shows that adult-child alliances and institutional strategies aimed at scaffolding children to participate in research and have a voice were not only important in building child safe practices at the organisational level but were also important to some extent in mitigating the vulnerabilities of the children involved (Moore, 2017).

In respect to the two examples discussed in this article, we posit that global injustice in the tourism industry is not only a problem of children's (lack of) power but, in addition, "one of deep reliance on others for standing together with children in their struggles" (Josefsson & Wall, 2020, p. 1050). A child rights informed theorisation of social justice must go beyond thinking of children as "dependant" or "independent"; children are relational and interdependent with others in their communities of practice (i.e. families, schools, and communities) (Graham & Fitzgerald, 2010), as well as, interdependent with others in the global tourism system. As such, it is important to work in partnership with children living in situations of vulnerability in the tourism industry in order to advance relevant, responsive, and ethical research, policy, and planning.

While working with children and young people in inclusive and participatory ways may be perceived as challenging and resource-intensive, there is now a wealth of knowledge in other disciplines and professions for tourism scholars to draw on. Various typologies and models of participation have emerged over the past two decades to guide research, policy and practice efforts and to help ensure these uphold children's human rights. One of the most widely recognised is the "Lundy model" that seeks to operationalise children's right to participation in matters that affect them (UNCRC Article 12) in the education sector (Lundy, 2007). Further developed into a "checklist for participation," it was included in Ireland's National Strategy on Children and Young People's Participation in Decision-Making 2015–2020. Lundy (2007) argues that "voice isn't enough" in conceptualising Article 12 of the UNCRC but that children must be given a safe, inclusive space to form and express their views *("space");* facilitated to express their views *("voice");* be listened to *("audience")* and have their views acted upon *("influence").*

Likewise, research that explored the links between participation and wellbeing of students in the education sector in Australia (Graham et al., 2018), reveals that along with student "*voice*", there needs to be recognition of their identity as human persons with agency and experience who can effect change at school (having '*influence*'). Having "*choice*" and "*working together*" were also identified as important elements of student participation positively associated with wellbeing, which require intergenerational dialogue and ongoing collaboration with significant adults in children's lives (Graham et al., 2018). Another well-known participation typology often cited in policy documents is Hart's (1992) "ladder", which locates children's participation in terms of rungs – the lowest rungs being tokenistic efforts at consulting with children, through to middle rungs where children are scaffolded and informed, and the top rungs representing children's participation in decision-making with adults. A key insight from these and other frameworks is that different approaches to participation might be engaged for different purposes. They also draw attention to the fact that what might be called "participation" may not constitute a child rights approach.

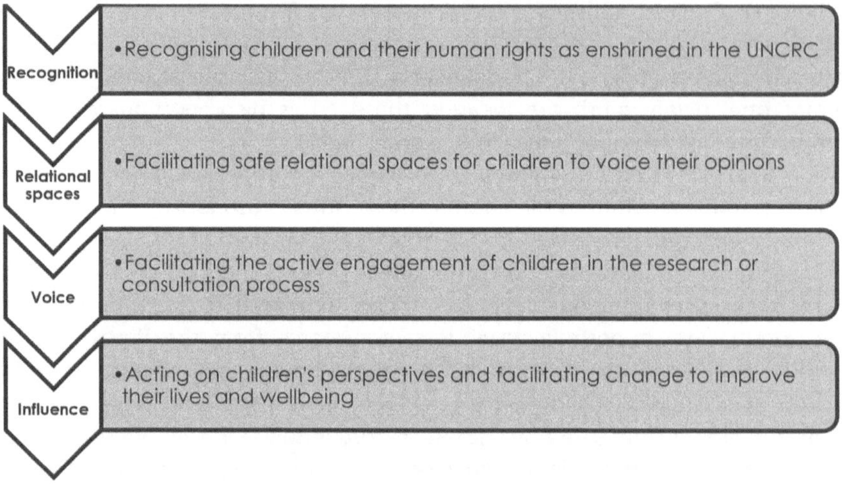

Figure 1. Child rights informed approach to children's participation in tourism.

Based on these frameworks and principles, we propose a child rights informed approach to children's participation in tourism research, policy, and planning, which might go some way in addressing social injustices experienced by children in the tourism industry, including those potentially exacerbated by global crises. Building on a growing tourism social justice scholarship and consideration of the importance of recognising and promoting the rights of vulnerable groups at the local level (Boluk & Carnicelli, 2019; Higgins-Desbiolles, 2020; Jamal & Higham, 2021; Rastegar et al., 2021; Weaver et al., 2022), as well as the understandings of participation discussed above, we consider four important dimensions in implementing a child rights informed approach to participation - recognition, relational spaces, voice and influence (see Figure 1). Below we briefly explain these dimensions and provide examples of what these might mean for advancing participation in the two tourism contexts we discussed earlier.

The first aspect in addressing children's vulnerability and lack of participation in the tourism industry is "recognition" of the indivisibility of children's rights to participation, protection, and provision as enshrined in the UNCRC. Considering the recognition of children in the contexts of child trafficking and orphanage tourism, for example, requires closer attention to the ways in which these children are socially constructed, usually as objects of despair and in need of being 'saved'. Such constructions limit the possibilities for children to be viewed as agentic and capable of contributing to research that might inform decision-making in complex matters that affect them (see also Beazley et al., 2009; Ursin & Skålevik, 2018). As previously discussed, one of the few studies which directly involved children in orphanage tourism research highlighted children's resilience despite their lived experiences of extreme family poverty and injustice (Miller & Beazley, 2022). Recognising children's rights to protection, provision but also participation in research about their lives may contribute to important child safe policies and safeguarding measures which are less about "saving" children and more about standing with children as they seek to influence positive change and improvement in their lives.

The second aspect involves setting up and facilitating safe relational spaces for children to voice their opinions, whether during the research process or in consultative processes for tourism policy and planning. This requires an ethical and reflexive approach to remain attuned to children's needs and uphold their dignity. The Ethical Research Involved Children (ERIC) initiative provides useful guidance for researchers as well as policymakers and organisations wanting to consult with children and young people in meaningful ways (Canosa & Graham, 2016; Graham et al., 2013). This includes a core emphasis on relational ethics as an ongoing social practice which nudges us beyond prescribed rules or codes of conduct to highlight the importance of in-situ ethical decision-making or "mindfulness" and reflexivity in the research process (Abebe & Bessell, 2014; Canosa, Graham & Wilson, 2018; Truscott et al., 2019). Doing research with vulnerable street children selling souvenirs to tourists, for example, may pose a series of complexities that arise when in the field related to matters such as informed consent, harms and benefits, confidentiality and payment (see, for example, Beazley & Miller, 2015; Huberman, 2012). Careful planning at the outset of the research/consultation, as well as, in-situ ethical decision-making, are important in keeping children safe but also in facilitating their active engagement in the process: What would be the likely benefits for the individual child participating in the research? What form will children's involvement in the research take? Can children be involved in the planning of the proposed research? If so, what strategies can we use to facilitate this?

The third aspect of the proposed child rights informed approach to participation concerns closer attention to what having "voice" means in tourism contexts. We propose that voice needs to go beyond "having a say" to encompass a stronger emphasis on *purpose*. Whose voices will be heard? Will it be those who are marginalised or those who are privileged? What will happen to the views that are expressed? Will adults help to facilitate expressing these views, and how will they do this? Facilitating relational spaces where children feel safe voicing their opinions is more likely to open up opportunities for dialogical encounters and intergenerational learning – adults learning from children as well as children learning from adults (Graham & Fitzgerald, 2010; Kemmis, 2008). Facilitating "voice" requires that children are informed (as enshrined in Article 17 of UNCRC - access to information) and able to express their opinions freely (Article 13 - freedom of expression). This necessitates careful planning to communicate information and consent material that is age-appropriate, visually appealing, and culturally safe for the children involved. In addition, research or consultative methods need to be responsive to the specific needs of children (i.e. suitably adapted methods for particular contexts/needs/ages of children) (Canosa et al., 2018; Graham et al., 2013).

Such a participatory approach is also more likely to engage both the researcher and young people in a dialogical learning process which better reflects ethically sound processes. According to one of the "fathers" of critical pedagogy, dialogical spaces for learning are necessary to foster change which is initiated at a grassroots level by those who are "oppressed" (Freire, 1970). Hence facilitating children's participation should not simply be about having "voice" but rather "voice with influence" that requires we not only listen but also respond (in legislative, policy, and practice terms) in ways that reflect their views about how best to improve their lives and well-being. When doing research with children in orphanage tourism contexts, for example, how do we plan for our findings to make a difference in their lives? What outcomes do we hope for? What obligations do we have to promote the consideration and use of the research findings in safeguarding programs, child safe tourism practices, and policy-making? These are important questions that need to be considered if a rights informed approach to social justice in tourism is to be achieved.

Conclusion

A child rights informed social justice approach to tourism recovery post-pandemic would provide avenues to explore children's lived experiences through meaningful inclusion in research (for example, regarding child labour in tourism and orphanage tourism) that is sensitive to their vulnerabilities but also fosters their active engagement in issues that affect their lives. There is a clear need to challenge the idea that children and young people are too vulnerable to participate in tourism research, policy, and planning and to overcome the ethical and methodological challenges that pose barriers to their participation. Human rights discourses and interventions can only contribute to solving the social injustices experienced by children in the tourism industry if children themselves "are understood not only as vulnerable people in need of protection, but also as full social beings with agency and dignity" (Wall, 2017, p. 283). Only when children's lived experiences of oppression and exclusion are recognised and acknowledged as valid and important can we promote more socially just and equitable forms of tourism. To rethink tourism post-pandemic and work towards achieving the Sustainable Development Goals (SDGs), systems must be in place to promote and enact children's rights not only to protection (i.e. freedom from forms of exploitation caused by tourism) but also to participation (i.e. participation in the governance of tourism).

Note

1. Including among others SDG12 (responsible consumption and production), SDG10 (reduced inequalities); SDG4 (quality education); SDG5 (gender equality); and SDG8 (decent work and economic growth).

Disclosure statement

No potential conflict of interest was reported by the authors.

ORCID

Antonia Canosa http://orcid.org/0000-0002-3422-9095
Anne Graham http://orcid.org/0000-0002-9308-8536

References

Abebe, T., & Bessell, S. (2014). Advancing ethical research with children: Critical reflections on ethical guidelines. *Children's Geographies, 12*(1), 126–133. https://doi.org/10.1080/14733285.2013.856077

Bakas, F. E. (2018). The political economy of tourism: Children's neglected role. *Tourism Analysis, 23*(2), 215–225. https://doi.org/10.3727/108354218X15210313504562

Balagopalan, S. (2019). Childhood, culture, history: Redeploying' multiple childhoods. In S. Spyrou, R. Rosen, & D. T. Cook (Eds.), *Reimagining childhood studies* (pp. 23–39.). Bloomsbury Academic.

Baum, T., & Hai, N. T. T. (2020). Hospitality, tourism, human rights and the impact of covid-19. *International Journal of Contemporary Hospitality Management, 32*(7), 2397–2407. https://doi.org/10.1108/IJCHM-03-2020-0242

Beazley, H. (2018). Community-based strategies to combat child trafficking in indonesia. In L. Shevellar & P. Westoby (Eds.), *The Routledge Handbook of Community Development Research* (pp. 84–98). Routledge.

Beazley, H., Bessell, S., Ennew, J., & Waterson, R. (2009). The right to be properly researched: Research with children in a messy, real world. *Children's Geographies, 7*(4), 365–378. https://doi.org/10.1080/14733280903234428

Beazley, H., & Miller, A. (2015). *The art of not been governed: Street children and youth in Siem Reap*. Springer Singapore.

Benali, A., & Oris, M. (2019). Consuming poverty: Volunteer tourism in an orphanage in Nepal. In J. Cheer, L. Matthews, K. E. van Doore, & K. Flanagan (Eds.), *Modern Day Slavery and Orphanage Tourism* (pp. 110–122). CABI.

Bessell, S. (2021). The impacts of COVID-19 on children in Australia: Deepening poverty and inequality. *Children's Geographies*, 1–11. https://doi.org/10.1080/14733285.2021.1902943

Boluk, K., & Carnicelli, S. (2019). Tourism for the emancipation of the oppressed: Towards a critical tourism education drawing on Freirean philosophy. *Annals of Tourism Research*, *76*, 168–179. https://doi.org/10.1016/j.annals.2019.04.002

Canosa, A., & Graham, A. (2016). Ethical Tourism Research Involving Children. *Annals of Tourism Research*, *61*, 219–221. http://doi.org/10.1016/j.annals.2016.07.006

Canosa, A., Graham, A., & Wilson, E. (2018). Reflexivity and ethical mindfulness in participatory research with children: What does it really look like? *Childhood*, *25*(3), 400–415. https://doi.org/10.1177/0907568218769342

Canosa, A., Graham, A., & Wilson, E. (2019). Progressing a Child-Centred Research Agenda in Tourism Studies. *Tourism Analysis*, *24*(1), 95–100. Retrieved from https://doi.org/10.3727/108354219X15458295632007

Canosa, A., & Schänzel, H. (2021). The Role of Children in Tourism and Hospitality Family Entrepreneurship. *Sustainability*, *13*(22), 12801. https://doi.org/10.3390/su132212801

Canosa, A., Simmons, C., Cook, P. J., Apostolides, L., Wall, A., & Evington, R. (2021). 'Reclaiming place in a tourist town': Preliminary findings from a social theatre research project run by young people during COVID-19. *Annals of Tourism Research Empirical Insights*, *2*(1), 100008. https://doi.org/10.1016/j.annale.2020.100008

Canosa, A., & van Doore, K. E. (2022). Orphanage Tourism. In D. Buhalis (Ed.), *Encyclopaedia of Tourism Management and Marketing Cheltenham*, UK and Northampton, MA, USA: Edward Elgar Publishing.

Canosa, A., Van Doore, K. E., Beazley, H., & Graham, A. (in press). Children's Rights in the Tourism Industry. *International Journal of Children's Rights*.

Canosa, A., Wilson, E., & Graham, A. (2017). Empowering young people through participatory film: A postmethodological approach. *Current Issues in Tourism*, *20*(8), 894–907. https://doi.org/10.1080/13683500.2016.1179270

Carpenter, K. (2019). A 'nice, knock-down argument' about orphanage tourism, modern slavery and the power and peril of naming. In J. Cheer, L. Matthews, K. E. van Doore, & K. Flanagan (Eds.), *Modern Day Slavery and Orphanage Tourism* (pp. 125–137). CABI.

Cheer, J. M. (2018). Geographies of marginalization: Encountering modern slavery in tourism. *Tourism Geographies*, *20*(4), 728–732. https://doi.org/10.1080/14616688.2018.1477171

Cheer, J. M. (2019). Introduction. In J. Cheer, L. Matthews, K. E. van Doore, & K. Flanagan (Eds.), *Modern day slavery and orphanage tourism* (pp. 3–18). CABI.

Cheney, K. E., & Ucembe, S. (2019). The orphan industrial complex: The charitable commodification of children and its consequences for child protection. In K. Cheney & A. Sinervo (Eds.), *Disadvantaged childhoods and humanitarian intervention* (pp. 37–61). Springer.

Cole, S. (2017). Water worries: An intersectional feminist political ecology of tourism and water in Labuan Bajo, Indonesia. *Annals of Tourism Research*, *67*, 14–24. https://doi.org/10.1016/j.annals.2017.07.018

Cuevas-Parra, P. (2021). Thirty years after the uncrc: Children and young people's participation continues to struggle in a COVID-19 world. *Journal of Social Welfare and Family Law*, *43*(1), 81–98. https://doi.org/10.1080/09649069.2021.1876309

Cullingford, C. (1995). Children's attitudes to holidays overseas. *Tourism Management*, *16*(2), 121–127. https://doi.org/10.1016/0261-5177(94)00022-3

Everingham, P., & Chassagne, N. (2020). Post COVID-19 ecological and social reset: Moving away from capitalist growth models towards tourism as buen vivir. *Tourism Geographies*, *22*(3), 555–566. https://doi.org/10.1080/14616688.2020.1762119

Everingham, P., Peters, A., & Higgins-Desbiolles, F. (2021). The (im) possibilities of doing tourism otherwise: The case of settler colonial Australia and the closure of the climb at Uluru. *Annals of Tourism Research*, *88*, 103178. https://doi.org/10.1016/j.annals.2021.103178

Freire, P. (1970). *Pedagogy of the oppressed* (2000 ed.). Continuum International Publishing Group.

Global Kids Online. (2020). How are children doing under lockdown #COVID19?. Retrieved from http://globalkidsonline.net/covidunder19/

Goldman, P. S., Bakermans-Kranenburg, M. J., Bradford, B., Christopoulos, A., Ken, P. L. A., Cuthbert, C., Duchinsky, R., Fox, N. A., Grigoras, S., Gunnar, M. R., Ibrahim, R. W., Johnson, D., Kusumaningrum, S., Agastya, N. L. P. M., Mwangangi, F. M., Nelson, C. A., Ott, E. M., Reijman, S., van IJzendoorn, M. H., Zeanah, C. H., Zhang, Y., & Sonuga-Barke, E. J. S. (2020). Institutionalisation and deinstitutionalisation of children 2: Policy and practice recommendations for global, national, and local actors. *The Lancet. Child & Adolescent Health*, *4*(8), 606–633.

Graham, A., & Fitzgerald, R. (2010). Progressing children's participation: Exploring the potential of a dialogical turn. *Childhood*, *17*(3), 343–359. https://doi.org/10.1177/0907568210369219

Graham, A., Powell, M. A., Taylor, N., Anderson, D. L., & Fitzgerald, R. (2013). Ethical research involving children (ERIC): Compendium. Retrieved from http://childethics.com/

Graham, A., Truscott, J., Simmons, C., Anderson, D., & Thomas, N. (2018). Exploring student participation across different arenas of school life. *British Educational Research Journal*, *44*(6), 1029–1046. https://doi.org/10.1002/berj.3477

Guiney, T., & Mostafanezhad, M. (2015). The political economy of orphanage tourism in Cambodia. *Tourist Studies*, *15*(2), 132–155. https://doi.org/10.1177/1468797614563387

Hall, C. M. (2010). Crisis events in tourism: Subjects of crisis in tourism. *Current Issues in Tourism, 13*(5), 401–417. https://doi.org/10.1080/13683500.2010.491900

Hart, R. A. (1992). *Children's participation: From tokenism to citizenship*. UNICEF Innocenti Research Centre.

Higgins-Desbiolles, F. (2020). Socialising tourism for social and ecological justice after COVID-19. *Tourism Geographies, 22*(3), 610–614. https://doi.org/10.1080/14616688.2020.1757748

Higgins-Desbiolles, F., Scheyvens, R. A., & Bhatia, B. (2022). Decolonising tourism and development: From orphanage tourism to community empowerment in Cambodia. *Journal of Sustainable Tourism*, 1–21. https://doi.org/10.1080/09669582.2022.2039678

Huberman, J. (2012). *Ambivalent encounters: Childhood, tourism and social change in Banaras, India*. Rutgers University Press.

International Labour Organization (2020). Child labour: Global estimates 2020, trends and the road forward.

Israfilova, F., & Khoo-Lattimore, C. (2019). Sad and violent but I enjoy it: Children's engagement with dark tourism as an educational tool. *Tourism and Hospitality Research, 19*(4), 478–487. https://doi.org/10.1177/1467358418782736

Jamal, T., & Higham, J. (2021). Justice and ethics: Towards a new platform for tourism and sustainability. *Journal of Sustainable Tourism, 29*(2–3), 143–157. https://doi.org/10.1080/09669582.2020.1835933

Jiménez, E. (2021). Decolonising concepts of participation and protection in sensitive research with young people: Local perspectives and decolonial strategies of Palestinian research advisors. *Childhood, 28*(3), 346–362. https://doi.org/10.1177/09075682211025587

Josefsson, J., & Wall, J. (2020). Empowered inclusion: Theorizing global justice for children and youth. *Globalizations, 17*(6), 1043–1060. https://doi.org/10.1080/14747731.2020.1736853

Jourdan, D., & Wertin, J. (2020). Intergenerational rights to a sustainable future: Insights for climate justice and tourism. *Journal of Sustainable Tourism, 28*(8), 1245–1254. https://doi.org/10.1080/09669582.2020.1732992

Kemmis, S. (2008). Critical theory and participatory action research. In P. Reason & H. Bradbury (Eds.), *The Sage Handbook of action research* (pp. 121–138). SAGE.

Kerr, M. M., & Price, M. (2018). "I know the plane crashed": Children's perspectives in dark tourism. In P. R. Stone, T. Seaton, R. Sharpley, & L. White (Eds.), *The Palgrave Macmillian handbook of dark tourism studies*. Palgrave Macmillian.

Laydjiev, I. (2014). Understanding crisis-induced change in global governance: The diplomatic management of the global financial crisis of 2008-2009. *QEH Working Paper Series - QEHWPS202*, 1–34.

Lew, A. A., Cheer, J. M., Haywood, M., Brouder, P., & Salazar, N. B. (2020). Visions of travel and tourism after the global COVID-19 transformation of 2020. *Tourism Geographies, 22*(3), 455–466. https://doi.org/10.1080/14616688.2020.1770326

Lundy, L. (2007). "Voice" is not enough: Conceptualizing Article 12 of the United Nations Convention on the Rights of the Child. *British Educational Research Journal, 33*(6), 927–942. https://doi.org/10.1080/01411920701657033

Lundy, L. (2019). A lexicon for research on international children's rights in troubled times. *The International Journal of Children's Rights, 27*(4), 595–601. https://doi.org/10.1163/15718182-02704013

Lyneham, S., & Facchini, L. (2019). Benevolent harm: Orphanages, voluntourism and child sexual exploitation in South-East Asia. *Trends and Issues in Crime and Criminal Justice*, (574), 1.

McMellon, C., & Tisdall, E. K. M. (2020). Children and young people's participation rights: Looking backwards and moving forwards. *The International Journal of Children's Rights, 28*(1), 157–182. https://doi.org/10.1163/15718182-02801002

Miller, A., & Beazley, H. (2022). We have to make the tourists happy'; orphanage tourism in Siem Reap, Cambodia through the children's own voices. *Children's Geographies, 20*(1), 51–63. https://doi.org/10.1080/14733285.2021.1913481

Moore, T. P. (2017). Children and young people's views on institutional safety: It's not just because we're little. *Child Abuse & Neglect, 74*, 73–85. https://doi.org/10.1016/j.chiabu.2017.08.026

Peleg, N. (2019). *The child's right to development*. Cambridge University Press.

Powell, M. A., Graham, A., Canosa, A., Anderson, D., Taylor, N., Robinson, S., Moore, T., & Thomas, N. P. (2021). Children and safety in Australian policy: Implications for organisations and practitioners. *Australian Journal of Social Issues, 56*(1), 17–41. https://doi.org/10.1002/ajs4.134

Prout, A., & James, A. (1997). A new paradigm for the sociology of childhood? Provenance, promise and problems. In A. James & A. Prout (Eds.), *Constructing and reconstructing childhood: Contemporary issues in the sociological study of childhood* (pp. 7–32). Farmer Press, Taylor and Francis.

Rastegar, R., Higgins-Desbiolles, F., & Ruhanen, L. (2021). Covid-19 and a justice framework to guide tourism recovery. *Annals of Tourism Research, 91*, 103161. https://doi.org/10.1016/j.annals.2021.103161

Richter, L., & Norman, A. (2010). Aids orphan tourism: A threat to young children in residential care. *Vulnerable Children and Youth Studies, 5*(3), 217–229. https://doi.org/10.1080/17450128.2010.487124

Riddell, S., & Tisdall, E. K. M. (2021). Transforming children's rights? Dilemmas, challenges and implementation. *Journal of Social Welfare and Family Law*, *43*(1), 1–7. https://doi.org/10.1080/09649069.2021.1876304

Rhoden, S., Hunter-Jones, P., & Miller, A. (2016). Tourism experiences through the eyes of a child. *Annals of Leisure Research*, *19*(4), 424–443. https://doi.org/10.1080/11745398.2015.1134337

Roche, S. (2020). Conceptualising children's life histories and reasons for entry into residential care in the Philippines: Social contexts, instabilities and safeguarding. *Children and Youth Services Review*, *110*, 104820. https://doi.org/10.1016/j.childyouth.2020.104820

Schänzel, H. A., & Smith, K. A. (2011). Photography and children: Auto-driven photo-elicitation. *Tourism Recreation Research*, *36*(1), 81–85. https://doi.org/10.1080/02508281.2011.11081664

Sedgley, D., Pritchard, A., & Morgan, N. (2011). Tourism and ageing: A transformative research agenda. *Annals of Tourism Research*, *38*(2), 422–436. https://doi.org/10.1016/j.annals.2010.09.002

Seymour, J. (2015). More than putting on a performance in commercial homes: Merging family practices and critical hospitality studies. *Annals of Leisure Research*, *18*(3), 414–430. https://doi.org/10.1080/11745398.2015.1078247

Seraphin, H., & Green, S. (2019). The significance of the contribution of children to conceptualising the destination of the future. *International Journal of Tourism Cities*, *5*(4), 544–559. https://doi.org/10.1108/IJTC-12-2018-0097

Stinson, M. J., Grimwood, B. S., & Caton, K. (2021). Becoming common plantain: Metaphor, settler responsibility, and decolonizing tourism. *Journal of Sustainable Tourism*, *29*(2–3), 234–252. https://doi.org/10.1080/09669582.2020.1734605

Truscott, J., Graham, A., & Powell, M. A. (2019). Ethical considerations in participatory research with young children. In *Participatory research with young children* (pp. 21–38). Springer.

UNICEF. (n.d). UNICEF and the sustainable development goals. Retrieved from https://www.unicef.org/sdgs

United Nations (1989). *Convention on the Rights of the Child (UNCRC)*. Office of the High Commisioner for Human Rights. Retrieved from http://www2.ohchr.org/english/law/crc.htm

United Nations Sustainable Development Group. (2020). Policy brief: The impact of COVID-19 on children. Retrieved from https://unsdg.un.org/resources/policy-brief-impact-covid-19-children

United Nations World Tourism Organisation (UNWTO). (n.d.). Global code of ethics for tourism. Retrieved from https://www.unwto.org/global-code-of-ethics-for-tourism

Ursin, M., & Skålevik, M. L. (2018). Volunteer tourism in Cambodian residential care facilities—A child rights-based approach. *The International Journal of Children's Rights*, *26*(4), 808–836. https://doi.org/10.1163/15718182-02604006

van Doore, K. E. (2016). Paper orphans: Exploring child trafficking for the purpose of orphanages. *The International Journal of Children's Rights*, *24*(2), 378–407. https://doi.org/10.1163/15718182-02402006

van Doore, K. E. (2019). Orphanages as sites of modern slavery. In J. Cheer, L. Matthews, K. E. van Doore, & K. Flanagan (Eds.), *Modern day slavery and orphanage tourism* (pp. 19–30). CABI.

Wall, J. (2017). The global human rights of modern child slaves. In A. M. Duane (Ed.), *Child Slavery before and after Emancipation: An argument for child-centered slavery studies* (pp. 270–286). Cambridge University Press.

Wall, J. (2019). From childhood studies to childism: Reconstructing the scholarly and social imaginations. *Children's Geographies*, 1–14. https://doi.org/10.1080/14733285.2019.1668912

Weaver, D. B., Moyle, B., & McLennan, C-l J. (2022). The citizen within: Positioning local residents for sustainable tourism. *Journal of Sustainable Tourism*, *30*(4), 897–914. https://doi.org/10.1080/09669582.2021.190

World Vision. (2021). *5 Global Crises the world can't ignore in 2021*. Retrieved from https://www.wvi.org/fragile-context/context/5-crises-the-world-can%27t-ignore-in-2021

Yang, M. J. H., Yang, E., & Khoo-Lattimore, C. (2020). Host-children of tourism destinations: Systematic quantitative literature review. *Tourism Recreation Research*, *45*(2), 231–246. https://doi.org/10.1080/02508281.2019.1662213

OPEN ACCESS

Tourism and refugee-crisis intersections: co-creating tour guide experiences in Leeds, England

Elisa Burrai, Dorina-Maria Buda and Emily Stevenson

ABSTRACT
Forcibly displaced people like refugees and asylum-seekers can be socio-culturally integrated in receiving societies through tourism. Our poststructuralist approach to ethics explains this potential as we combine three concepts: Zygmunt Bauman's "strangerhood", Emmanuel Levinas' "ethical responsibility", and Jacques Derrida's "hostipitality". We draw on the initiative of a social enterprise in the English city of Leeds which encourages displaced people to contribute as tour-guides. During 2017 and 2019 we conducted in-depth individual and focus-group interviews with refugees, asylum-seekers, and public sector stakeholders to examine integration of displaced people via tourism. Findings highlight that contributing to the tourism sector, individual refugees and asylum-seekers are provided with a meaningful platform for self-representation moving beyond tokenistic notions of participation, and become co-creators of diverse and inclusive societies.

1. Introduction

There were 82.4 million forcibly displaced people worldwide in 2021 as a result of persecution, conflict, violence, and human rights violations. In 2019 there were 79.5 million forcibly displaced people around the world, an increase of 8.7 million people from 2018; this includes 26.0 million refugees, 45.7 million internally displaced people, and 4.2 million asylum-seekers (UNHCR, 2020). Worldwide, one person becomes displaced every three seconds. The Institute for Economics and Peace (2020) estimates that climate change crisis will displace 1.2 billion people by 2050. Additionally, major geopolitical changes across the globe, such as the Syrian war, or the more recent Ukrainian crisis,lead to an exponential increase in forcibly displaced people migrating to Europe. In their vulnerable situation, forcibly displaced people are most at risk during times of political, environmental and health crises (Darling, 2017). Since 2015, images and narratives about distressed refugees, especially those fleeing the Syrian civil war continue to make headline news in what is considered to be "the worst humanitarian crisis in Europe since 1945" (Karakoulaki et al., 2018, p. iv). Similar images reoccurred in February 2022 when Russia invaded Ukraine. Since then, more than 2 million refugees fled Ukraine with an expectation that there will be more than 4 million displaced Ukrainians by July 2022 (UNHCR, 2022). Thus, in this paper we examine the intersection between this major challenge, the so-called "refugee-crisis" with the tourism phenomenon, and offer insights about meaningful integration of refugees in receiving countries.

Refugee status is defined by the 1951 Convention relating to the Status of Refugees as a person who flees their country "owing to well-founded fear of being persecuted for reasons of race, religion, nationality, membership of a particular social group or political opinion" [Chapter 1, Article 1, A (2)]. During the refugee status determination which includes registration, and admissibility processes, as well as *prima facie* and group determination, the intermediary step of asylum-seeking is given until the refugee claim is decided by the receiving country. Thus, not all asylum-seekers will eventually be recognised as refugees, but every refugee was initially an asylum-seeker (UNHCR, 2019, p. 64). For the UNHCR, these two categories of forcibly displaced people are different to "migrants", who move from their native country to another, unrelated to the refugee definition (UNHCR, 2019, p. 64). It should be noted that there is not universally or legally accepted definition of the migrant. The International Organisation for Migration (para. 1) defines migrants as persons "who move away from their place of usual residence, whether within a country or across an international border, temporarily or permanently, and for a variety of reasons". The UNHCR regard international migrants as those crossing borders to improve their life via work or study, or family reunification, or other such reasons. There are, however, some recent debates in refugee and migration studies that "refugees are migrants" (Refmig.org, para. 1; Costello et al., 2020). For clarity and purpose of this study we use the legally accepted UNHCR definitions for refugees and asylum-seekers since the United Kingdom is signatory of the 1951 Refugee Convention.

The inextricable connections between tourism and forced migration have been signalled within schools of thought in tourism studies such as the critical turn and new mobilities paradigm (Hannam et al., 2006; McRae, 2003; Russell, 2003; Sheller & Urry, 2006). In a century of "increased movement of people both in tourism and through less attractive social contexts like those involved in the relocation of refugees" (McRae, 2003, p. 243) it is crucial to meaningfully unpack the complex interrelation between "obligatory as well as voluntary forms of travel" (Hannam et al., 2006, p. 10). This also resonates with questions posed by scholars in forced migration studies about the scope of their research field, namely "how (or indeed whether) the contours of this field can be defined in light of the continuum of "forced" to "voluntary" forms of movement, mobility, and immobility" (Fiddian-Qasmiyeh, 2014, p. 1). As such there is recent increased academic attention to the link between tourism, refugees, and forced displacement (Alrawadieh et al., 2019; Ormond & Vietti, 2021), with recent research focusing mainly on Greece (Trihas & Tsilimpokos, 2018; Pappas & Papatheodorou, 2017).

To contribute to this body of knowledge on tourism and forced displacement, we analyse intersections between the refugee-crisis and tour-guiding through the involvement of refugees in the tourism sector offering tour-guide experiences. We do so via the lens of poststructuralist ethics which enables us to reflect upon contemporary geopolitical changes, and to deconstruct ethics and responsibility in the hope of opening up possibilities for more support of refugees in the tourism sector. Thus, we propose a novel theoretical approach which combines three concepts – strangerhood (Bauman, 1991), ethical responsibility (Levinas, 1991) and hostipitality (Derrida, 2001)[1] – so as to offer a more humanistic approach to the socio-cultural integration and welcome of refugees and asylum-seekers via tourism. We argue that actively involving refugees and asylum-seekers in creating and delivering tourism services such as guiding city-tours can contribute to fostering societal change and acceptance for the *other-than-us*.

Since 2015, about 400+ tours have been conducted by one unique tour-operator in Berlin *Refugee Voices Tours (*2021) which offers walking tours guided by a Syrian refugee, and presents parallels between the history of Berlin and the current situation in Syria. In the Netherlands, *Pocket Stories* invites visitors to engage with this question when touring around the country: "how are you connected to migration and what are your roots?" (Pocket Stories, 2021). At a European level, the European Union co-funds Migrantour in 16 European cities from Valencia to Bologna, Paris to Ljubljana, and Lisbon to Cagliari to "offer intercultural urban walks, a form of responsible tourism at kilometre zero, which sees as protagonists fellow citizens also coming from distant worlds" (Migrantour, 2021, para. 1). The only organisation of this type in England is *Tales of a City Tours* which provides a unique perspective on the city of Leeds through cultural walking tours led by tour guides with a refugee background (Do It Org, 2021, para. 1).

Tales of a City Tours social entrepreneurship is at the centre of our study for which we gathered research material in Leeds between 2017-2019 from 20 individual in-depth interviews and one group interview, with refugees, asylum-seekers, and support workers in the public sector who work with refugees. In 2018 one of the authors founded the *Tales of City Tours* with the aim to involve refugees and asylum-seekers in designing and leading walking-tours around the city centre of Leeds. Refugees and asylum-seekers are first approached by the social entrepreneur to understand whether they are interested in getting involved in guiding city-tours. Those interested meet regularly with the founder to design the walking tours, and to share personal stories related to parts of the city. Financially, the social enterprise had different source of funding. These were European Union training funds which aimed specifically at the development of the business idea to support vulnerable people, a small start-up grant from UnLtd the Foundation for Social Entrepreneurs (a charity in the United Kingdom) that supports social entrepreneurship (UnLtd, 2020), and a levy £10 per person paid by each visitor taking the guided city tours.

By designing and leading their own city-tours of Leeds, refugees have both complete autonomy in designing the tours, as well as the chance to create their own narratives during the tours, based upon their own background, interpretation and personal connection to their receiving city. In such city-tour experiences, the tour-guide with a refugee background facilitates connections between visitors, places and refugees. This allows for a deconstruction of hostipitality –hospitality/hostility – attitudes and stereotypes that perceive refugees as "strangers" within our societies (Bauman, 1991; Derrida, 2000). Additionally, this offers further opportunity to redefine and reorient tourism through more just and inclusive ways of designing and experiencing it (Higgins Desbiolles, 2020).

2. Conceptual borromean connections: strangerhood, ethical responbility, and hostipitality in tourism

Borromean connections refer to intersections found in a Borromean Knot whereby three elements are tied together topologically – that is, if one element is removed the other two are left unconnected. The Borromean Knot is not a Venn diagram, even though a representation in plane resembles one, because the three simple closed curves are knotted in three-dimensional space that are topologically linked. Originating in the Italian aristocratic House of Borromeo in the 1300s, the Borromean Knot has been used in sciences from mathematics to Lacanian psychoanalysis, but not in tourism studies.

We use the Borromean Knot image to bring together, explore and expand three philosophical and socio-cultural concepts: Zygmunt Bauman's notion of "strangerhood", Emmanuel Levinas's ideas on "ethical responsibility", and Jacques Derrida's theorisations of "hostipitality". In this way we bring a novel lens to analyse tourism in relation to the refugee crisis. These three concepts are held together by the background of forced displacement and tourism, and complement one another as if tied in a Borromean knot (Figure 1). Understanding how strangerhood is framed around strangers as "refugees and chaos", we identify a need for ethical responsibility towards "the other" which becomes meaningful only if we engage with and dismantle established socio-political structures of unwelcome and hostility in order to recreate more inclusive ones. Interlinking the concepts allows us to understand refugees in relation to travel and tourism. Thus, through the interconnected pattern of the rings, integration and participation in tourism can develop.

Bauman's Bauman (1991) notion of 'strangerhood' offers insights into uncertainty and chaos as part of the social world. Chaos is symbolised by 'the other', the stranger who dismantles the

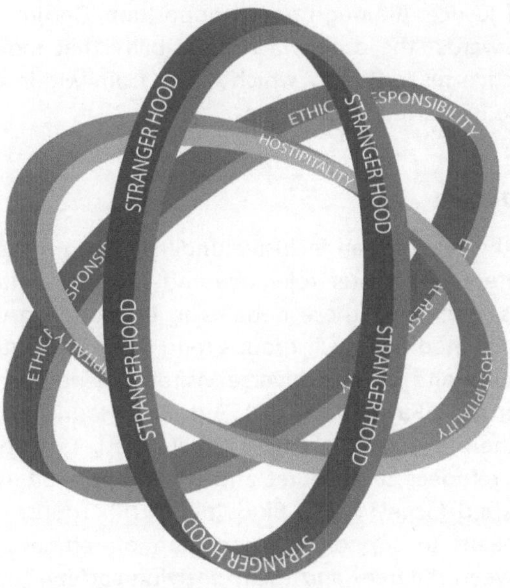

Figure 1. Borromean rings showcasing the three concepts knotted together.

"stable and fixed boundaries modernity has established" (Marotta, 2002, p. 39). Within Bauman's liquid conceptualisation of modern societies, social relationships and identities are subject to fluidity and changes too. At this point, the explanation of Levinas (1991) on the responsibility towards 'the other' assumes its relevance in regard to refugees considered as 'others' via their negotiated identities and relationships within the new societies they enter. The stranger becomes an ethical problem because "the stranger is ... someone who refuses to remain confined to the 'far away' land or go away from own and hence ... defies the easy expedient of spatial or temporal segregation" (Bauman, 1991, p. 59). Hence, the spatial boundaries which frame the identities of 'the self" and 'the other' become an ethical problem (Bauman, 1991). In this regard, Levinas understands 'the self" in relation to 'the other' both as social interaction, and also through a role of responsibility towards 'the other'. Ethics, therefore, takes shape of the individual responsibility for 'the other'. This ethical responsibility is unconditional and unbounded from specific political and socio-geographical contexts (Levinas & Kearney, 1986).

Drawing on Levinas' understanding of ethical responsibility towards "the other – the stranger" we acknowledge the limitations of Levinas' ethics. Levinas proposes an ethical approach to "the other" confined within the relationship and interaction of two individuals. This fails to acknowledge a broader, communal sense of ethical responsibility mediated by "the political world of the impersonal "third" – the world of government, institutions, tribunals, schools, committees" (Levinas & Kearney, 1986, p. 29). Thus, we interject Derrida's thinking on hospitality to address an individual's responsibility towards "the other" within well-defined political boundaries of welcome.

In this context, hospitality refers to "the right of a stranger not to be treated with hostility" when they arrive on someone else's territory (Derrida, 2000, p. 4). In line with a postmodern view of "the other", Derrida, like Bauman and Levinas, proposes that "the other" "the stranger" "the foreigner" is someone different from "the self" or "the guest". Thus, our responsibility towards "the other" implies sacrifices towards the "other others", such as the wider socio-political context in which we operate. Ethical responsibility towards a "collective other" can open up inclusive and more just avenues otherwise suppressed by totalitarian and populist regimes. Via a hospitality-hostility/hostipitality lens we deconstruct the very meaning of hospitality as welcome to reveal new meanings of justice. Although not in opposition, Derrida adds to Levinas' thinking on ethical responsibility towards "the other", a responsibility that moves away from individual encounters. This is a collective responsibility which works from within established socio-political structures to re-create more inclusive and just ones.

2.1. Strangerhood and tourism

"Strangerhood" highlights the need for an inclusive understanding of "refugee-ness", which relating to tourism has the potential to foster tolerance and openness within increasingly complex and unwelcoming societies. Anchoring these debates in literature on tourism and marginalised communities, it can be highlighted that such groups tend to be "exploited" for their cultural capital, used to promote an area and provide diverse visitor experiences, yet often without them benefitting directly (Mackenzie & Raymond, 2020). Although, tourism has been regarded as the "interpretive vehicle of "othering" par excellence" (Hollinshead, 1998, p. 121; the importance of encouraging migrants and refugees to interpret and utilise their cultural heritage through tourism experiences is emphasised (Scheyvens & Biddulph, 2018). Tourism can foster "participatory cultural self-expression" meant to support the inclusion of refugees, by helping to develop mutual understanding between refugees and their receiving society (Ntaltagianni & Dimas, 2016, p. 24). Encounters between different cultures generate hybrid spaces which can become centres of cultural production and "emergent belonging" for marginalised people (Amoamo, 2011, p. 1255).

To harmoniously integrate diverse cultural identities of refugees it is essential we take into account their intangible aspects such as their traditions and cultural heritage which they carry with them from their places of origin (Mackenzie & Raymond, 2020). Engaging in tourism and hospitality sectors, these forcibly displaced migrants can become an integrative part of the receiving societies instead of further suffering from socio-cultural exclusions (Scheyvens & Biddulph, 2018). Indeed, it is argued that in order to forge more inclusive and just tourism, migrants and such marginalised groups should be involved in the production and consumption of tourism. Inclusivity entails self-representation, it challenges prevalent power relations, involves a wider range of people in the decision-making process and provides opportunities to include new places on tourism maps (Scheyvens & Biddulph, 2018). Inclusivity of those deemed "strangers" means also learning from different cultures and meaningful exchanges between hosts and guests (Scheyvens & Biddulph, 2018). Strangerhood in tourism, means addressing inequality; challenging stereotypes or generalised histories and fostering awareness of situation of minorities. Nonetheless, inclusion/exclusion of migrants or "strangers" in tourism occurs differently depending on the scale considered. This means that although at a macro scale (i.e. national level), those with a migrant background are more likely to face exclusion and hostility, at a micro level (i.e. local or regional) inclusive (bottom-up, grassroots) practices are more likely to develop (Darling & Bauder, 2019).

In contemporary western European societies, immigrants are often subject to exclusionary and oppressive practices which dominate public discourses on migration (Darling, 2017). To unpack the challenges for forcibly displaced people, the concept of "strangerhood" proves helpful (Bauman, 1991). We are offered a postmodern explanation on how the political, socio-cultural changes in western European societies have impacted the way that "the stranger comes to symbolise the very ambivalence that modernity's will-to-order is attempting to suppress" (Pietsch & Marotta, 2009, p. 188). The stranger is socially constructed as a permanent other. The stranger cannot ever become an insider in the host country, and they will always be the persons to blame for disrupting the social system of order they enter. It is maintained that the relation between "us and them" is socially constructed underlying a form of collective identity (Bauman, 1991). Identity is as it acquires meaning through its opposition to another. This opposition takes shape in formation of boundaries where collective identities are bounded in convergent "we-ness" and "they-ness" (Bauman, 1992). This opposition can also be read in terms of *friends – us, the host*, and *enemies – them, the strangers* (Bauman, 1992; Derrida, 2000).

In acknowledging Bauman's view on socially constructed identities that are often shaped into collectivities, we contest the idea of a single refugee community. Instead, we recognise that the diverse nature of ethnic minority groups in many tourism destinations creates rich, multifaceted and individualised narratives (Higgins Desbiolles, 2020). We, therefore, acknowledge this variety of subjectivities, rather than assuming the existence of a monolithic collective identity held together by a uniform "refugee-ness".

Additionally, strangers, within the context of our project, are mainly refugees who escape persecution, war and violence. Tourism literature remains sporadic on the conceptualisation of "strangers" as a socio-culturally distant "Other" as it is the case of refugees. There is, however, acknowledgement of the importance of "the Other" for sustainable planning (Lacey et al., 2012), and to partake in encounters between hosts (others) and guests (strangers) (Aramberri, 2001).

2.2. Ethical responsibility in tourism

Responsibility towards "the other", according to Levinas, specifically towards refugees, explains "that there is not circumstance under which we could declare that it was not our concern" (Campbell, 1994, p. 462). The ways most western European countries treat refugees show a lack of ethical humanism (Bauman, 2016). Taking this ethical dimension of responsibility into consideration allows us to challenge contemporary discourses on refugees. "The other", we contend, is the stranger identified by Bauman (1991) who disrupts the harmonious being of western European societies.

In tourism there is recent attention on voluntary and forced mobilities in connection with justice in ways that shared urban spaces of inclusion and justice are being reconsidered to make the transition to more environmentally sustainable and socially just societies (Sheller, 2018). Tourism is characterised by a landscape of uneven (im)mobilities. Privileged mobilities include tourists, expatriates and second homeowners whereas migrants, refugees and asylum-seekers are part of precarious mobilities. Unequal movements are not only the result of global economic disparities, but they are also products of socio-political discourses on inclusion/exclusion through which different types of mobility are legitimised (Bianchi et al., 2020).

Bringing together mobility, ethical responsibility and justice makes poignant the need to "develop a deeper understanding of how uneven mobility relates not only to how we move around cities, but also gendered and racialized colonial histories and neocolonial presents" (Sheller, 2018, p. 18; Tomassini et al., 2021). Thus, tourism literature on justice, fairness and ethics is becoming more prominent in spite of the argument that "theoretical links to justice and ethics have been slow to emerge at the destination level, especially about fairness, equity and justice for disadvantaged local groups, including poor, minority and indigenous populations" (Jamal & Camargo, 2014, p. 11).

Ethics is "the putting into question of my spontaneity by the presence of the Other (Autrui)" (Levinas, 1991, p. 43). The ethical is the identification of a point of alterity or exteriority, in opposition to the Same. This exterior being is named "face" and it is defined as: "the way in which the other [l'Autre] presents himself (sic), exceeding the idea of the other in me" (Levinas, 1991, p. 50). The face is the condition of possibility for ethics and "the ethical relation is one in which I am related to the face of the Other, where the French word "autrui" refers to the other human being, whom I cannot evade, comprehend, or kill and before whom I am called to justice, to justify myself" (Critchley, 1999, p. 5). In simpler terms, ethics is responsibility towards the other, and this responsibility is pre-ontological and unconditional (Levinas, 1991). Ethics is a requisite for the self to exist. Without recognition of the other person and their ethical responsibility, there cannot be a self. Hence, "the self cannot be defined without the other and it is in ethical relationship to this other that the self is bestowed" (Goodman, 2012, pp. 16–17).

Responsibility is not impacted by any political, cultural or social factors. It is, instead, natural and independent from our geographical location. As Levinas (1991, p. 96) explains: "the other is not simply close to me in space, or close like a parent, but he [sic] approaches me essentially insofar as I feel myself – insofar as I am – responsible for him [sic]". Such an approach presents some limitations regarding the lack of acknowledgment of the complex surrounding of social interactions. Indeed, Levinas (1991) interprets ethics as responsibility for "the other" without problematising the role of justice and the State. The ethical relationship between two subjects is defined as "the political world of the impersonal "third" – the world of government, institutions, tribunals, schools, committees" (Levinas & Kearney, 1986, p. 29). Responsibility for the other is, also, part of "a deconstructionist politics that would open us to the other"s appeal" (Popke, 2003, p. 308). Such a view complements Levinas' thinking adding scrutiny of the role and the importance of "the impersonal third". The acceptance and involvement of migrants and refugees in receiving countries is part of the ethical responsibility individuals have for others (Levinas, 1991).

Engaging in the tourism industry, such as becoming tour-guides, represents an opportunity for refugees to connect with their receiving country. This enables refugees to cultivate a sense of belonging in a new society, through gradual comprehension of a new socio-cultural environment (Ntaltagianni & Dimas, 2016). Being involved as tour-guides, becomes a means for cultural creativity and new configurations of diversity (Amoamo, 2011). Refugees can retain their identities through provisions of tourism services, such as guiding city-tours, whilst the host society becomes connected to refugees through an understanding of their origins.

Through refugees' involvement in tourism, we argue that mutual respect, acceptance and inclusivity can be achieved. Involvement unfolds through a process of co-creation, although perhaps even more so when working with individuals with different ethnic backgrounds. The importance of bonding and bridging social networks should therefore not be underestimated. This is particularly pertinent when working with refugees who are less likely to trust due to previous traumatic experiences. Furthermore, when interacting with newly arrived refugees whose access to social capital is minimal, the collaborative process can be a way of facilitating inclusion by creating intercultural connections (Zetter et al., 2006). Refugees' participation in tourism can be a positive force to redefine the boundaries set by mainstream patriotic discourses favouring hybrid communities.

2.3. Hostipitality and deconstruction in the creation of city-tours guided by refugees

For Derrida, the concept of hospitality needs to be deconstructed to be understood and exercised. Hospitality is "at once timeless, archaic, modern, current, and future ... the single word "hospitality" magnetizes – the historical, ethical, juridical, political, and economic questions" that societies face (Derrida, 2000, p. 3). Deriving from Latin *hospitalitas*, the concept of hospitality is argued to be a human right rather than philanthropy, and carries its own opposition "hostility" incorporated into it. Hospitality "is parasitized by its opposite, "hostility"" (Derrida, 2000, p. 3) shown towards the undesirable guest, the other, the stranger, the foreigner. Within the context of welcome, this coexistence of *hospitality-hostility* generates for Derrida a portmanteau word combining their meaning and sounds into *hostipitality*.

We anchor debates about Derrida's hospitality in literature on tour-guiding and migration. Tour guides can enhance visitors' knowledge and experiences, yet their role goes beyond their knowledge and skills in that they play a crucial part in creating safe, memorable and hospitable environments for their guests (Alazaizeh et al., 2019; Farkić et al., 2020). When discussing tour-guiding in relation to migration, it is argued that in their guiding work, migrants have the potential to "transcend strangerhood" and "acquire localhood" breaking through those oppressive walls of cultural and political not acceptance of the Other (Mellino & Vietti, 2019; Ormond & Vietti, 2021, p. 6). Migrants through their tour narratives and routes can give a more just and even shape to the relationships with their guests. In 2010, Migrantour was founded to provide guided tours of different European cities by intercultural guides with a migration background (Ormond & Vietti, 2021). Ormond and Vietti (2021) discuss how two European-based initiatives, Migrantour and the Roots Guide, can become pedagogical tools to combat xenophobia and exploitation. In particular, the focus on co-created guided walking tours by first- and second-generation migrants resonates with our work. Thus, tourism researchers begin to interrogate the possibility for tourism to forge another kind of ethics necessary to "temper growing xenophobia and anti-immigrant sentiment" (Ormond & Vietti, 2021, p. 2).

The refugee crisis painfully highlights that societies live in times whereby hostility rather than hospitality is extended to others, strangers, foreigners such as refugees and asylum-seekers. Hostipitality stems from a positive response to calls for change which then becomes openness to others. It refers to dismantling and transforming the law of the state which governs cities. In this view of reformed laws that make states and cities more welcoming for refugees, Derrida refers to a new cosmopolitics that takes shape into autonomous cities supportive of each other, yet independent from the state. Cosmopolitics and hospitality are about opening cities to refugees reorienting the politics of the state through transformation of:

> membership by which the city belongs to the state, as in a developing Europe or in international juridical structures still dominated by the inviolable rule of the state sovereignty - an intangible rule, or one at least supposed such, which is becoming increasingly precarious and problematic nonetheless. (Derrida, 2001, p. 4)

Cosmopolitics depends on open spaces in which hospitality towards the other can exist. Within these open spaces, tolerance and inclusivity can be fostered through the development of tourist experiences which bring together refugees, local visitors, and international tourists.

Yet, socio-cultural barriers exist when working with displaced people and other marginalised communities in tourism, which often prevents any sort of meaningful engagement with these groups considered unknowledgeable and unskilled to develop tourism strategies (Ormond & Vietti, 2021). Yet actively involving refugees and asylum-seekers in tourism services via creation of city-tours guided by refugees, a process of social and institutional deconstruction can unfold. This deconstructive process helps to create open, interactive and meaningful spaces where different social values merge.

3. Interwiewing forcibly displaced people and local support workers

Research material for this project was collected between June 2017 and September 2019 in the city of Leeds where the social enterprise *Tales of a City Tours* is located. We conducted 20 semi-structured interviews with seven refugees, three asylum seekers (see Table 1), and 10 local support workers from the public, non-governmental, and not-for-profit sectors (see Table 2). In the 10 interviews conducted with both refugee and asylum-seekers in Leeds (Table 1), participants shared their understanding of tourism and its potential to foster inclusivity, tolerance and participation in the socio-economic fabric of Leeds. To complement the narratives collected from asylum-seekers and refugees, we interviewed 10 support workers who have direct contact with refugees and asylum-seekers, in their roles as volunteers, administrators, head or curator of cultural events in Leeds (Table 2). Interviewees were recruited mainly using snowball technique or through the researchers' personal contacts.

Additionally, a group interview with refugees and asylum-seekers was organised shortly after the initial individual interviews conducted in June 2017 whereby 4 out of 10 participants attended. This provided an opportunity to share ideas, knowledge and perspectives, facilitating

Table 1. Refugees and asylum-seekers[a].

	Pseudonym	Status	Ethnicity	Age	Length of stay in England
1.	Sandra	Refugee	Lebanese	42	17 years
2.	Salma	Refugee	Syrian	Not disclosed	2 years
3.	Gaani	Refugee	Somalian	30	6 years
4.	Kael	Asylum seeker	Iranian	32	10 months
5.	Royar	Asylum seeker	Kurdish	27	7 months
6.	Konjit	Refugee	Ethiopian	35	2 years
7.	Sajan	Refugee	Indian	48	8 years
8.	Leila	Refugee	South African	47	16 years
9.	Marius	Asylum seker	Guinean	31	10 years
10.	Mabel	Refugee	Iranian	33	2 years

Source: Authors.
[a]Details as at September 2019 when last interviews were conducted.

Table 2. Local support workers in public and non-governmental sectors.

No.	Name	Role
1.	Laura	Curator of exhibitions at a Leeds museum
2.	Jim	Charity founder helping the homeless integrate into society through guided tours
3.	Chloe	Charity trustee & volunteer helping refugees/asylum-seekers settle in Leeds
4.	Malia	Charity administrator helping refugees/asylum-seekers settle in Leeds
5.	Emma	Account Manager for governmental organisation
6.	Marie	Charity worker helping refugees/asylum seekers settle in Leeds
7.	Vera	Head of Creative Engagement at a theatre
8.	Luke	Volunteer at a charity for refugees/asylum-seekers
9.	Fiona	Charity worker for refugees/asylum-seekers settling in Leeds
10.	Louise	Head of community partnership in a Leeds theatre

Source: Authors.

an environment of collective learning between individuals of different ethnic backgrounds in a community-based setting. The group interview lasted 90 minutes, and collaborative activities in the tourism industry were discussed, with a focus on guided city-tours in Leeds.

Individual interviews took place at participants' homes, workplaces, having food together or while meeting their families. During the interviews, and indeed throughout the research process we remained aware of our positionalities of two migrant academics and one local practitioner, all three of us women. There are similarities in our migrant backgrounds with some of our participants, yet significant differences between the forced migrants' precarious socio-economic conditions and our privileged academic or practitioner status. Throughout the process, one of the authors, both a practitioner and a researcher, reflected on her perceived position of "power". She reflected on issues such as conducting interviews in English her first language, the setting of these interviews, as sometimes these were conducted on the University campus which could have been intimidating for some participants.

Reflecting on our positionalities and how we had to mediate between our own beliefs, values and socio-cultural backgrounds, we acknowledge that "ethical research is produced through negotiated spaces and practices of reflexivity that is crucial about issues of positionality and power relations at multiple scales" (Sultana, 2007, p. 375). Our responsibility throughout the research process was to be particularly aware of the context in which we moved and of the knowledge which we were representing and co-creating with our participants. Thus, during this project we reflected on how our words and actions could result in establishing power hierarchies rather than deconstructing them; and, therefore, we emphasised to our participants the co-creation nature of this project.

Discussions with participants revolved around three key thematic areas: involvement in the broader socio-cultural fabric of Leeds; participation in activities around the city; and perceptions of tourism experiences (as tourists and/or as tour-guides) as potential tools to deconstruct the socio-political discourses on refugee-ness. The topics discussed in the individual and group interviews included: heritage, identity, participation, inclusion, sense of belonging and tourism. All individual interviews lasted between 22 and 81 minutes, were tape-recorded, and then transcribed. Pseudonyms are used to ensure anonymity and confidentiality of participants, and throughout the interviewing process strict ethical guidelines were followed.

To analyse our material, we draw on critical qualitative methodological approaches whereby we take an interpretivist epistemological stance to investigate the multiple subjectivities at play in such sensitive matters pertaining to forced migrants. Thus, the interview transcriptions were deductively analysed through theoretical links and thematic coding with the concepts discussed in previous sections. The themes identified were grouped under the broad categories of strangerhood (Bauman, 1991) ethical responsibility (Levinas, 1991) and hostipitality (Derrida, 2001) which enabled to identify the informants' articulations of *participative strangerhood*, *inclusion through ethical responsibility* and *deconstructive tourist experiences*. These are discussed in the following sections of the paper.

To better understand and analyse the emotional space formed between the interviewer and the interviewee the concepts of empathy and identification are useful as they explain the intersubjective relationship between researchers and participants. Empathy allows space to manoeuvre while oscillating between observation, characterised by an awareness of the differences and similarities within the "researcher-researched" relationship (Bondi, 2003). Emotions of care, compassion, and empathy are known to be emotional pre-requisites for engaging in positive dialogue with "the other". This empathetic space of sharing thoughts, ideas, and emotions was achieved by being open with participants about the researcher's background.

By striving for "affective interchanges" and "caring encounters", it was not intended to remain "objective" towards our participants. Instead, our aim was to create an empathetic space whereby each participant's worldview could be safely discussed. Yet, we were aware of the sort of empathic pain which serves to "draw attention to social difference, and the ways some bodies remain privileged" over others (Frazer & Waitt, 2016, p. 177). It was important, in situations like this, to avoid indulging in what is called "narcissistic emotion" (Chouliaraki, 2013, p. 23), and simply easing our discomfort by overstating our ability to help the participants through this research, allowing ourselves some form of "moral redemption" (Frazer & Waitt, 2016, p. 184).

3.1. The context: English city of leeds

We collected research material in Leeds, a city in the North of England. Leeds was chosen for three reasons. First, the city in 2017 hosted 29.01 million tourists and tourism is worth around 1.75 billion to the local economy (Visit Leeds, 2021). In 2019, Leeds hosted 30.42 million day and night visitors worth £1.89 billion to the local economy (Leeds City Council, 2020). Second, Leeds has over 170 different ethnic groups and the number of refugees and asylum-seekers in the Yorkshire and Humber region as a whole was 5,693 in March 2018 (Uppal & Bhabra, 2018). The diverse cultural landscape of Leeds makes it a vibrant and rich destination that aims at being a compassionate city with a strong economy. Migrant communities are valuable contributors to the city's economy, culture and diversity, and Leeds aspires to ensure people are welcomed and supported. Yet, this local narrative seems to be in tension with increasingly hostile national policy discourses towards forced migrants and against local responses of compassion, tolerance and inclusivity proposed at localised grassroots levels. Third, the authors are familiar with and emotionally connected to Leeds as they work and reside in the city.

Regarding the city's refugee population, at the beginning of April 2019, Home Office figures show that 3,087 people were supported in Leeds City Region while awaiting a decision on their claims (Mahmood et al., 2019). The Local Migration Profile report (Mahmood et al., 2019) shows that in 2019 there were also around 125 unaccompanied asylum-seeking children being looked after by Leeds City Region local authorities. The high number of refugee support organisations operating in Leeds indicates the need for hands-on technical and emotional support for refugees needing to integrate.

To challenge the increasingly hostile narrative and policy debates at national level towards refugees and asylum-seekers, one of the authors, a British citizen from Leeds, started in January 2018 the *Tales of a City Tours* social enterprise. As part of this tourism venture, refugees are involved as tour-guides in the process of creating, designing and leading walking tours in Leeds. Refugees and asylum-seekers have complete autonomy designing these city-tours, but only refugees are legally allowed to guide as such, given their legal status and permission for employment. Asylum seekers equally have total autonomy to decide to which extent they want to be involved designing the tours, and have the opportunity to voice their preferences in regards to points of interest of the tours as well as to develop their knowledge on tourism-related issues. Once asylum-seekers become recognised refugees they can be then legally employed. For example, Leila, a South African refugee tour-guide, decided to include in her tours the main post-office building in the city centre as, when she first arrived in Leeds in 2000, she went there once per week to collect her cash allowance to which refugees are entitled. The tours welcome a variety of regional and international tourists, as well as local visitors. Tour groups have a refugee tour-guide, and lead between 8 and 12 visitors. The tours last an hour and take place in Leeds city centre. The refugee tour-guide explains why certain parts of the city centre were chosen to visit and how those relate to their personal socio-cultural background.

4. Refugee-crisis and tourism intersections

4.1. Tourism experiences of "participative strangerhood"

The "strangerhood" concept (Bauman, 1991) helps us unpack ideals of socio-cultural integration of asylum-seekers and refugees within the spatio-political boundaries of unfamiliar, often unwelcoming, countries. Drawing on "strangerhood" and the fluidity of human identities we better understand the ways in which asylum-seekers and refugees can be socio-culturally integrated. Providing tourism services such as tour-guiding, the ethnically diverse refugees in our project enrich tourism provisions through unique stories to share with visitors. Yet, this variety can also be perceived as a threat to the "stable and fixed boundaries modernity has established" (Marotta, 2002, p. 39). The concept of "strangerhood" addresses the intersubjective and relational nature of identities.

Some participants, despite being forcibly removed from their native countries, claim a collective "territorial identity" based on their geographical origins (Ntaltagianni & Dimas, 2016). Participants in our project focused on the importance of their nationality or ethnicity: *"being Ethiopian"*, or *"being Arabic"*. They demonstrate pride in their cultural heritage, particularly intangible aspects like cultural and religious festivals, traditional dances, history and cultural systems. Leila, a South Africa woman in her forties, and a refugee tour-guide refers to the collective memories of an annual traditional ethnic ceremony in South Africa: "it's something that we all shared, and all grew up looking forward to" (Interview, July 17, 2017).

Often, the self"s identity is opposed to the other"s, and shapes a collective identity characterised by "us/them" or "familiar/unfamiliar". Identity is socially and discursively constructed, thus it gains meaning in opposition to "the other". Having refugee/asylum-seeker status defines our informants as "strangers" because of their exposure to social exclusion and suppression (Bauman, 2016). To illustrate this, Sandra, a Lebanese woman in her forties, admits that being a refugee is "something which makes you feel smaller" (Interview, July 3, 2017). Mabel, too, an Iranian woman in her early thirties, explains: "I had to leave my country, my family, my home, my job … So, I'm really not happy to be a refugee. It was not my choice" (Interview, July 17, 2017). Those still awaiting their asylum-seeker status spoke about having a "temporary identity" (Marius, interview, July 17, 2017), since their future is entirely in the hands of the authorities. As Royar, a Kurdish man in his late twenties explains:

> People always ask me, what are your plans for the future, will you be an engineer again? ... But because I am waiting, I cannot have any plans, maybe this country will throw me out... so I am not sure of what kind of life I will live and what I will do. (Interview, July 11, 2017)

Leaving home determines being strangers. Refugees are "the ones who, in leaving the home of their nation, are the bodies out of place in the everyday world they inhabit, and in the communities they come to live" (Ahmed, 2000, p. 78). This becomes clear in Chloe's words who states that, at first the refugees she knows liked UK, but after a while they experience a less pleasant reality, such as racism or complicated bureaucratic processes:

> There's always a pattern, they come over and, 'the U.K.'s great, I love the U.K.', 'the U.K.'s lovely, I'm here, I'm home', you know, 'praise Allah', they're here, they're happy. Then there's always a dip where they encounter racism, a little bit of racism, they will encounter the bureaucracy of the Home Office. (Interview, June 13, 2019)

Being forcibly displaced, involves travel, movement, and crossing borders, amongst other aspects. It constrains those forcibly displaced to question their "self". These travels shape the fluid constructions of identities throughout their journeys. For some refugees crossing borders and moving away from home reinforces a sense of belonging. Gaani and Royar, for example, strongly believe that their status is part of their history, and that it constitutes their identities. Gaani, a Somali man in his thirties, says: "Once you deny where you come from, and your history, you also deny where you are" (Interview, July 5, 2017). Involving refugees as tour-guides is intended to allow individuals to assert and reform their identity by choosing which parts of their story they wish to tell. This means that there will be some who want to avoid any stigma associated with being a "refugee" and have no desire to provide such details.

For others moving away from home reinforces a sense of freedom. Being a refugee in a new country has empowered some to assert their identity without fear of oppression. Sajan, in his forties, argues that in England, he feels valued as a human being and is able to express his true identity, unlike in India: "They did not respect me and my gender... that was my big problem, there they put me down" (Interview, July 14, 2017).

Refugees' narratives help us understand the importance of individual and collective identities, and the potential to deconstruct socio-cultural practices (Derrida, 2001) so as to enable the development of an ethical humanism (Levinas, 1991) towards strangers (Bauman, 2016). In explaining her involvement with helping refugees, Marie, a British charity worker of Pakistani heritage, says:

> I started [working in charities] because I always wanted to help people but I always was going to help people in the U.K. and Pakistan, because I didn't [...] know anything about it, but you know, one of my brothers was killed by a bunch of drunken youths. My faith also tells me that you help everyone, regardless of religion, culture... " (Interview, September 12, 2019).

For Marie, ethical humanism is unconditional and goes beyond fixed socio-cultural boundaries (Levinas, 1991). Some refugees are motivated to get involved in the creation of city-tours by the opportunity to connect with people from all over the world suggesting that, for refugees who may lack support networks, building social capital is important (Zetter et al., 2006; Scheyvens & Biddulph, 2018). Leila, the South African mother, also notes that working in tourism would increase a refugee's sense of belonging to a place, by showing there are people who are "kind and genuine and show interest to you as a person" (Interview, July 17, 2017).

Potential barriers for the development and delivery of city-tours were discussed in the interviews. These include knowledge of English to guide tourists around Leeds, and the worry of not feeling emotionally ready to share personal experiences throughout the tours. During the interviews, our participants believe that tourism is a vehicle to encourage differences to coexist. Luke, academic and volunteer for a charity helping forcibly displaced people to settle in Leeds, having taken a refugee-led city-tour says:

> because I do the walks for the students and for friends who come to visit Leeds for the first time [...], I didn't hear anything about Leeds that I didn't already know, but what was interesting for me was what different bits in the centre meant to the guy leading the walk, you know. So he [the refugee tour-guide] says: ≪*you might wonder why we've stopped here, that's because over the road there, is the Post Office where I had to come and get my money each week, I didn't know what a Post Office was before I came here, there are no Post Offices in Iraq*≫, it never occurred to me (Interview, July 15, 2019).

People's identities shape the concept of strangerhood. This concept is understood differently depending on the context and whether it is used by individuals or collectivities. Often, strangers are linked to fear, chaos, and instability (Bauman, 1991). They are usually people who do not share common or familiar socio-cultural patterns to locals already living in a place. For some refugees, being strangers is feeling excluded from forms of belonging and identity (Ahmed, 2000). Yet, tourism can enable the reconciliation of differences. It is possible to balance the tensions between "order, strangerhood and freedom when we accept our moral condition" (Marotta, 2002, p. 52). Yet, while balancing this tension we should still be responsible towards the other.

4.2. Meaningful integration through ethical responsibility in tourism

In our interviews, participants referred to the challenges of integrating in Leeds, explaining that they were victims of discrimination and experienced isolation. These challenges of strangerhood in digitalised societies symbolise chaos and uncertainty. Being perceived as a "disturbing" stranger affects the ability of refugees to express their true identities (Bauman, 2016). Thus, being a tour-guide could facilitate socio-cultural connections between ethnically, culturally or religiously different groups. Such connections help address problems around integration experienced by displaced people living in the U.K. (Zetter et al., 2006). Displaced people's involvement in delivering tourism experiences can create positive societal changes through communicating with tourists' different cultures.

Interviewees explain that connecting to other refugees with similar experiences made them feel part of a community. Gaani confirms: "Leeds is such a diverse city... walking side by side with other migrant communities helped me to feel at home" (Interview, July 5, 2017). Leila explains how she relates to newly arrived refugees, and uses her experiences to support and empower them, something which she believes is much needed in a society which becomes increasingly individualised. All participants agreed that interaction with the other, can help refugees cope in the outside world, where different ethnicities often co-exist.

Participants referred to barriers when working with people from different ethnic backgrounds, including having to "concentrate to understand other speaking styles" (Sajan, interview, July 14, 2017). The sense of ethical responsibility is addressed by Gaani who states that "humanity can only be upheld by humans". Luke echoes similar thoughts:

> I believe we all have a responsibility to do our best to make people's lives better if we can, and so, if people who have come to live here, they should have the right to expect that when they are here [England], they're going to be treated with respect and made to feel comfortable, so they do feel able to belong and to contribute, and so I want them to know about the place that they're in (Interview, July 15, 2019).

As Levinas explains: "we will say that since the Other looks at me, I am responsible for him [sic], without even having taken on responsibilities in his regard; his [sic] responsibility is incumbent on me" (1986, p. 96). Hence, Luke's individual felt responsibility towards refugees as "the other" shapes the ethical relation within a society.

Involving refugees in tourism can be instrumental in improving cross-cultural understanding. This cultural understanding and acceptance of diversity is achieved through structural deconstruction at a societal level (Derrida, 1997). That is, connecting to "the other" through tourism experiences shaped by a multitude of voices and backgrounds. Fiona, a charity worker, explains that through tourism, Leeds welcomes:

> people from all over the world … we know that Leeds is really cool and welcoming city, we're a very multi-cultural city. We're such a diverse city […] all being different cultures together and it benefits us. It's such a lovely thing to have so many wonderful, interesting cultures side by side, but I think bringing in people who potentially need support and love and have come from various different backgrounds … is a good city for that (Interview, June 28, 2019).

Sandra, Marius and Gaani all suggest that their multicultural background, multilingual ability and open-mindedness would make them more approachable to different people, who may want to connect with them on city-tours. This is in line with the idea of a tour-guide's unique identity, in which their intercultural competence allows for a greater understanding of others (Ormond & Vietti, 2021).

Furthermore, Marius, resonating with Levinas's ethical responsibility, highlights how societal and cultural connections can be created by emphasising a common humanity amongst the tour-guide and tour-guests: "We can talk about our differences, but I believe what we have in common is far larger" (Interview, July 17, 2017). Indeed, a city-tour, according to Marius, could spark a debate on universal topics like music, or love, bringing refugees and tourists together. As Marius says: "music doesn't have any boundaries" and "love is universal … nobody rejects love" (Interview, July 17, 2017). Emphasising commonalities was something which both Konjit and Royar believe helps with the inclusion process: "For me as a Christian, I believe that England is blessed … when I see the crosses, I feel at home" (Konjit, interview, July 11, 2017).

4.3. Co-creative hospitable tourist experiences

Discussing with Sajan about tourism being culturally defined, and how each participants' social reality differs due their diverse geographical origins, amongst others, he observes that tourism in England is pervasive and welcoming, whilst in India tourism is reserved only for the rich. Participants in our project stressed that city-tours must be "affordable", visiting the city's less expensive sites. The elitist character of tourism, often profit driven, is highlighted by Chloe, a volunteer for a charity which helps forcibly displaced people to settle in Leeds:

> tourism is worth £9 Billion to Yorkshire economy, it's the second biggest industry in Yorkshire … Why do people do tourism? To make money. People don't get involved in tourism to give asylum-seekers nice days out. They get involved in tourism to make money, whether it's running a café, running a tea shop, running an event, why do they do it? To make money … Tourism is not a charity, tourism is a business, a massive business. (Interview, June 13, 2019)

Chloe explains that asylum-seekers live on £7 per day, hence they lack the financial resources to experience tourism. However, for refugees being involved in the delivery of tourism experiences in Leeds means raising awareness of different cultures and cultural exchanges between tourists and tour-guides. Meaningful city-tours can, according to Salma, a Syrian mother of two, positively impact all parties involved by facilitating deeper welcoming connections between tour groups and refugees, who have a genuine desire to learn through offering authentic hospitable engagements with their guests (Ormond & Vietti, 2021). The development of city-tours led by refugees could produce meaningful societal change by deconstructing the apparent socio-political order of Leeds.

This is echoed by Laura, an exhibitions curator at Leeds City Museum, who emphasises the importance of creating refugee-led tours to have positive impacts on citizens' attitudes towards refugees and asylum-seekers:

> I'm in awe of people and the things that they've experienced and kind of their story and how they've come to be in Leeds and it really makes me reflect on things that I just never consider… about how I can move around the world and the choices that are open to me that aren't to other people or the difficult decisions that people have had to make that I've been in the fortunate position of never having to make those. And then personal and sort of professional level, I'm just really gaining so much, learning so much about people and the world from just talking to people, which sounds stupid, but again, just looking beyond… my own experience and what I know and think (Interview, June 04, 2019).

There are good intentions coming from Leeds regarding integration and welcoming of refugees. Particularly, the non-governmental and not-for-profit organisations engage in alternative ways of deconstructing less welcoming approaches to hosting refugees. Malia, a charity administrator who helps refugees and asylum-seekers in Leeds, explains:

> the way you might promote [events/activities] to engage refugee communities would be very different to the way you might promote to engage British communities, other communities. So, there's a real challenge in that, like we've had some organisations try and do boxing, a great initiative I think. I think it's fabulous, great stress relief, all of this stuff, but for some in a refugee community, fighting is a trigger and so actually that's not a helpful thing. So, what we would see as a great stress release, something to get rid of built-up anxiety and whatever, anger or all that kind of stuff, but how do you cross that? (Interview, June 28, 2019).

Certain activities sincerely aiming to integrate refugees, like boxing, are institutionally bounded and regarded as being "right". Destabilising fixed and accepted categories of understanding would be part of deconstruction as a process of questioning how ideas and arguments come into being. This is particularly relevant concerning some governmental discourses on migration and refugee-ness which are often disparaging. It is not suggested that state institutions be rejected or dismantled, but to deconstruct structures within institutions that have become too rigid or that remain unquestioned and unchallenged (Derrida, 2001). This is aligned with Marie's efforts to make refugees feel welcome in the receiving country:

> they [refugees/asylum seekers] talk about their home that they have left and they are looking for a home, so this is how I see refugees, but they're on the road and they always live temporary lives and they tell me, we always live temporary lives, we never feel settled, so for us, the biggest thing is to feel settled. So, like in Europe, the least we can do is make them feel settled (Interview, September 12, 2019)

Marie's welcoming hospitality ought to become a mode of socio-political action and a means for ongoing questioning to open alternative views on the meaning of ethics and justice. This adds to previous hopeful calls to reframe travel, tourism and hospitality sectors, their hierarchies and practices which should become more resistant to mainstream socio-political attacks on diversity and inclusivity.

5. Conclusion

The refugee-crisis despite its hideousness, has generated opportunities for refugees to partake in provision of tourism services by offering guided city-tours in Leeds. Thus, ideas of strangerhood and ethical responsibility are brought to the fore so as to understand how forcibly displaced people can be meaningfully integrated in the receiving country through ways of rethinking, reorienting and redesigning tourism. Engaging with the tourism sector facilitates the emergence of cosmopolitics – as a tool to deconstruct socio-cultural barriers often existent in current societies between "westerners"/locals and "strangers"/refugees (Derrida, 2001). Tourism experiences in cities like Leeds can give "rise to a place (lieu) for reflection on the questions of asylum and hospitality – and for a new order of law and democracy to come to be put to the test" (Derrida, 2001, p. 23).

Here we contributed to tourism debates on the intersection between the global refugee-crises, responsibility in tourism, and injustices that refugees and asylum-seekers face. Our contribution stems from the need of interlinking "research and praxis" to pave just and inclusive tourism futures (Jamal & Higham, 2021). We do so by combining ideas of stangerhood (Bauman, 1991), ethical responsibility (Levinas, 1991) and hostipitality (Derrida, 2001). This enables us to further critical understandings on ways for tourism to overcome societal challenges, like exclusion, hostility, and intolerance towards forcibly displaced people. Interviews conducted for this project highlight sensitivities around refugee identities to be considered when integrating forcibly displaced people via tourism. A refugee guided city-tour mirrors the dynamism of ever-shifting refugee identities, thus allowing individual stories to become a new part of the (shared) intangible cultural heritage of the city (Ntaltagianni & Dimas, 2016). A re-identification with this intangible heritage in the host city has the potential to bridge the gap between "us" and "them". Our study chips away at hegemonic tourism sector structures to strive towards collaboration and meaningful integration of forced migrants in receiving countries via the tourism sector. Presenting ways in which forced migrants can contribute to the tourism sector and be part of participatory spaces enables the development of stronger social ties with other ethnic groups. To this end we stressed the importance of collaborating and co-creating with refugees spaces for sharing knowledge where each participant's worldview is valued.

However, an intervention such as the guided tours can only be effective if it is part of a long-term dynamic process where current socio-political discourses are challenged and dismantled by the emergence of powerful participatory (grassroots) approaches. It should be remembered that although this participatory approach is designed to be beneficial in other contexts too, it might not generate the exact same findings. Thus, further research would help to understand how refugees and asylum-seekers co-create tourism experiences in other destinations. Additionally, the focus on Leeds and on this tourism venture has to be acknowledged in relation to the wider socio-political landscape where tourism develops. Whereas sentiments of hostility towards migrants are more evident at a macro-scale/national government, there is an increasing devolution of mechanisms of security and border enforcement at a micro-scale of local councils (Darling, 2017). This fosters the growth of sanctuary movements across Leeds and other UK cities. Albeit the political dynamics that underpin the formation of certain tourism experiences is not the main focus on this paper, we recommend that future studies look into how political structures shape ways of doing tourism with and for ethnic minorities. There is great potential for further research on how refugee-led tours impact the lives and attitudes of tourists and citizens. Such a focus would complement the findings of our study by demonstrating how other stakeholders are impacted by the inclusion of ethnic minorities in tourism experiences.

Notes

1. British-Polish philosopher Zygmunt Bauman in his book *Modernity and Ambivalence* (1991) identifies strangerhood as the condition of individuals not statically being part of a single societal subsystem, instead becoming part of a mobile, complex socially displaced system. Strangerhood, thus, intersects with ethical responsibility advocated by French-Lithuanian philosopher Emmanuel Levinas in *Totality and Infinity* (1991), who invokes ethics in relation to the other. Levinas explains ethics to be connected with moral responsibility, which becomes a political act through a form of semiotic analysis known as deconstruction, and advanced by Algerian-French philosopher Jacques Derrida in his thesis on *Hostipitality* first delivered at Bosphorus University, 9–10 May 1997 in Istanbul.

ORCID

Elisa Burrai http://orcid.org/0000-0002-1068-6304

References

Alazaizeh, M. M., Jamaliah, M. M., Mgonja, J. T., & Ababneh, A. (2019). Tour guide performance and sustainable visitor behavior at cultural heritage sites. *Journal of Sustainable Tourism*, *27*(11), 1708–1724. https://doi.org/10.1080/09669582.2019.1658766

Alrawadieh, Z., Karayilan, E., & Cetin, G. (2019). Understanding the challenges of refugee entrepreneurship in tourism and hospitality. *The Service Industries Journal, 39*(9–10), 717–740. https://doi.org/10.1080/02642069.2018.1440550

Ahmed, S. (2000). *Strange encounters: Embodied others in post-coloniality*. Routledge.

Amoamo, M. (2011). Tourism and hybridity: Re-visiting Bhabha's third space. *Annals of Tourism Research, 38*(4), 1254–1273. https://doi.org/10.1016/j.annals.2011.04.002

Aramberri, J. (2001). The host should get lost: Paradigms in the tourism theory. *Annals of Tourism Research, 28*(3), 738–761. https://doi.org/10.1016/S0160-7383(00)00075-X

Bauman, Z. (1991). *Modernity and ambivalence*. Polity Press.

Bauman, Z. (1992). Soil, blood and identity. *The Sociological Review, 40*(4), 675–701. https://doi.org/10.1111/j.1467-954X.1992.tb00407.x

Bauman, Z. (2016). *Strangers at our door*. Polity Press.

Bianchi, R. V., Stephenson, M. L., & Hannam, K. (2020). The contradictory politics of the right to travel: Mobilities, borders & tourism. *Mobilities, 15*(2), 290–306. https://doi.org/10.1080/17450101.2020.1723251

Bondi, L. (2003). Empathy and identification: Conceptual resources for feminist fieldwork. *ACME: An International E-Journal for Critical Geographies, 2*(1), 64–76.

Chouliaraki, L. (2013). *The ironic spectator: Solidarity in the age of post-humanitarianism*. John Wiley & Sons.

Campbell, D. (1994). The deterritorialization of responsibility: Levinas, Derrida, and ethics after the end of philosophy. *Alternatives: Global, Local, Political, 19*(4), 455–484. https://doi.org/10.1177/030437549401900402

Convention Relation to the Status of Refugees. (2022). Available https://www.ohchr.org/en/professionalinterest/pages/statusofrefugees.aspx

Costello, C., Nalule, C., & Ozkul, D. (2020). Recognising refugees: understanding the real routes to recognition. *Forced Migration Review*, (65), 4–7.

Critchley, S. (1999). *The ethics of deconstruction: Derrida and Levinas.*, Edinburgh University Press.

Darling, J. (2017). Forced migration and the city: Irregularity, informality, and the politics of presence. *Progress in Human Geography, 41*(2), 178–198. https://doi.org/10.1177/0309132516629004

Darling, J., & Bauder, H. (2019). Introduction–Sanctuary cities and urban struggles: Rescaling migration, citizenship, and rights. In *Sanctuary cities and urban struggles* (pp. 1–22). Manchester University Press.

Derrida, J. (1997). *Deconstruction in a nutshell: a conversation with Jacques Derrida.*, Fordham University Press.

Derrida, J. (2001). *On cosmopolitanism and forgiveness*. Psychology Press.

Derrida, J. (2000). Hostipitality. *Angelaki, 5*(3), 3–18. https://doi.org/10.1080/09697250020034706

Do It Org. (2021). *Tales of a city tours*. Available: https://do-it.org/organisations/tales-of-a-city-tours

European Commission. (2015). Refugee crisis: Takes decisive action. https://ec.europa.eu/commission/presscorner/detail/en/IP_15_5596

Farkić, J., Filep, S., & Taylor, S. (2020). Shaping tourists' wellbeing through guided slow adventures. *Journal of Sustainable Tourism, 28*(12), 2064–2080. https://doi.org/10.1080/09669582.2020.1789156

Fiddian-Qasmiyeh, E. (2014). Refugees helping refugees: How a Palestinian camp in Lebanon is welcoming Syrians. *The Conversation*. Retrieved from https://theconversation.com/refugees-helping-refugees-how-a-palestini-an-camp-in-lebanon-is-welcoming-syrians-48056.

Frazer, R., & Waitt, G. (2016). Pain, politics and volunteering in tourism studies. *Annals of Tourism Research, 57*, 176–189. https://doi.org/10.1016/j.annals.2016.01.001

Goodman, D. M. (2012). *The Demanded Self: Levinasian ethics and identity in psychology*. Duquesne University Press.

Hannam, K., Sheller, M., & Urry, J. (2006). Mobilities, immobilities and moorings. *Mobilities, 1*(1), 1–22. https://doi.org/10.1080/17450100500489189

Higgins Desbiolles, F. (2020). Diversity in tourism: A perspective article. *Tourism Review, 75*(1), 29–32. https://doi.org/10.1108/TR-06-2019-0226

Hollinshead, K. (1998). Tourism, hybridity and ambiguity: The relevance of Bhabha's 'third space' cultures. *Journal of Leisure Research, 30*(1), 121–156. https://doi.org/10.1080/00222216.1998.11949822

Institute for Economics & Peace. Ecological Threat Register (2020). Understanding Ecological Threats, Resilience and Peace, Sydney, September 2020. Retrieved from: http://visionofhumanity.org/reports

International Organisation for Migration. (2022). .Available at https://www.iom.int/about-migration

Karakoulaki, M., Southgate, L., & Steiner, J. (2018). Observations on migration in the twenty-first century: Where to from here? *E-International Relations*, Retrieved from https://www.e-ir.info/2018/08/29/observations-on-migration-in-the-twenty-first-century-where-to-from-here/

Jamal, T., & Camargo, B. A. (2014). Sustainable tourism, justice and an ethic of care: Toward the just destination. *Journal of Sustainable Tourism, 22*(1), 11–30. https://doi.org/10.1080/09669582.2013.786084

Jamal, T., & Higham, J. (2021). Justice and ethics: Towards a new platform for tourism and sustainability. *Journal of Sustainable Tourism, 29*(2–3), 143–157. https://doi.org/10.1080/09669582.2020.1835933

Leeds City Council. (2020). *Rediscover Leeds launched in bid to help city's key industries recover*. Retrieved from https://news.leeds.gov.uk/news/rediscover-leeds-launched-in-bid-to-help-citys-key-industries-recover.

Levinas, E. (1991). *Totality and infinity: an essay on exteriority* (4th ed). Kluwer Academic Publishers. [1961].

Levinas, E., & Kearney, R. (1986). Dialogue with Emmanuel Levinas. In Cohen, R. (Ed), *Face to face with Levinas* (pp. 13–33). State University of New York Press.
Mackenzie, S. H., & Raymond, E. (2020). A conceptual model of adventure tour guide well-being. *Annals of Tourism Research*, *84*, 102977. https://doi.org/10.1016/j.annals.2020.102977
Mahmood, S., Uppal, B., & Bhabra, P. (2019). *Migration in Leeds – Annual update*. Retrieved from https://democracy.leeds.gov.uk/documents/s185667/Scrutiny%20Report%20-%20Migration%20update%20-%20March%202019.docx.pdf
Marotta, V. (2002). Zygmunt Bauman: Order, strangerhood and freedom. *Thesis Eleven*, *70*(1), 36–54. https://doi.org/10.1177/0725513602070001005
McRae, L. (2003). Rethinking tourism: Edward Said and a politics of meeting and movement. *Tourist Studies*, *3*(3), 235–251. https://doi.org/10.1177/1468797603049658
Mellino, M., & Vietti, F. (2019). L'antropologia applicata tra 'tecniche di mercato' e 'pratiche politiche. Riflessioni Sui Migranti, Migrantour e Noi. *Antropologia Pubblica*, *5*(1), 123–132.
Migrantour. (2021). Retrieved from http://www.mygrantour.org/en
Ntaltagianni, C., & Dimas, P. (2016). *Heritage Interpretation as a Means of Social Integration of Refugees in the EU*. Retrieved from https://www.herita.be/sites/default/files/2019-10/proceedings_interpret_confpapers.pdf#page=19
Ormond, M., & Vietti, F. (2021). Beyond multicultural 'tolerance': Guided tours and guidebooks as transformative tools for civic learning. *Journal of Sustainable Tourism*, *30*,1–17.
Lacey, G., Peel, V., & Weiler, B. (2012). Disseminating the voice of the other: A case study of philanthropic tourism. *Annals of Tourism Research*, *39*(2), 1199–1220. https://doi.org/10.1016/j.annals.2012.01.004
Pietsch, J. & Marotta, V. (2009). Bauman, strangerhood and attitudes towards immigrants among the Australian population, *Journal of Sociology*, *45*, 187–200.
Pappas, N., & Papatheodorou, A. (2017). Tourism and the refugee crisis in Greece: Perceptions and decision-making of accommodation providers. *Tourism Management*, *63*, 31–41. https://doi.org/10.1016/j.tourman.2017.06.005
Pocket Stories. (2021). Available: https://www.ourpocketstories.org/our-vision
Popke, E. J. (2003). Poststructuralist ethics: subjectivity, responsibility and the space of community. *Progress in Human Geography*, *27*(3), 298–316. https://doi.org/10.1191/0309132503ph429oa
RefMig Refugees are Migrants. (2022). Available https://www.refmig.org/
Refugee Voices Tours. (2021). Retrieved from https://refugeevoicestours.org/
Sheller, M. (2018). *Mobility justice: The politics of movement in an age of extremes*. Verso Books.
Sheller, M., & Urry, J. (2006). The new mobilities paradigm. *Environment and Planning A: Economy and Space*, *38*(2), 207–226. https://doi.org/10.1068/a37268
Scheyvens, R., & Biddulph, R. (2018). Inclusive tourism development. *Tourism Geographies*, *20*(4), 589–609. https://doi.org/10.1080/14616688.2017.1381985
Sultana, F. (2007). Reflexivity, positionality and participatory ethics: negotiating fieldwork dilemmas in international research. *ACME: An International E-Journal for Critical Geographies*, *6*(3), 374–385.
Tomassini, L., Lamond, I., & Burrai, E. (2021). Global citizenship & Parrhesia in small values-based tourism firms. *Leisure Sciences*, 1–19. https://doi.org/10.1080/01490400.2021.1874574
Trihas, N., & Tsilimpokos, K. (2018). Refugee crisis and volunteer tourism in Lesvos. *Greece. Journal of Tourism & Sustainability*, *2*(1), 42–56.
United Nations High Commissioner for Refugees. (2019). *Global trends*. Retrieved from https://www.unhcr.org/5ee200e37.pdf
United Nations High Commissioner for Refugees. (2020). *Refugee statistics*. Retrieved from: https://www.unrefugees.org/refugee-facts/statistics/#:~:text=By%20the%20end%20of%202019,remained%20at%20a%20record%20high.
United Nations High Commissioner for Refugees. (2021). *UNHCR global report 2020*. Retrieved from. https://www.unhcr.org/flagship-reports/globalreport/
UNHCR. (2022). *Ukraine Situation: Regional Refugee Response Plan*. Retrieved from https://data2.unhcr.org/en/documents/details/91114
UnLtd. (2020). Retrieved from: https://www.unltd.org.uk.
Uppal, B., & Bhabra, P. (2018). *A strategic approach to migration in Leeds*. Retrieved from: https://democracy.leeds.gov.uk/documents/s178855/Migration%20Cover%20Report%20130718.pdf
Visit Leeds. (2021). *Full steam ahead for Leeds' visitor economy performance*. Leeds City Council. Retrieved from: https://www.visitleeds.co.uk/news/its-more-than-temperatures-rising-in-leeds-as-visitor-numbers-hit-record-levels/.
Zetter, R., Griffiths, D., Sigona, N., Flynn, D., Pasha, T., & Beynon, R. (2006). *Immigration, social cohesion and social capital*. Joseph Rowntree Foundation.

Seeking justice beyond the platform economy: migrant workers navigating precarious lives

Tyler Riordan, Richard N.S. Robinson and Gerhard Hoffstaedter

ABSTRACT

Recent literature focuses on the inherent challenges of food delivery work. Less is known about how these injustices impact workers and their lives more broadly, or how workers navigate them. This empirical article is based on a 12-month ethnographic theory-relevant case study and includes an innovative shadowing method focused on migrant food delivery workers in Brisbane, Australia. We found that temporary migrant workers face intersectional injustices, exacerbated by the COVID-19 crisis, both within and beyond the platform economy. Nevertheless, the workers agentically enacted their own justices to navigate their precarity, in the absence of institutional supports. This study's theoretical contribution is the development of a model explaining the agentic and structural underpinnings of the injustices that migrant platform workers experience. Moreover, our contribution reveals that the unique attributes that migrants bring to bear on their platform work provide them with the affordances to navigate the injustices they experience. In so doing they mitigate some of the negative impacts of platform work, and indeed derive benefits that non-migrant platform workers might not.

Introduction

Digital platforms have disrupted traditional service models across the hospitality and tourism sectors. In food delivery, platforms offer new markets to restaurants, convenience to customers, and work opportunities for people with inequitable employment access (Pigatto et al., 2017). Workers join platforms such as Uber Eats and Deliveroo because of limited barriers to entry, the inability to find work in other fields, and the precarious nature of the broader labour market (Barratt et al., 2020). Many are attracted to platform work by claims of its flexibility and autonomy affordances (Vieira, 2021). Contrarily, there is another side to platform work, which is incompatible with sustainable employment in tourism (Mooney et al., 2022).

Despite their status as "essential" workers, conditions for platform workers are increasingly precarious. Negative conditions were exacerbated by the COVID-19 pandemic (Aloisi & De Stefano, 2022; Figueroa et al., 2021). Migrant workers, who conduct most food delivery in Australia (McDonald et al., 2020), face increased risks including death (Zhou, 2021). When work-related injustices occur, workers receive little support from platforms, governments, or other stakeholders (McDonald et al., 2020). Most research on the platform economy focuses on labour aspects such as precarity, efficiency, and algorithmic control (Schor et al., 2020), with insufficient examination of how workers navigate these intersecting injustices across other life domains (see Schüßler et al., 2021).

Furthermore, Campbell and Price (2016) suggest there is a lack of clarity in the literature about the relationship between those who are precariously employed and precarious work. Thus, Robinson et al. (2019) propose a fuller understanding is required vis-à-vis the interconnectedness between precarious workers, precarious work, and precarious lives. Following Campbell and Price (2016) we adopt a "theory-relevant" case specifically focusing on migrant food delivery workers to avoid confounding the messiness between precarious workers, precarious work, and precarious lives across incomparable contexts. Theory development from specific relevant cases has a long tradition in qualitative research, which builds on the conventions of grounded theory and case study research (Eisenhardt, 1989). Burawoy (2009) further articulated the process of bringing theory, in this case precarity more broadly, "to the field" via ethnography. This "theory-relevant" approach is congruent with the Mertonian tradition of mid-range theory, which bridges so-called grand theories with specific empirical contexts (Joo et al., 2020). Thus framed, this article sets out to address these research gaps by investigating migrant food delivery workers' circumstances within and beyond the platform economy. To do so, we applied a justice lens to answer the research question: *how do temporary migrant food delivery workers navigate the injustices that they face within and outside the platform economy?*

We developed a case study using ethnographic research with migrants who deliver food in Brisbane, Australia. This methodology, in keeping with theory-relevant approaches (Burawoy, 2009; Campbell & Price, 2016), included an innovative shadowing method, participant observation, and semi-structured interviews with delivery workers and other key stakeholders. Specifically, we wanted to investigate how migrant food delivery workers attempt to find justice and dignity in an unjust system from an experiential perspective.

The article begins with a review of precarious work. We then present an overview of conditions for (migrant) workers in the platform economy. After providing more details on our ethnographic approach, we then critically present empirical data evidencing how workers experience injustices. Next, we discuss how workers turn to each other to agentically navigate the intersecting injustices they face within and outside the platform economy. We then expound our theoretical contribution by proposing a comprehensive injustice framing of migrant work/er precarity which accounts for a constellation, or ecosystem (see Lusch & Vargo, 2014 as applied to service work) of factors. In this study we make two theoretical contributions. First, we develop a model that explains the agentic and structural underpinnings of the injustices migrant platform workers experience. Second, we reveal that the unique attributes that migrants bring to bear on their platform work enables them to navigate the injustices they experience. In so doing they mitigate some negative impacts of the work, and indeed find benefits that non-migrant platform workers might not.

Precarious work, platforms and migrant workers

Work has become increasingly precarious for many in the global North. Precarious work is characterised by lack of regulatory protections, low wages and wage theft, employment insecurity, limited benefits and entitlements, and limited worker control over wages and conditions (Campbell & Price, 2016; Vosko, 2010). In precarious work, risks are passed from businesses or governments to workers (Kalleberg & Vallas, 2017). The increase of precarious work in the global North impacts other areas of workers' lives including career progression, relationships, and subjective wellbeing (Kalleberg, 2018). According to the International Labour Organization (2016) precarious workers are denied benefits including fair treatment, voice, safe and healthy work workplaces, enhanced productivity, employment and income security, equality of access, and social protections. In tourism and hospitality, precarious work contributes to social cleavages and economic inequalities (Robinson et al., 2019) and is particularly concerning for migrant workers (Anderson, 2010). Resultant are calls for more sustainable workforces (Ioannides et al., 2021).

Conditions in the tourism and hospitality industries have worsened during the COVID-19 pandemic, associated lockdowns, and border closures. While precarious work has long been an issue, the COVID-19 crisis brought underlying concerns to wider attention (Baum et al., 2020). Further, negative working conditions disproportionately affect vulnerable workers, such as migrants who we focus on in this article (Cook et al., 2021; Sun et al., 2021). Restaurants increasingly outsourced responsibilities to food delivery platforms to remain open and meet consumer demands (Ma et al., 2021). For food delivery workers, conditions have become more precarious as workers face increased risks with limited support (Cano et al., 2021). However, in response to the broader intersecting challenges of precarious work and the COVID-19 crisis, many workers still turned to the platform economy to survive. For some, food delivery work offers an outlet to leave the broader labour market which is characterised by temporary contracts, lows wages and lack of flexibility (Vieira, 2021).

Conditions in the platform economy can be precarious due to management arrangements. Food delivery workers in Australia are independent contractors, paid per piece, and not compensated for waiting time (Goods et al., 2019). Algorithmic management has replaced human supervision (Duggan et al., 2020) and platforms treat workers as interchangeable and replaceable (Riordan et al., 2023). Platforms issue tasks and instructions to workers through smartphone applications (Veen et al., 2020) and workers lack clarity on decision-making processes and pay rates (Gregory & Maldonado, 2020) which can negatively affect them (Wood et al., 2019). While algorithms are sometimes portrayed as neutral, design decisions and algorithmic management are shaped by pre-existing power dynamics between managers and workers (Jarrahi et al., 2021). Algorithmic management can constrain choice and lead to increased precarity for food delivery workers (Veen et al., 2020). Indirect control is maintained through customer feedback whereby workers conduct emotional labour in exchange for favourable ratings (Gandini, 2019). Algorithmic management can also lead to discriminatory practices, surveillance, and control (Galière, 2020; Shapiro, 2018). These issues are compounded for migrant workers who may lack local knowledge and the social capital to navigate these circumstances—and during the pandemic many were entrapped due to their inability to return home easily (Martin & Bergmann, 2021).

The circumstances of temporary migrants in Australia are unique (Berg & Farbenblum, 2020). For decades Australian governments have politicised migration such that many migrants, especially refugees and asylum seekers, become suspended in a visa-status-less (or uncertain) limbo (Ratnam, 2019). Temporary migrants thus often end up in precarious work circumstances (Farbenblum & Berg, 2021), including the platform economy. According to historical patterns in the global North, racial and ethnic minorities are overrepresented in precarious roles (Kalleberg & Vallas, 2017). Some food delivery platforms target migrant workers who they perceive to be exploitable (Altenried, 2021). In Australia, young temporary migrants are three times more likely to be platform workers than citizens (McDonald et al., 2020). Migrant workers in the platform economy experience higher job insecurity than non-migrant workers (Holtum et al., 2022). Migrant labour has increasingly become a feature of global research on food delivery (van Doorn & Vijay, 2021). However, the COVID crisis has exacerbated structural issues for migrant workers.

Concepts such as crisis are imbued with meaning. Crisis is a narrative construction which is often used to refer to moments of truth or turning points in history (Roitman, 2013). As Roitman (2021, p. 2) argues, our efforts to solve a crisis can only progress if we "agree about what exactly is in crisis and for whom". Therefore, people experience crisis in numerous ways, according to their individual and collective experiences. Further, times of crisis expose the fragility of existing economic structures (Narotzky & Besnier, 2014). For this article, the COVID-19 crisis refers to the global health pandemic and associated socio-economic consequences (Roitman, 2021); focused on experiences of migrant food delivery workers.

Justice and dignity

Rastegar et al. (2021, p. 1) argue that COVID-19 has become "a justice issue following years of unsustainable growth, growing conflict, inequity and injustice". Therefore, more just approaches are required in the tourism industry for marginalised groups such as women, children, minorities, illegal workers, Indigenous peoples (Jamal, 2019)—and migrants. Principles of justice and just approaches also intertwine with human dignity.

Dignity relates to both basic livelihood and social value; and a sense that one feels equal to their peers (Kant, 1997; Narotzky, 2016). Dignity is a human right (United Nations, 1948) and should entitle everyone to the same rights and duties which we ask for ourselves (Benhabib, 2013). Thus, dignity gives our lives value (Rawls, 1971) and an inner worth that goes beyond one's market value (Kant, 1997). In the labour context, dignity relates to meaningful work (Hodson, 2001) and is dependent on factors such as access to well-paid, safe and secure work conditions (Bolton, 2007). Crucially, in industries such as hospitality and tourism, dignity is central to decent work and sustainable workforces (Winchenbach et al., 2019). Having dignity is a requirement to developing a sense of justice as without understanding one's worth it is hard to recognise that one has been treated unjustly (Pritchard, 1972). However, models of justice which prioritise group identity over social institutions and interactions can be problematic (Fraser, 2001). Instead, Fraser (2001, p. 24) proposes a status model to redress injustices through a politics of recognition that overcomes "subordination by establishing the misrecognized party as a full member of society, capable of participating on a par with other members". This speaks to a moral obligation, and a global political responsibility to address systemic injustices across borders (Jamal & Higham, 2021).

Injustices are rooted in processes and practices that systematically disadvantage some groups (or individuals) in relation to others (Fraser, 2020). Injustices often stem from the failure of the state to protect its citizens or the failure of institutions to protect workers and when cries of injustice become collective they point to systematic failings (Narotzky, 2016). Injustices can be remedied through processes such as redistribution, recognition, affirmation, and transformation (Fraser, 2020). When facing uncertainty, vulnerable people seek workplace justice as a way to protect themselves (Jamal & Higham, 2021); sometimes through deviant and nefarious means (Lugosi, 2019; Robinson, 2008). Leading scholars argue for fairer and more just conditions for platform workers (Graham & Woodcock, 2018). Thus, justice has become a crucial lens through which we investigate migrant workers' experiences in the platform economy.

Materials and methods

We chose a case study utilising an ethnographic approach to explore and better understand the experiences of migrants who conduct food delivery in the platform economy. Ethnographic fieldwork investigates small case numbers to immersively explore social phenomena (Atkinson & Hammersley, 1994). We designed a one-year multi-sited ethnography to fully capture the workers' "social universe" (Bourdieu, 1985). Fieldwork included traditional ethnographic methods, such as interviewing, participant observation and simply being in the field for prolonged periods of time, as well as novel tools, such as shadowing delivery drivers to capture their experiences more closely.

Food delivery work takes place at multiple temporal and spatial intervals which presented unique fieldwork challenges. In response, we developed innovative methods to capture worker experiences. The first, involved participant-observation in public places where food delivery

Table 1. Food delivery worker profiles.

Pseudonym	Country	Professional background	Visa	Vehicle
Carlos	Peru	Biomedical Science (Student)	Student	Bicycle
Li	Panama	Family Business (Service)	Student	E-bike
Anita	Colombia	Lawyer	Student	E-bike
Juliana	Colombia	Doctor	Bridging	E-bike
Jose	Colombia	Electrician	Bridging	Moped
Diego	Colombia	Chemical Engineer	Student	E-bike
Ricardo	Colombia	Underwriter	Student	E-bike/Car
Leticia	Brazil	Biologist	Student	Motorbike
Fernanda	Brazil	Dentist	Partner	Car
Oscar	Mexico	Mechanical Engineer	Student	Bicycle/Moped
Sara	Colombia	Financial Engineer	Student	Bicycle/E-bike
Manuel	Colombia	Business Administration	Student	E-bike
Peter	Macau	Communications (Student)	Graduate	E-bike
Catalina	Colombia	Business Administration	Student	E-bike
Omar	Colombia	Engineer	Bridging	E-bike
Maria	Colombia	Accountant	Student	E-bike
Juan	Mexico	Bank Chain Manager	Student	Motorbike
Ashfin	Iran	International Business Management	Student	E-bike
Olivia	Brazil	Lawyer	Student	Car
Yang	Burma	Carpentry (student)	Unknown	E-bike
Hiro	Japan	Economist	Partner	Car
Anthony	Colombia	Lawyer	Student	E-bike
Carolina	Colombia	Industrial Engineer	Student	E-bike
Vijay	India	Business Administration	Student	Car
Pamela	Chile	Transport Engineer	Partner	Car

interactions take place—including restaurants, malls, shopping centres, and footpaths. The second, was based on "shadowing" which is typically used in business settings and allows researchers to "look over the shoulder" of people while they work (McDonald, 2005). Shadowing—a form of one-on-one ethnography—is interpretive and incorporates a reflexive approach which involves critical awareness of how the researcher views themselves and what impact this has on the research (and data) (Gill, 2011). Shadowing also offers unique insight into the messiness of workers' lives, their interactions, and the ways they experience phenomena (van der Weele & Bredewold, 2021).

Practically, shadowing involved following (cycling alongside or driving with) workers while they delivered meals. Shadowing also involved the researcher hanging out with delivery people beyond the context of delivering food. This included time waiting for orders, catching up after shifts, sharing meals, and so on. This allowed the researcher to develop rapport, observe behaviour, and engage in deep conversations based on trust built over extended periods. Shadowing by bicycle provided further insights into the physical and mental challenges of the role and life in Australia more generally. Ingold and Vergunst (2008) argue that "walking with" participants can facilitate understandings of embodied experiences. "Ride-alongs" have been utilised with ride-share drivers (Dubal, 2019; Holtum et al., 2022), and ethnographic methods have been applied by researchers who worked as food delivery riders (Cant, 2019; Richardson, 2020; van Doorn & Badger, 2020). As workers use phones for work and in-between jobs, we also incorporated elements of digital ethnography to capture the online interactions which entangle their daily lives (Sinanan & McDonald, 2018). Finally, we conducted semi-structured interviews with workers and other stakeholders that regularly interact with this cohort. To the best of our knowledge, this study is the first to access the wider "social universe" of food delivery workers. Institutional approval was granted following risk assessment and human research ethics applications.

Data were collected between September 2020 and October 2021 in Brisbane, Australia. The project includes 25 current or former food delivery workers (see Table 1). Migrant workers were deliberately sought out due to their prevalence in food delivery roles (McDonald et al., 2020), as

Table 2. Stakeholder interview profiles.

Pseudonym	Role	Institution
Paul	Managing Director	Charity
Thomas	Councillor	Local Government
Naomi	Mechanic	Bicycle shop
Joseph	Moderator	Facebook group
Theresa	Counsellor	Migration agency
Will	General Manager	Food delivery platform
Michael	Safety Manager	Education institution
Harry	Coordinator	Union
David	Operations Manager	Food delivery platform
Elliot	Owner	E-bike lender

well as the scarcity of research relating to their unique circumstances. The first author conducted over 50 shadowing shifts—with nine workers. 19 interviews were also conducted with 21 food delivery workers.[1] Workers in this study primarily engaged with Uber Eats, and less frequently with Deliveroo, Menulog and DoorDash. A further 10 interviews were conducted with stakeholders (see Table 2). Researchers held additional discussions with other delivery workers and stakeholders throughout fieldwork. Data include reflective fieldwork journals, interview transcripts, video recordings, GPS tracking displays, and social media interactions.

As is common in qualitative research, preliminary sense-making began during fieldwork and continued throughout the research process. The researchers employed data reduction to select, focus, simplify, abstract, and transform the raw data and guide the final stages of fieldwork (Miles & Huberman, 1984). All data were indexed chronologically at the completion of fieldwork, and interviews were professionally transcribed. Data were manually coded and relationships were explored and identified (Hammersley & Atkinson, 2007). Data were then grouped into larger themes through a process of inductive reasoning and theorisation mobilised an analytic induction process (Hammersley & Atkinson, 2007). Finally, empirical and theoretical findings were critically reviewed iteratively via interpretative engagement with the relevant literature.

Results

Our findings demonstrate the challenges that migrant food delivery workers face within and beyond the platform economy. These injustices encroach their lives as precarious workers, platform workers and temporary migrants in Australia. However, this cohort do not suffer in silence. Instead, they use their skillsets to navigate intersecting injustices through agency, resistance, collaboration and alongside external support.

Precarious workers

Working circumstances for temporary migrants have become more precarious since the beginning of the COVID-19 crisis. The COVID-19 pandemic and associated lockdowns have highlighted the precarious nature of non-standard work. Most informants are highly educated but face barriers to employment in Australia and believe that they do not have opportunities to work in their fields. Many previously worked in industries such as hospitality and cleaning which, as Sara explains, is where she was bullied, exploited, and "lost my dignity in this job". With the onset of the pandemic lockdowns, some lost their jobs. Carlos, a 20-year-old university student, explains why he no longer works in a hotel:

> I got fired … because of COVID and since I was casual, they just like got rid of me like (snaps fingers).

For Maria, who previously did food delivery occasionally, "the pandemic hit, and I had to do Uber [Eats] … because there was no more work". In summary, many workers start delivering food to make ends meet or to escape precarious conditions in the broader labour market.

Precarious platform work

Food delivery work can also be precarious. Orders and earnings are irregular and informants believe they are unfairly compensated. Juliana feels "the company makes a lot of money from us, but never pays us back" while her partner Jose added "there is no adequate remuneration". During COVID-19 lockdowns informants described "parking lots" of riders waiting outside restaurants. Anita, who had never ridden a bicycle before coming to Australia, recalls the early days of the pandemic:

> We went out almost a full day, that is [the] twelve hours that [the platform] allows you to be connected to the app. And we barely made thirty dollars a day, and we had to go out almost every day for ten or twelve hours to be able to at least make the rent.

Regular changes to processes and payments frustrate workers. Diego, like others, does not "know how the algorithm of Uber works". The lack of transparency and inconsistency of orders (and earnings) makes it difficult for workers to control time or plan for the future. Some feel they are manipulated into making decisions, without knowledge of the consequences. Workers also sense that the rating system is unfair and customers have too much control. Workers fear that their accounts may be suspended or "deactivated" without warning. As independent contractors, workers received little support from platforms and were ineligible for government financial support (including COVID-19 crisis payments) which was allocated to employees.

Food delivery workers face further challenges related to equipment, exposure to traffic, and other risks. Naomi, a bicycle mechanic, explains some difficulties:

> The bikes that these guys ride just fall into disrepair … it's from excessive use and not being able to afford maintenance … one of my services will cost effectively their entire night's wage.

Most migrants rent electric bicycles from companies that Thomas describes as "parasitic". Some are "informal" arrangements and repairs are difficult to access as, according to Catalina, "that wasn't included in the rent". David, former operations manager for a food delivery platform, explained that as workers are contractors "you don't have to tell them how to do it" or give them safety training "because they are 'experts' in their field". However, safety training does not remove the key risks of cycling for work. Some workers feel unprotected and unsupported by platforms as Peter opines: "they just send the message to remind you please wear the helmet. And that's it". Informants were involved in accidents and injured, while researchers witnessed "near misses" during fieldwork. Many, like Juliana, dislike cycling on the road:

> It is dangerous here because sometimes there are areas where there is no sidewalk, and there is no bike path, so I have to take the road … and the cars go very fast.

Informants discussed deaths of workers in other parts of Australia. They felt safer and more protected from other road users as they are legally able to cycle on footpaths in Queensland. No deaths were reported in the state during fieldwork; however, stakeholder interviews and informal conversations with law enforcement representatives revealed that such data may not be recorded. When workers have accidents, they have limited or complicated avenues to recourse.

Food delivery workers face other challenges due to the location and nature of the work. Workers described attempts to steal food that they are delivering, phones, and bicycles. Interactions with customers can also be problematic. Some ask "why are you taking so long?" and others blame workers when something goes wrong—irrespective of who is at fault. Some workers like Li, a Panamanian of Chinese descent, have been abused and subjected to xenophobic incidents:

> When the customer came to me… [and] saw that I'm Asian, they started to say a lot of bad words about Chinese and about Coronavirus. So they took the food and threw it in my face.

Some avoid night work due to the risks associated with "drunk people" who harass them or "yell things at you from [passing] cars". Some females stopped working at night due to unwanted attention. After an incident of sexual abuse, Sara, a finance engineer, wore a ring and told customers she was married to avoid being accosted. Inadequate lighting is another issue. Migrant workers perceive that they are targeted by police and council workers. During fieldwork, after Anthony was reprimanded for cycling in a pedestrian mall, he and Manuel both commented that "they don't stop Australians, they only stop foreigners". Some workers have been excluded from places that they normally access because they are wearing a delivery backpack—irrespective of whether they are working at the time.

The COVID crisis amplified problems that migrant food delivery workers face. Delivery workers encountered increased risk of exposure to the virus as they regularly entered restaurants and apartment buildings. In response, platforms implemented safety measures such as no-contact deliveries. However, there was simultaneously an increase in deliveries to hospitals, hotels used for quarantine, and other high-risk sites. Informants reported that customers, restaurants, and hotel staff did not always follow safety procedures. Some felt pressure to comply with requests that they felt were unsafe. Temporary migrants were also excluded from government support payments due to their visa status. This was especially challenging for international students whose visa condition limits them to 40 hours work each fortnight. While these conditions were recently relaxed for the tourism and hospitality sector, food delivery workers were unsure if they qualified. Some workers were aggravated as waiting time is not compensated by platforms, yet still counts towards their allocated working time according to their visa conditions.

Lack of support as independent workers and temporary migrants

Migrant food delivery workers do not always know how to access assistance when things go wrong. In the first instance, workers describe a lack of support from platforms. For example, Uber closed their office in Brisbane in March 2020. According to Peter their "support call line NEVER [Peter's emphasis] pick up the phone". All issues and complaints are submitted through the app, which can be complicated and time consuming for workers. When a worker has an accident or is deactivated by the platform, they do not know what to do and are unsatisfied or angered by inadequate or unrelated automated responses from platforms. Ricardo explains his frustration after an accident:

> You have to pay the ambulance… the X-rays… it's expensive… and you tell Uber… "I have had an accident" and [they] say… "Communicate with the insurance department".

However, the insurance department requested additional paperwork that was difficult for him to complete. Instead, Ricardo, like others, relied on claiming from his compulsory student health insurance. Platform workers have limited access to sick pay. Informants could not afford to take days off, and some recommenced work before injuries had healed. In Ricardo's case, he needed to work "to pay my visa", and he could not "pay rent". After an accident, Ashfin felt "powerless", while for Anthony it negatively "changed his perception of food delivery work". When workers approached government bodies, like the Fair Work Commission, for help they were told that they "cannot do anything" (Joseph) as they are not employees.

Due to their temporary status in Australia, migrant workers typically lack social capital or do not see it within their scope to make complaints. Some informants did not know where to look for external support and processes were complicated to navigate for English language learners. Further, workers were not aware of the protections they deserve or the resources available to them. Joseph, a stakeholder and former international student, details some challenges:

> We don't know our rights. We don't know we can actually say, "no", we cannot fight back. So, we assume that, whatever they're telling us, because they are the boss and they are the company… they're doing the legal… the right things.

Their status as non-citizens and non-voters also mean limited avenues for recourse and little political will to improve access. Thomas, a local government councillor, elaborates:

> We know that these people are working in a very dangerous job. We know they're being exploited yet it doesn't come up as an election issue at all because they're not voters.

This illustrates how migrant food delivery worker's industrial disempowerment is mirrored in, and by, a general disenfranchisement in other life domains.

Impacts of precarious circumstances

The cumulative impacts of injustices (and fighting them) affect the relationships, physical, and mental health of informants. Oscar, an engineer, found "it was very stressing to be working five or six days per week and cycling a lot". For Pamela, "the most difficult part over money, or the conditions… is the emotional burden". Many find the job "lonely" and aspects of the work are dehumanising. Some reported relationships issues due to perceived pressures including Juliana and Jose who "almost broke up". Leticia, a biologist from Brazil, wished to "feel more worth" and requested that platforms "treat me as a person, not as a number". She continued that "we are so used to being not seen… [we are] invisible". Diego explains how some restaurant workers "don't even look at you". Carlos shared his thoughts:

> People should be more conscious that we are also people, we are not like robots who deliver food. We have problems, issues, and so these people treat us as if we are nothing.

Some informants feel shame while doing delivery. While waiting for a storm to pass, Omar explained that he does not work when it is raining as even he "feels sorry for himself". Carlos does not tell Australian friends he does delivery as "I don't want them to look at me different". Joseph, a moderator of Facebook groups for migrants, suggests "Australian society doesn't even know… what we're going through". Migrant workers face a range of injustices in Australia that permeate their lives. While some feel that they have no support that is not always the case. In response to their precarious circumstances, food delivery workers need to create their own pathways to justice.

Navigating injustices

Migrant food delivery workers use their skills and wherewithal to navigate the precarious circumstances and injustices they face in Australia. Workers achieve this through agency, resistance, and collaboration. However, not all injustices are able to be overcome. In such instances, they seek support from others to minimise negatives and extract benefits where possible.

Agency

Informants choose to work in food delivery. For some, it offers an escape from the broader labour market as "it is better to do Uber Eats" than alternatives available to them. This is due to the perceived autonomy, flexibility to choose hours, and potential to earn quickly. The flexibility was appealing to Diego as he could "go online anytime, whenever you want, if you don't want to work, you don't work". Diego also found the "freedom" beneficial compared to his friends who could not attend social occasions as they "have to work". Ricardo explains why he prefers food delivery:

> I don't say Uber is a bad job. Actually. It's a good job because you can … be your own boss and you can get good money.

Migrant workers also choose platform work to resist aspects of the migration system that they believe are unfair by working beyond the legal hours stipulated by their visa conditions.

Workers also draw on their past experiences and skills to respond to challenges. For example, when a worker was involved in an accident they were unable to access insurance from the platform due to their nationality. In response, they used their agency to gain financial support as described in the following field note:

> Using his previous experience working in the military. He wrote letters to the [platform] company. And they ended up paying for his hospital bills, and giving him compensation of 150 dollars a day … for two weeks. So he mentioned that he knew how to do the paperwork that was required, that he got his friends from England to check the grammar.

As well as individual acts of agency, workers also come together to resist platforms and other stakeholders.

Resistance[2]

Migrant food delivery workers also seek justice through acts of resistance. Some workers resist platforms by individually and collectively rejecting low paying orders; lobbying for safety equipment; joining unions; participating in media coverage; starting court proceedings; using multiple platforms; evading work restrictions; renting accounts; and accepting payment for food that they do not deliver. They also used "tricks" and "hacks" to make their work more profitable and enjoyable. Others respond to restaurants that they believe have treated them poorly by cancelling orders or leaving negative online reviews. Such acts target platforms and their algorithms, restaurants, and customers. While these acts make workers feel better, they have little impact on their broader circumstances.

Collaboration

Migrant food delivery workers take it upon themselves to look after one another. In the absence of support from platforms, workers provide guidance to each other. Workers felt that such acts were necessary due to the inconsistent nature of communication with platforms. Pamela elaborates:

> The support … the last year and a half was terrible. If we have any problems, we kind of have to solve it by ourselves. We cannot call anyone. If we write to someone, they [can] take, minutes, hours, days to answer.

Workers also support each other during or after damage or theft of work equipment. Anita, a Colombian lawyer, explains why she helped a worker when she witnessed their bicycle being stolen:

> There is a lot of collaboration and even more so now because we are outraged by the theft of bicycles, we try to help each other because we know that the police are not going to do anything.

Migrant food delivery workers also offer each other support at times of crisis. In the context of this study, workers raised funds for peers who were injured, or to cover funeral and repatriation costs. Some workers relied on peers and other friends when they could not afford to buy food.

Navigating precarious circumstances with external support

While migrants used their own initiative to overcome issues, some required assistance to navigate structural injustices. During the COVID crisis, workers sought food and financial support from community organisations, religious groups, and educational institutions. Paul, a managing director of a charity, explains how they began to help delivery workers:

> Initially, we were for homeless and disadvantaged people and people at risk of homelessness, that's been our primary focus, but when we had the COVID-19, we found that, we had a lot of international students who were in need of food.

Community members also assisted informants in other ways. Joseph—who moderates Facebook groups for people who speak Spanish—was motivated to help as he "found a lot of issues, a lot of gaps between the Australian society and where we were as immigrants". Diego gave an example of how an Australian friend helped him:

> She has an electric bike and she's not using it. I told her I am going to buy an electric bike and she told me "no, don't buy anything. I will lend it to you".

When Sara arrived in Brisbane without accommodation—a week before a COVID lockdown—she was taken in by a person she calls "a little angel and my Australian mother". Finally, customers offer tips or other incentives to workers; bystanders intervene to protect workers from abuse; and restaurant staff sometimes gift free food or drinks. These examples of hardship that food delivery workers experience and the ways various stakeholders mobilised to administer justice speak to the spill-overs of the exacerbated impacts of the pandemic from precarious work into precarious lives.

Discussion

Our research shows that food delivery workers face complex injustices within and beyond the confines of platform work. Such injustices can be attributed to the intersectional challenges of precarious work, the platform economy, and the broader circumstances of temporary migrant workers in Australia. Intersectionality is useful for our analysis as it captures the overlapping and multiple dimensions of social identities, such as race, gender, migration status, class domination and other discriminating factors (Crenshaw, 1991). In the absence of formal avenues to support, migrant workers take it upon themselves to enact their own forms of restitution for the injustices they face. Workers seek individual and collective justice through acts of agency, resistance, collaboration, and with the assistance of external support.

Our findings demonstrate that some temporary migrants see food delivery as a chance to escape precarious employment circumstances. Workers use their agency to leave industries such as cleaning and hospitality—which are inherently precarious (Robinson et al., 2019). This is consistent with other research that found food delivery workers choose the role to both take advantage of the opportunities that labour platforms offer (van Doorn et al., 2022) and escape precarity in the broader labour market (Vieira, 2021). However, others turn to food delivery as it is one of few opportunities available to them. While many workers have suffered during COVID-19, the most severe impacts were felt along pre-existing lines of marginality putting migrant workers at greater risk than their peers (Cook et al., 2021). Irrespective of such issues, food delivery work offered temporary migrants a lifeline during the COVID crisis which was important as many suffered critical losses of income and were unable to meet basic needs (Berg & Farbenblum, 2020). Rather than being hapless victims, migrant workers use their agency to respond to their exploitative circumstances (Rydzik & Anitha, 2020). These results are significant as they show that some migrant workers choose food delivery as it is more appealing than alternatives available to them.

Our research also indicates that migrants feel exploited while working for platforms. The opportunity to "be your own boss" is crucial for these businesses to recruit workers. While workers perceive flexibility as advantageous, they also recognise limitations. For example, workers are flexible to choose when and where to work however most work at mealtimes and weekends when customer demand is high (Goods et al., 2019). Our results validate other research which found platform workers have limited autonomy as they are not aware of processes such as how work is issued and managed (Veen et al., 2020). Further, platforms change processes without notification, transparency, or justification to workers (Goods et al., 2019). Workers in this study believe that payments are unfair. Some argue that the rating system is problematic and customers have too much control (Rosenblat & Stark, 2016). Platforms are not the only stakeholders to disappoint workers thus providing a nuanced extension of their related experiences from precarious workers to precarious work (Campbell & Price, 2016). Informants feel unsupported due to their circumstances as platform workers—such inequalities and precarisation of exploitable migrants are inherent features of platform labour (van Doorn et al., 2022). Workers in our study feel invisible at times; as generations of workers have done before the existence of the "gig" economy (Alkhatib et al., 2017; Anderson, 2010). Some are taken advantage of by predatory bike lenders, others are concerned by theft and believe they lack support from police because they are migrants. As a result, some workers feel trapped in precarious work.

Safety was a concern for workers and responses from platforms and governments were deemed inadequate. Algorithmic management puts pressure on workers who are incentivised to work quickly (Gregory, 2021). While cycling infrastructure is inappropriate for delivery workers, they were grateful for unique laws in Queensland that allow them to ride on footpaths as they believed cycling on busy roads contributed to interstate worker deaths. Controversially, platforms did not report all worker deaths as they were not engaged with deliveries at the time of incidents and therefore deemed not to be working (Begley, 2021). Consequently, families were denied death benefit claims (Begley, 2021). Some federal and state government inquiries were launched in response resulting in platforms introducing new safety features and checklists (Zhou, 2021). However, procedures are not always enforced, or understood by workers. In addition, not all delivery workers follow traffic rules (Gregory, 2021) and some have created ways to agentically avoid safety procedures (Riordan et al., 2023). Many informants were injured which correlates with US research which found over 50% of food delivery workers have had an accident or crash (Figueroa et al., 2021). When things go wrong, migrant workers are reluctant to complain for fear of deactivation by platforms. This is in line with other research which shows that temporary migrants who are exploited do not make claims as they fear deportation or job loss (Farbenblum & Berg, 2021).

Workers also lack other support due to their status as temporary migrants. Workers were disappointed by their exclusion from government support and what was available was inadequate to meet financial needs (Berg & Farbenblum, 2020). When things went wrong in their work or living circumstances, informants did not always know how or where to seek help. Due to their temporary status, migrant workers may not see it within their scope to make complaints; or they may not be aware of the resources available or protections they deserve. This matches other research, which found temporary migrants face barriers to external support complicated by difficulty navigating processes in the English language (Berg & Farbenblum, 2020). Consistent with contemporary hospitality research (Robinson et al., 2022) some workers were abused by customers, restaurant workers, and general bystanders. Incidences of racism have increased in Australia since the onset of COVID-19 (Berg & Farbenblum, 2020). Some find the work lonely and describe feelings of shame. These circumstances can affect migrant workers in diverse ways including negative health impacts (Berg & Farbenblum, 2020). These results demonstrate that a lack of social capital and unfamiliarity with local systems can be a barrier to justice for temporary migrant workers, and reveal how precarious workers and precarious work (Campbell & Price, 2016) morph into precarious lives (Robinson et al., 2019). Accordingly, we propose a framework

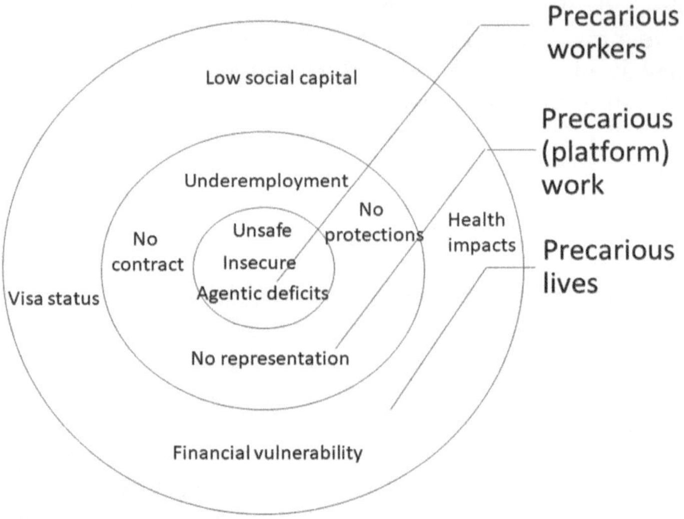

Figure 1. Injustices ecosystem framework.

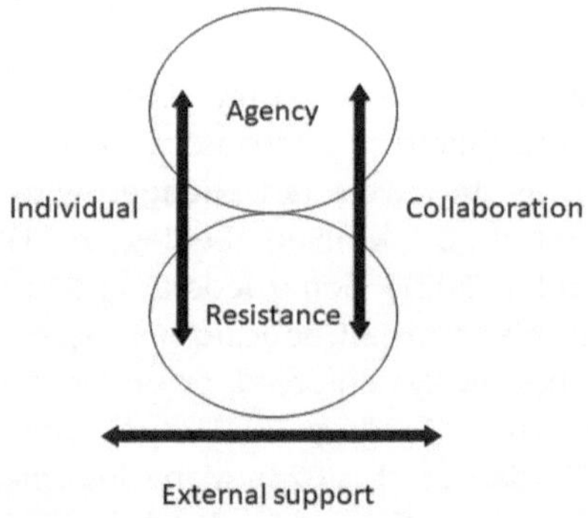

Figure 2. Navigation of injustices framework.

of the migrant food delivery worker's ecosystem, highlighting the relationships between these three domains and thus grounding the spill-over of injustices they endure and agentically navigate (see Figure 1).

The spill-overs between these layers of the unjust ecosystem occur due to an intersection of their deficits as migrants. Unable to access governmental supports, a form of institutionalised racism (Krieger, 2021), migrants mainly have access to the low entry and low skilled jobs—which are often characterised by unsafe and insecure work. Platform work at least enables migrants to work on their own terms to some degree; for example, when, where, and how often they work. Yet they are still subjected to the structural precariousness of platform work, inter alia a lack of protection and representation. In addition, their ignorance, reluctance to engage with, and even fear of systems of support, amplify their vulnerability. These deficits compound already precarious lives creating vulnerability not just in their work, but also financially and health-wise, hindering the development of social capital outside their own migrant networks. Structurally, myriad stakeholders are complicit in migrant food delivery workers' unjust precarities across these domains, including the businesses engaging with the platforms, customers, and government policies and agencies (e.g. immigration, police, medical, welfare).

Nonetheless, migrant workers find ways to engage and oppose the cross-cutting and complex injustices that they face. Migrant food delivery workers are able to draw on their past experiences to overcome injustices—and inform their professional decisions (Lugosi & Ndiuini, 2022). They navigate injustices through agency and resistance as shown in Figure 2. These actions are implemented by individuals, through collaboration, and with external support. Migrant food delivery workers need to enact justice for themselves as they believe formal institutions cannot or will not respond (Sykes, 1986). As elsewhere, the absence of transparent information and other provisions mean platform workers turn to one another for material assistance and emotional support (Gregory, 2021; Qadri, 2021) and to seek racial and economic justice (Dubal, 2021). While workers can resist some aspects of platform work, such actions do not materially change their relationships with platforms (Barratt et al., 2020). Instead, workers' everyday acts of resistance are informal and mostly take place offstage rather than publicly challenging platforms (Anwar & Graham, 2020). Further, when this cohort are unable to overcome injustices themselves, they also seek and receive unsolicited support from other community members. These actions are important for migrant food delivery workers who have higher barriers to justice than their peers (Figueroa et al., 2021).

Migrant platform workers lack parity with local workers due to their intersectional challenges. On the one hand, as independent contractors they are not entitled to the full benefits of employees. As Vieira (2021) argues, attempts to improve platform workers' conditions must also coincide with general improvements to labour legislation. On the other, as temporary migrants (largely from the global South) workers are denied the rights and privileges of citizens. While local platform workers may face injustices as precarious and platform workers, temporary migrant workers need to navigate unique challenges in their precarious lives which deny them parity. According to Fraser (2001), participatory parity whereby all adult members of a society are granted equal socio-economic entitlements is a requirement for justice. A lack of justice is also connected to a lack of dignity—which is central to decent work and sustainable workforces in industries such as hospitality and tourism (Winchenbach et al., 2019). Denial of fair and just conditions also questions the dominant discourse around the benefits of platforms and platform work. Migrants who are reliant on income from platforms carry out most platform work in Australia (McDonald et al., 2020), which contests the "sharing economy" and "side hustle" discourses.

Finally, justice and dignity for platform workers is an urgent matter for society and governments to act on. At the farthest end of the scale, this is exemplified by the unreported deaths of workers. Informants have been in accidents, denied insurance, robbed, and abused, and as such have been denied recognition or affirmation (Fraser, 2020). They are dissatisfied with responses and feel entitled to fairer compensation and material improvements to work and living conditions in Australia, which would necessitate a redistribution of available resources. The injustices migrant food delivery workers experience cascades into structurally precarious work injustices and concomitantly into precarious lives, as illustrated in the injustices ecosystem framework we propose (see Figure 1). This framework seeks to explain the agentic and structural underpinnings of the injustices we have documented in this article.

Conclusion

Researchers and policy-makers fail to consider the unique circumstances of migrants who conduct most platform work (for exceptions, see Altenried (2021), van Doorn et al. (2022), and van Doorn and Vijay (2021)). We argue that stakeholders can no longer ignore this key aspect of platform workers profiles. This case, taking a "theory-relevant" approach (after Burawoy, 2009; Campbell & Price, 2016), into a unique cohort of platform workers, addresses shortcomings of shoehorning specific phenomena into grand (theory) narratives (Joo et al., 2021). We make two theoretical contributions. First, we developed a model that explains the agentic and structural underpinnings of injustices that migrant platform workers experience. Second, we reveal that the unique affordances that migrants bring to bear on their platform work enable them to navigate the injustices they experience. In so doing, they mitigate some of the work's negative impacts, and indeed find benefits that non-migrant workers might not.

The evidence is clear: migrant platform workers encounter inter-connected injustices due to (in)actions of platforms, consumers, businesses, governments, institutions, other stakeholders, and the public. While the COVID crisis has highlighted some justice issues, migrant platform workers will continue to need to seek their own justice long after the pandemic passes. As such, we join calls for fairer platforms (Fairwork, 2022) and a radical rethink of the integration of migrants into labour markets (van Doorn et al., 2022). To achieve such ends, we advocate for a broader justice-based framework which considers the intersecting precarity of migrant (platform) workers—bound by participatory parity for all workers. Envisioned here is a transformative process such that migrant platform workers can access justice (and dignity) across their work and life domains. This research is important as industries such as tourism and hospitality further embrace precarious work models in the "sharing" economy. Future research should focus on the wider experiences of the migrants who conduct platform work and how it affects other areas of their lives. Furthermore, experiences and (in)justices of women in the platform economy warrant focused attention. We posit it would also be valuable to further explore the relationship between a sustainable workforce and precarity.

Notes

1. Two joint interviews were held.
2. Detailed examples of resistance have been omitted to avoid revealing current workers' tactics.

Acknowledgements

We thank the anonymous reviewers for their suggestions and advice. We also thank the informants who shared their time and expertise throughout the project.

Declaration of interest statement

No potential conflict of interest was reported by the authors.

ORCID

Tyler Riordan http://orcid.org/0000-0002-7538-9212
Richard N.S. Robinson http://orcid.org/0000-0001-9737-3812
Gerhard Hoffstaedter http://orcid.org/0000-0001-6083-3141

References

Alkhatib, A., Bernstein, M. S., & Levi, M. (2017). *Examining crowd work and gig work through the historical lens of piecework* [Paper presentation]. Paper presented at the CHI Conference on Human Factors in Computing Systems, Colorado (pp. 4599-4616). https://doi.org/10.1145/3025453.3025974

Aloisi, A., & De Stefano, V. (2022). Essential jobs, remote work and digital surveillance: Addressing the COVID-19 pandemic panopticon. *International Labour Review*, *161*(2), 289–314. https://doi.org/10.1111/ilr.12219

Altenried, M. (2021). Mobile workers, contingent labour: Migration, the gig economy and the multiplication of labour. *Environment and Planning A: Economy and Space*, *0*(0), 1–16. https://journals.sagepub.com/doi/full/10.1177/0308518X211054846

Anderson, B. (2010). Migration, immigration controls and the fashioning of precarious workers. *Work, Employment and Society*, *24*(2), 300–317. https://doi.org/10.1177/0950017010362141

Anwar, M. A., & Graham, M. (2020). Hidden transcripts of the gig economy: Labour agency and the new art of resistance among African gig workers. *Environment and Planning A: Economy and Space*, *52*(7), 1269–1291. https://doi.org/10.1177/0308518X19894584

Atkinson, P., & Hammersley, M. (1994). Ethnography and participant observation. In N. Denzin & Y. Lincoln (Eds.), *Handbook of qualitative research* (pp. 248–261). Sage.

Barratt, T., Goods, C., & Veen, A. (2020). 'I'm my own boss…': Active intermediation and 'entrepreneurial' worker agency in the Australian gig-economy. *Environment and Planning A: Economy and Space*, *52*(8), 1643–1661. https://doi.org/10.1177/0308518X20914346

Baum, T., Mooney, S., Robinson, R. N. S., & Solnet, D. (2020). COVID-19's impact on the hospitality workforce—New crisis or amplification of the norm? *International Journal of Contemporary Hospitality Management*, *32*(9), 2813–2829. https://doi.org/10.1108/IJCHM-04-2020-0314

Begley, P. (2021, 25 June). Food delivery driver's death wasn't recognised by Uber Eats and his family are still fighting for the insurance. *ABC News*. https://www.abc.net.au/news/2021-06-25/background-briefing-uber-eats-delivery-drivers-death/100239920

Benhabib, S. (2013). *Dignity in adversity: Human rights in troubled times*. Polity.

Berg, L., & Farbenblum, B. (2020). As if we weren't humans: The abandonment of temporary migrants in Australia during COVID-19. Available at SSRN 3709527.

Bolton, S. (2007). Dignity in and at work: Why it matters. In S. Bolton (Ed.), *Dimensions of dignity at work* (pp. 3–16). Routledge.

Bourdieu, P. (1985). The social space and the genesis of groups. *Theory and Society*, *14*(6), 723–744. https://doi.org/10.1007/BF00174048

Burawoy, M. (2009). *The extended case method: Four countries, four decades, four great transformations, and one theoretical tradition*. University of California Press.

Campbell, I., & Price, R. (2016). Precarious work and precarious workers: Towards an improved conceptualisation. *The Economic and Labour Relations Review*, *27*(3), 314–332. https://doi.org/10.1177/1035304616652074

Cano, M. R., Espelt, R., & Morell, M. F. (2021). Flexibility and freedom for whom? Precarity, freedom and flexibility in on-demand food delivery. *Work Organisation, Labour & Globalisation*, *15*(1), 46–68. https://doi.org/10.13169/workorgalaboglob.15.1.0046

Cant, C. (2019). *Riding for Deliveroo: Resistance in the new economy*. John Wiley & Sons.

Cook, J., Threadgold, S., Farrugia, D., & Coffey, J. (2021). Youth, precarious work and the pandemic. *YOUNG*, *29*(4), 331–348. https://doi.org/10.1177/11033088211018964

Crenshaw, K. (1991). Mapping the margins: Intersectionality, identity politics, and violence against women of color. *Stanford Law Review*, *43*(6), 1241–1300. https://doi.org/10.2307/1229039

Dubal, V. B. (2019). *An Uber ambivalence: Employee status, worker perspectives, & regulation in the gig economy (November 15, 2019)*. https://ssrn.com/abstract=3488009

Dubal, V. B. (2021). The new racial wage code. *Harvard Journal of Law and Policy, 14*, 511-550.

Duggan, J., Sherman, U., Carbery, R., & McDonnell, A. (2020). Algorithmic management and app-work in the gig economy: A research agenda for employment relations and HRM. *Human Resource Management Journal*, *30*(1), 114–132. https://doi.org/10.1111/1748-8583.12258

Eisenhardt, K. M. (1989). Building theories from case study research. *The Academy of Management Review*, *14*(4), 532–550. https://doi.org/10.2307/258557

Fairwork. (2022). *Principles*. https://fair.work/en/fw/principles/

Farbenblum, B., & Berg, L. (2021). *Migrant workers' access to justice for wage theft: A global study of promising initiatives*. https://www.migrantjustice.org/publications

Figueroa, M., Guallpa, L., Wolf, A., Tsitouras, G., & Hernández, H. (2021). *Essential but unprotected: App-based food couriers in New York City*.

Fraser, N. (2001). Recognition without ethics? *Theory, Culture & Society*, *18*(2-3), 21–42. https://doi.org/10.1177/02632760122051760

Fraser, N. (2020). From redistribution to recognition?: Dilemmas of justice in a 'postsocialist' age. In S. Seidman & J. C. Alexander (Eds.), *The new social theory reader* (pp. 188–196). Routledge.

Galière, S. (2020). When food-delivery platform workers consent to algorithmic management: A Foucauldian perspective. *New Technology, Work and Employment*, *35*(3), 357–370. https://doi.org/10.1111/ntwe.12177

Gandini, A. (2019). Labour process theory and the gig economy. *Human Relations*, *72*(6), 1039–1056. https://doi.org/10.1177/0018726718790002

Gill, R. (2011). The shadow in organizational ethnography: Moving beyond shadowing to spect-acting. *Qualitative Research in Organizations and Management: An International Journal*, *6*(2), 115–133. https://doi.org/10.1108/17465641111159116

Goods, C., Veen, A., & Barratt, T. (2019). "Is your gig any good?" Analysing job quality in the Australian platform-based food-delivery sector. *Journal of Industrial Relations*, *61*(4), 502–527. https://doi.org/10.1177/0022185618817069

Graham, M., & Woodcock, J. (2018). Towards a fairer platform economy: Introducing the Fairwork Foundation. *Alternate Routes, 29*, 242–253.

Gregory, K. (2021). 'My life is more valuable than this': Understanding risk among on-demand food couriers in Edinburgh. *Work, Employment and Society, 35*(2), 316–331. https://doi.org/10.1177/0950017020969593

Gregory, K., & Maldonado, M. P. (2020). Delivering Edinburgh: Uncovering the digital geography of platform labour in the city. *Information, Communication & Society, 23*(8), 1187–1202. https://doi.org/10.1080/1369118X.2020.1748087

Hammersley, M., & Atkinson, P. (2007). *Ethnography: Principles in practice*. Routledge.

Hodson, R. (2001). *Dignity at work*. Cambridge University Press.

Holtum, P. J., Irannezhad, E., Marston, G., & Mahadevan, R. (2022). Business or pleasure?: A comparison of migrant and non-migrant Uber drivers in Australia. *Work, Employment and Society, 36*(2), 290-309.

Ingold, T., & Vergunst, J. L. (2008). *Ways of walking: Ethnography and practice on foot*. Ashgate Publishing, Ltd.

International Labour Organization. (2016). *Non-standard employment around the world: Understanding challenges, shaping prospects*. https://www.ilo.org/global/publications/books/WCMS_534326/lang–en/index.htm

Ioannides, D., Gyimóthy, S., & James, L. (2021). From liminal labor to decent work: A human-centered perspective on sustainable tourism employment. *Sustainability, 13*(2), 851. https://doi.org/10.3390/su13020851

Jamal, T. (2019). *Justice and ethics in tourism*. Routledge.

Jamal, T., & Higham, J. (2021). Justice and ethics: Towards a new platform for tourism and sustainability. *Journal of Sustainable Tourism, 29*(2-3), 143–157. https://doi.org/10.1080/09669582.2020.1835933

Jarrahi, M. H., Newlands, G., Lee, M. K., Wolf, C. T., Kinder, E., & Sutherland, W. (2021). Algorithmic management in a work context. *Big Data & Society, 8*(2), 205395172110203–205395172110214. https://doi.org/10.1177/20539517211020332

Joo, D., Cho, H., Woosnam, K. M., & Suess, C. (2020). Re-theorizing social emotions in tourism: Applying the theory of interaction ritual in tourism research. *Journal of Sustainable Tourism, 0*(0), 1–16. https://www.tandfonline.com/doi/full/10.1080/09669582.2020.1849237

Kalleberg, A. L. (2018). *Precarious lives: Job insecurity and well-being in rich democracies*. Polity Press.

Kalleberg, A. L., & Vallas, S. P. (2017). Probing precarious work: Theory, research, and politics. In A. L. Kalleberg & S. P. Vallas (Eds.), *Precarious work* (Vol. 31, pp. 1–30). Emerald.

Kant, I. (1997). *Groundwork of the metaphysics of morals* (M. Gregor, Trans.). Cambridge University Press.

Krieger, N. (2021). Structural racism, health inequities, and the two-edged sword of data: Structural problems require structural solutions. *Frontiers in Public Health, 9*, 655447. https://doi.org/10.3389/fpubh.2021.655447

Lugosi, P. (2019). Deviance, deviant behaviour and hospitality management: Sources, forms and drivers. *Tourism Management, 74*, 81–98. https://doi.org/10.1016/j.tourman.2019.02.014

Lugosi, P., & Ndiuini, A. (2022). Migrant mobility and value creation in hospitality labour. *Annals of Tourism Research, 95*, 103429–103412. https://doi.org/10.1016/j.annals.2022.103429

Lusch, R. F., & Vargo, S. L. (2014). *The service-dominant logic of marketing: Dialog, debate, and directions*. Routledge.

Ma, J., Webb, T., & Schwartz, Z. (2021). A blended model of restaurant deliveries, dine-in demand and capacity constraints. *International Journal of Hospitality Management, 96*, 102981. https://doi.org/10.1016/j.ijhm.2021.102981

Martin, S., & Bergmann, J. (2021). (Im)mobility in the Age of COVID-19. *International Migration Review, 55*(3), 660–687. https://doi.org/10.1177/0197918320984104

McDonald, P., Williams, P., Stewart, A., Mayes, R., & Oliver, D. (2020). *Digital platform work in Australia: Prevalence, nature and impact*. https://eprints.qut.edu.au/203119/

McDonald, S. (2005). Studying actions in context: A qualitative shadowing method for organizational research. *Qualitative Research, 5*(4), 455–473. https://doi.org/10.1177/1468794105056923

Miles, M. B., & Huberman, A. M. (1984). Drawing valid meaning from qualitative data: Toward a shared craft. *Educational Researcher, 13*(5), 20–30. https://doi.org/10.3102/0013189X013005020

Mooney, S., Robinson, R. N. S., Solnet, D., & Baum, T. (2022). Rethinking tourism's definition, scope and future of sustainable work and employment. *Journal of Sustainable Tourism, 0*(0), 1–19. https://www.tandfonline.com/doi/full/10.1080/09669582.2022.2078338

Narotzky, S. (2016). between inequality and injustice: Dignity as a motive for mobilization during the crisis. *History and Anthropology, 27*(1), 74–92. https://doi.org/10.1080/02757206.2015.1111209

Narotzky, S., & Besnier, N. (2014). Crisis, value, and hope: Rethinking the economy: An introduction to supplement 9. *Current Anthropology, 55*(S9), S4–S16. https://doi.org/10.1086/676327

Pigatto, G., Machado, J., Negreti, A., & Miranda Machado, L. (2017). Have you chosen your request? Analysis of online food delivery companies in Brazil. *British Food Journal, 119*(3), 639–657. https://doi.org/10.1108/BFJ-05-2016-0207

Pritchard, M. S. (1972). Human dignity and justice. *Ethics, 82*(4), 299–313. https://doi.org/10.1086/291853

Qadri, R. (2021). What's in a network? Infrastructures of mutual aid for digital platform workers during COVID-19. *Proceedings of the ACM on Human-Computer Interaction, 5*(CSCW2), 1–20. https://doi.org/10.1145/3479563

Rastegar, R., Higgins-Desbiolles, F., & Ruhanen, L. (2021). COVID-19 and a justice framework to guide tourism recovery. *Annals of Tourism Research, 91*, 103161. https://doi.org/10.1016/j.annals.2021.103161

Ratnam, C. (2019). Fitting refugees into the normative narrative of Australian multiculturalism. *Urban Geography*, *40*(8), 1198–1209. https://doi.org/10.1080/02723638.2019.1653136

Rawls, J. (1971). *A theory of justice*. Oxford University Press.

Richardson, L. (2020). Platforms, markets, and contingent calculation: The flexible arrangement of the delivered meal. *Antipode*, *52*(3), 619–636. https://doi.org/10.1111/anti.12546

Riordan, T., Robinson, R. N. S., & Hoffstaedter, G. (2023). 'Contactless' delivery: Migrant experiences in the platform economy. In I. Ness (Ed.), *Platform labour and global logistics: A research companion*. Routledge.

Robinson, R. N. S. (2008). Revisiting hospitality's marginal worker thesis: A mono-occupational perspective. *International Journal of Hospitality Management*, *27*(3), 403–413. https://doi.org/10.1016/j.ijhm.2007.09.003

Robinson, R. N. S., Martins, A., Solnet, D., & Baum, T. (2019). Sustaining precarity: Critically examining tourism and employment. *Journal of Sustainable Tourism*, *27*(7), 1008–1025. https://doi.org/10.1080/09669582.2018.1538230

Robinson, R. N. S., Oren, O., & Riordan, T. (2022). *Serving up a fair go: Surfacing cultural issues in hospitality employment*. https://business.uq.edu.au/files/82074/Fair-Go-Surfacing-cultural-issues-hospitality-employment.pdf

Roitman, J. (2013). *Anti-crisis*. Duke University Press.

Roitman, J. (2021). *Framing the crisis: COVID-19*. https://ssrn.com/abstract=3952226

Rosenblat, A., & Stark, L. (2016). Algorithmic labor and information asymmetries: A case study of Uber's drivers. *International Journal of Communication*, *10*, 3758–3784.

Rydzik, A., & Anitha, S. (2020). Conceptualising the agency of migrant women workers: Resilience, reworking and resistance. *Work, Employment and Society*, *34*(5), 883–899. https://doi.org/10.1177/0950017019881939

Schor, J. B., Attwood-Charles, W., Cansoy, M., Ladegaard, I., & Wengronowitz, R. (2020). Dependence and precarity in the platform economy. *Theory and Society*, *49*(5-6), 833–861.

Schüßler, E., Attwood-Charles, W., Kirchner, S., & Schor, J. B. (2021). Between mutuality, autonomy and domination: Rethinking digital platforms as contested relational structures. *Socio-Economic Review*, *19*(4), 1217–1243. https://doi.org/10.1093/ser/mwab038

Shapiro, A. (2018). Between autonomy and control: Strategies of arbitrage in the "on-demand" economy. *New Media & Society*, *20*(8), 2954–2971. https://doi.org/10.1177/1461444817738236

Sinanan, J., & McDonald, T. (2018). Ethnography. In J. Burgess, A. Marwick, & T. Poell (Eds.), *The SAGE handbook of social media* (pp. 179–195). SAGE.

Sun, Y.-Y., Sie, L., Faturay, F., Auwalin, I., & Wang, J. (2021). Who are vulnerable in a tourism crisis? A tourism employment vulnerability analysis for the COVID-19 management. *Journal of Hospitality and Tourism Management*, *49*, 304–308. https://doi.org/10.1016/j.jhtm.2021.08.014

Sykes, G. W. (1986). Street justice: A moral defense of order maintenance policing. *Justice Quarterly*, *3*(4), 497–512. https://doi.org/10.1080/07418828600089081

United Nations. (1948). Universal declaration of human rights. *UN General Assembly*, *302*(2), 14–25.

van der Weele, S., & Bredewold, F. (2021). Shadowing as a qualitative research method for intellectual disability research: Opportunities and challenges. *Journal of Intellectual & Developmental Disability*, *46*(4), 340–350. https://doi.org/10.3109/13668250.2021.1873752

van Doorn, N., & Badger, A. (2020). Platform capitalism's hidden abode: Producing data assets in the gig economy. *Antipode*, *52*(5), 1475–1495. https://doi.org/10.1111/anti.12641

van Doorn, N., Ferrari, F., & Graham, M. (2022). Migration and migrant labour in the gig economy: An intervention. *Work, Employment and Society*, *0*(0), 1–13.

van Doorn, N., & Vijay, D. (2021). Gig work as migrant work: The platformization of migration infrastructure. *Environment and Planning A: Economy and Space*, *0*(0), 1–21.

Veen, A., Barratt, T., & Goods, C. (2020). Platform-capital's 'app-etite' for control: A labour process analysis of food-delivery work in Australia. *Work, Employment and Society*, *34*(3), 388–406. https://doi.org/10.1177/0950017019836911

Vieira, T. (2021). The unbearable precarity of Pursuing Freedom: A critical overview of the Spanish sí soy autónomo Movement. *Sociological Research Online*, *0*(0), 1–17. https://doi.org/10.1177/13607804211040090

Vosko, L. F. (2010). *Managing the margins: Gender, citizenship, and the international regulation of precarious employment*. Oxford University Press.

Winchenbach, A., Hanna, P., & Miller, G. (2019). Rethinking decent work: The value of dignity in tourism employment. *Journal of Sustainable Tourism*, *27*(7), 1026–1043. https://doi.org/10.1080/09669582.2019.1566346

Wood, A., Graham, M., Lehdonvirta, V., & Hjorth, I. (2019). Good gig, bad gig: Autonomy and algorithmic control in the global gig economy. *Work, Employment & Society*, *33*(1), 56–75.

Zhou, N. (2021, 27 June). Call for federal regulator for Australia's gig economy after sixth delivery rider death revealed. *The Guardian*. https://www.theguardian.com/australia-news/2021/jun/27/call-for-federal-regulator-for-australias-gig-economy-after-sixth-delivery-rider-death-revealed

Part III

Tools and processes that build (in)justice in tourism

Do international sanctions help or inhibit justice and sustainability in tourism?

Siamak Seyfi, Colin Michael Hall, Jarkko Saarinen and Tan Vo-Thanh

ABSTRACT

Despite the expanded use of sanctions as a soft foreign policy tool in the post-Cold War era, there is yet little knowledge on the implications of this coercive tool in relation to justice, ethics and sustainability in destinations to which sanctions are applied. Using Iran as a case study and grounded in international relations and political science literature, this study used semi-structured interviews with tourism actors to assess the direct and indirect effects of sanctions on tourism with respect to justice, rights and sustainability. The informants suggested that sanctions have worsened mobility rights, rights to communication, and the economic and financial rights of tourism actors, thereby limited their capacity to contribute to inclusive and sustainable development. Distributive pressures within Iran arising from the sanction-driven economic disruption have clearly undermined the empowerment capacity of tourism to contribute to improve gender justice, thereby standing in opposition to the principles of justice and the Sustainable Development Goals (SDGs). However, notions of justice and rights are uneven in space and time, with their application inherently dependent on its definition in particular contexts. By portraying new insights from the restrictions emanating from sanctions, this study contributes to a deeper understanding of a very popular instrument of foreign policy and its humanitarian and justice implications in destinations affected by sanctions.

Sanctions can create severe and undue suffering for individuals who have neither perpetrated crimes nor otherwise borne responsibility for improper conduct.

(Michelle Bachelet, United Nations High Commissioner for Human Rights, 2021).

Introduction

Economic sanctions are increasingly used in international relations and they have become integral to the repertoire of coercive geopolitical tools of statecraft (Cortright & Lopez, 2018) and championed as 'the humane alternative to war' (Boomen, 2014). In the post-Cold War era, sanctions have become a 'go-to' policy and are imposed at the multilateral level (e.g., at the UN), regional (e.g., the EU in particular), and unilaterally by a single state (Hufbauer et al., 2008). They range from broad economic and trade sanctions to more targeted measures (so-called smart sanctions) such as asset freezes, travel bans, financial or commodity restrictions, and arms embargo (United Nations Security Council, 2020). Sanctions are frequently used in conjunction with other foreign policy measures to affect a state's or group's policies or actions when such policies or actions are deemed as a threat to world peace and security (Cortright & Lopez, 2018).

Sanctions have been used for decades and have been widely discussed in international relations, political economy and political science (Gowlland-Debbas, 2021; McDowell, 2021). Nonetheless, recent reviews of tourism and geopolitics (e.g., Mostafanezhad & Norum, 2016; Hall, 2017) overlooked sanctions as a significant topic, potentially because for the most part sanctions have not been applied to major tourist generating regions or destinations until relatively recently (Seyfi & Hall, 2020a, 2020b), and there has therefore been a substantial lag in the literature. Additionally, despite the burgeoning interest in studies of crises in tourism, existing research is predominantly centred on issues of terrorism and political security and has not framed sanctions as a crisis event for affected destinations (Hall, 2010; Ritchie & Jiang, 2019; Wut et al., 2021). As a result, until recently, sanctions had become an ignored element in studies of the politics of tourism and crisis. However, while any crisis may exacerbate existing inequalities, the crises precipitated by sanctions often compound and complicate existing issues of economic, social and environmental injustice for individuals, communities and specific populations (Drury & Peksen, 2014; Özdamar & Shahin, 2021). Research also indicates that, because of the social and political instability caused by sanctions, individuals' political and social rights may be violated more frequently (Fathollah-Nejad, 2014), thereby limiting their empowerment and capacity to contribute to inclusive and sustainable development (Tahmasebi, 2018; Seyfi et al., 2020a).

While scholars have stressed that planning for sustainable tourism must be driven by the rights, interests, and social needs of vulnerable groups in local communities (Boluk et al., 2019; Rastegar et al., 2021), little is known about how tourism continues to operate in sanctioned countries and how tourism actors respond to the sustainability and justice challenges sanctions pose. This is particularly important in sanctions-ridden destinations, which are often low-income countries with significant justice and rights issues and pre-existing heterogeneity in employment generation, income distribution and poor working conditions for women and marginalised groups (Gutmann et al., 2020). It is also unknown whether justice related sustainability issues are only a concern of academic commentators and interest groups, for example, or whether they are actively considered by tourism actors such as policy makers, business interests and members of the wider community. Hence, these issues point to the gaps that this study attempts to fill.

In the light of these research gaps and applying a justice lens and located in an international relations and political science disciplinary base, the research objective central to this study include investigating the hitherto unheard voices of key informants in Iranian tourism and hospitality sector regarding the politico-economic crises emanating from international sanctions. This study also attempts to cultivate an improved understanding of the impacts of sanctions on Iranian tourism and to explore how sanctions affect notions of justice and the Sustainable Development Goals (SDGs). Taken collectively, this paper responds to the recent calls by scholars to transcend disciplinary boundaries for increased research related to justice and tourism (e.g., Jamal, 2019; Higgins-Desbiolles, 2018; Rastegar et al., 2021; Jamal & Higham, 2021) and extends understandings of sanctions-led crisis in economically and social fragile destinations.

Iran, which has been subjected to a steady stream of tough sanctions (Takeyh & Maloney, 2011), presents an intriguing case study for such an investigation. This study is also timely in the light of the United States' reimposition of sanctions against Iran following a brief period of relaxation of sanctions (2015–2018 resulting from the 2015 Iran nuclear deal) and contemporary debates over their continuation, as well as the possible consequences for the county's tourism industry.

In this study sanctions, justice and sustainability are framed within the context of tourism geopolitics, as they reflect the way in which the sanctions-tourism nexus has becomes a focus, or casualty, of foreign policy decision-making. As a result this research combines both traditional geopolitical theory which focuses on the nation-state as a geopolitical actor as well as more recent expansions of the notion of critical geopolitics which explores ways in which political space is experienced and practiced (Power & Campbell, 2010). As Seyfi and Hall (2020c) observed, research on sanctions provides a new direction in critical geopolitical and foreign policy studies in a tourism context because of the effects of the denial or constraint of mobility as well as the way in which some forms of tourism act as a means of resistance to sanctions, while also connecting to critical geopolitical research on sustainability and notions of justice (Power & Campbell, 2010). As a result, the study contributes to a greater understanding of a frequently used instrument of foreign policy on destinations and the experience of tourism actors through a justice, ethics and sustainability lens. Therefore, in responding to the need for more multidisciplinary empirical work related to justice and tourism, this study emphasizes sanctions as an important crisis event for tourism destinations and ethical challenge in relation to SDGs.

Literature review

Crisis precipitated by sanctions in relation to tourism and SDGs

The term 'sanctions' in international law refers to "coercive measures, taken by one State or in concert by several States (the sanctioner), which are intended to convince or compel another State (the sanctionee) to desist from engaging in acts violating international law" (Ilieva et al., 2018, p.201). Sanctions can be used to achieve a range of political goals, including restricting trade and economic development in targeted countries and having a negative influence on their GDP, bilateral trade, and financial services (Özdamar & Shahin, 2021). Indeed, it is the deliberate goal of sanctions to cause economic damage and create economic stress and pain so as to encourage policy change in the target country (Rarick, 2007). For instance, Neuenkirch and Neumeier (2015) illustrate that multilateral (e.g., UN sanctions) and unilateral sanctions (e.g., some US sanctions) can cut the targeted country's GDP by up to 25% and 13%, respectively. While economic sanctions are often used as indirect tools to achieve a political change, they primarily place severe constraints on the economic development of sanctioned countries, and, therefore, can significantly challenge their ability to achieve the SDGs (Afshari, 2019). Adopted by the United Nations in 2015, the SDGs are a set of 17 interlinked global goals with specific targets aimed at ending poverty, protecting the planet and ensuring that by 2030 all people enjoy peace and prosperity (United Nations, 2015). In their comprehensive analysis of unilateral sanctions imposed by US since 2000, Rosenberg et al. (2016) reported that sanctions reduced the human development index of targeted countries. This note is highly significant because finance and economic development, multilateral trade, and foreign investment are all required to achieve the SDGs, especially in the low and middle-income countries, which contradicts SDG17's emphasis on poor and developing countries' integration into the global economy and the strengthening of the multilateral trading system' (United Nations, 2015). Although the conceptualization of SDGs has been criticized for reinforcing contemporary neoliberal capitalism (Hall, 2019), sanctions are unmistakably hurdles to state-industry collaboration and the achievement of the SDGs in their current form.

Sanctions can also limit the international community's ability to cooperate, coordinate, and share obligations, as outlined in Goal 17 of the SDGs. This might result in a lack of access to knowledge and international scientific and educational expertise and, even, publishers (Mozafari, 2016). However, while the sanctions challenged business development and conduct, some studies suggest that a shortage of imported goods caused by sanctions might occasionally stimulate domestic production in countries affected by sanctions. For instance, the study of Farzin (2017) showed that local businesses in Iran have learned to withstand the impact of sanctions and established their own initiations to spur innovation.

Sanctions also impose financial constraints on a variety of industries in target nations (for example, foreign investment ban and technology transfers). Seyfi et al. (2019) argue that sanctions imposed on Iran have largely constrained technological support for cultural conservation in Iran, which is arguably a vital component of the country's sustainable development for international visitation (Rastegar et al., 2021). Such restrictions are in sharp contrast with target 11.4 of the SDG11, which calls for strengthening efforts to protect and safeguard the world's cultural heritage. This also reflects Jamal et al. (2010) observation on the significance of cultural justice for local people and fairness for the sustainable use and conservation of a destination's cultural heritage.

The legitimacy and effectiveness of sanctions is often criticized because of their devastating impacts on human rights and the well-being of civilian populations. Especially women and their rights are often seriously affected by sanctions. In their cross-national study spanning 146 nations from 1971 to 2005, Drury and Peksen (2014) reported that sanctions have a substantial detrimental effect on women's rights, their position and agency in society and labour force participation. A similar observation was noted in a more recent tourism-focused study by Seyfi et al. (2020a) which reported on the gendered impacts of sanctions and argue that sanctions have negatively affected and deteriorated economic, psychological, social and political aspects of Iranian women empowerment. They further note that sanctions contradict women's equality and empowerment as enshrined in the SDG5. Overall, as Rustler (2019) notes, sanctions have the potential to lock nations into unsustainable practices such as unintended consequences for the environment and the climate in the short, medium, and especially the long-term that undermine several of the SDGs and have long-term negative consequences for socio-economic and environmental quality of life, including health.

Theorizing the ethics of sanctions through a justice lens

Sanctions are incompatible with the principles of social justice proposed by Rawls (1971) and which draw upon ethical perspectives of the liberal enlightenment (Wood, 1999; Brown, 2006), the Kantian 'categorical imperative', which calls for people to act in a such a way that, if everyone did likewise, the world would be a better place (Ramet, 2019). Significantly, Kant's and Rawls' views are in stark opposition to those of Hobbes (Flikschuh, 2012), which saw the relationship between sovereignty and the individual as "nothing more than the fear of violent death" (Ramet, 2019, p.197) or Rousseau's notion of a general will (Steinberger, 2008). Instead, and significantly for notions of social justice, Kant also spelled out a 'practical imperative', itself drawn from the categorical imperative: "Act in such a way that you always treat humanity, whether in your own person or in the person of any other, never simply as a means, but always at the same time as an end" (Kant, 1948, p.91). Interestingly, from the perspective of both sustainability and justice, Kant argued that the responsibility of sovereign government was "to constrain the wealthy to provide the means of sustenance to those who are unable to provide for even their most necessary natural needs. The wealthy have acquired an obligation to the commonwealth, since they owe their existence to... its protection and care" (Kant, 1948, p.136). Kantian logic may be somewhat removed from the realpolitik imperatives of contemporary neoliberal capitalism or even the demands of authoritarian states. However, in policy terms such perspectives on justice may provide an ideological underpinning for some foreign policy positions on sanctions (e.g. smart sanctions) as well as a 'reasoned voice' to the applications of sanctions to entire regimes, e.g. following Russia's invasion of Ukraine (King, 1999; Scott, 2018).

According to Rawls's (1971) theory of justice, a situation of fairness exists when "all social primary goods - liberty and opportunity, income and wealth, and the bases of self-respect - are to be distributed equally unless an unequal distribution of any or all of these goods is to the advantage of the least favoured' (Rawls, 1971, p.303). Existing evidence suggests that the imposition of sanctions is clearly opposite to such principles of justice with most countries' economic sanctions disproportionately impacting the least advantaged groups of society (Drury & Peksen, 2014; Hatipoglu & Peksen, 2018). Thus, economic sanctions do not meet the purpose of allowing the greatest benefits to the least advantaged of society. Instead, sanctions are often approached under notions of rule-based international law, rather than specific concerns with particular notions of justice (Scott, 2018).

Rawls's (1971) notion of justice has been the subject of many debates and several scholars argue that that they do not consider differences in communities and their values (Daniels, 1989; Jamal, 2019) as well as the context in which they are applied. In their brief examinations of Rawls' theory of justice, Fennell (2006) and Duffy and Smith (2004) highlight the complexities of applying an ethics of justice to tourism. Rawls's theory has also been criticised for the extent to which it fails to acknowledge complexity, human irrationality, and notions of justice that are different from those of the Western philosophical tradition (Chan, 2005), as well as offering only a 'weak sustainability' perspective with respect to intergenerational rights (Mathis, 2009).

Fennell (2006) states that tourism is an inherent issue of justice. For Duffy and Smith (2004) social justice refers to the fair distribution of power, goods and other resources within and across societies. This echoes the view of Higgins-Desbiolles (2008) that the concept of justice in tourism encompasses more than a fair cost-benefit distribution. Jamal and Hales (2016) and Rastegar et al. (2021) also suggest a shift from distributive justice (e.g., Rawls's theory) to performative justice. They argue that where vulnerable populations are affected by a crisis, justice is done performatively through recognizing local rights, needs, and social spaces. Such a situation highlights the essentially contested nature of concepts of justice and sustainability (Gallie, 1956), particularly in light of the SDGs. While providing a universal (Kantian/Rawlsian) perspective on sustainability and justice, the implementation of the SDGs is performative given that it utilises the notion of common but differentiated governance to give local effect to the SDGs (Meuleman & Niestroy, 2015). However, the tension between these perspectives is perhaps irreconcilable given that it also reflects the relationship between universal and local rights, and competing concerns of justice that are inherently difficult for policy-makers to manage. While acknowledging the significance of performative and local rights, this paper primarily frames sustainability and justice issues in tourism within universal notions of justice and rights, however imperfect, because of concerns of the way local interpretations of rights may be misused to justify constraints on the empowerment of women and specific groups in society in particular (Bidegain Ponte & Enríquez, 2016; Sen, 2019).

Universal notions of rights and justice are also sometimes used to justify the application of sanctions (Schwendinger & Schwendinger, 2014; Teitel, 2015). However, while proponents of the use of sanctions contend that coercive tools such as economic sanctions are a cost-effective means for changing the behavior of targeted governments by avoiding the high costs of military intervention (Marinov, 2005), the most comprehensive study of the effectiveness of economic sanctions assessed that the measure works only about 35% of the time (Hufbauer et al., 2008), while having undesirable effects on the wider population. As such, there is considerable scholarship on the ethics of sanctions. The majority of this research stream employ 'just war' theory as a framework for evaluating their moral permissibility, focusing on sanctions aimed at changing undesirable political policies or conduct (Winkler, 1999; Ellis, 2021). For example, Damrosch (1994) argues that sanctions can be a morally acceptable tool of foreign policy if they are administered consistently to comparable situations, uphold international standards, and are not used only for political purposes. However, Ellis (2021) criticized the straightforward application of just war principles to sanctions and explained the distinction between war and economic sanctions:

> ... war is constituted by bombing, shooting or stabbing but economic sanctions are constituted by refusing to trade. While there is a strong *pro tanto* duty to not bomb, shoot or stab individuals, there is no comparable *pro tanto* duty to trade. That does not mean sanctions are always morally permissible, only that the moral issues involved are very different. We have no reason to believe that moral principles developed to govern war are also appropriate for governing sanctions. This approach to the ethics of economic sanctions ought to be abandoned (Ellis, 2021, pp.410-411).

From a tourism perspective, such ethical debates have been applied to sanctions on tourism to a destination (Seyfi & Hall, 2020a). For example, in the longstanding debate over sanctions on Myanmar tourism, some proponents of sanctions have suggested boycotting Myanmar travel and tourism to effect 'regime change'. The argument is that it is unethical for individuals to support the military dictatorship, which controls most of the country's productive capacity including tourism that attracts considerable foreign exchange. In contrast, those who argue against sanctions suggest that visiting such destinations might help to reduce the isolation of the wider population as well as generating needed employment and visitors can advocate for justice. From a justice perspective Rarick (2006) asks the fundamental question of: is it appropriate to destroy a country in order to save it? However, although contested, the most appropriate ethical course from a justice and sustainability perspective is ultimately decided upon at the national or the global scale. It is to these ethical concerns that the reminder of the paper will turn in the specific case of sanctions on Iran.

Method

The study context: Iran under four decades of crippling sanctions

Following the establishment of a theocratic Islamic Republic in 1979, Iran has been the target of one of the longest and toughest international economic sanctions' regimes in response to the country's nuclear activities, human rights record, and intervention in regional affairs (Takeyh & Maloney, 2011; Dizaji, 2019). In the aftermath of the seizure of the U.S. Embassy in Tehran in 1979, the United States, as one of the key actors in the imposition of unilateral sanctions has tried to curb Iran's geopolitical ambitions through a variety of foreign policy instruments, including economic sanctions (Takeyh & Maloney, 2011). Since then, these instruments have grown in their stringency. The question of whether or not these long-running sanctions have succeeded in changing Iran's behavior is extremely debatable (Madani, 2021). However, what is clear is that Iran's economy has struggled as a result of four decades of sanctions, which led to major economic problems, high inflation, and limited access to foreign markets (Pratt & Alizadeh, 2018). Due to the tense political relations with Western countries, and particularly the United States (which were major tourism markets for Iran before the revolution), the Iran–Iraq war (1980–1988) and steady imposition of sanctions, the country's tourism market experienced major changes and shifted to neighboring countries. According to the WTTC (2020), Iraq, Azerbaijan, Turkey, Pakistan and Kuwait account for over 70% of the inbound arrivals to Iran in 2019.

While Iran was the most popular tourist destination in the Middle East from 1967 to 1977 (Morakabati, 2011), following the revolution international travel to Iran was severely restricted by sanctions. After the 2015 nuclear agreement (formally known as the Joint Comprehensive Plan of Action), international sanctions were partially lifted, transformed Iran's political and economic atmosphere and created significant chances for the struggling tourist sector (Khodadadi, 2016; Seyfi & Hall, 2018). Importantly, the lifting of sanctions and Iran's integration into the international community transformed the country's image from that of a 'pariah state' to that of a 'booming destination' (Khodadadi, 2016) leading to substantial growth of incoming tourists, visitor spending, employment and tourism-related investment. For instance, in 2017 almost five million international tourists visited the country, nearly three times as many as in 2009 (UNWTO, 2019).

With an over 85 million population, Iran is the second-largest country in the Middle East followed by Saudi Arabia. Historically, Iran has been heavily dependent on oil revenues. However, the tourism industry plays an increasingly important role in the economic growth of the country. Iran's strategic position at the crossroads of major civilizations and trade routes (e.g., the Silk Road), along with its climate and landscapes provide significant opportunities for tourism development (Seyfi & Hall, 2018). As of February 2022, 26 sites are listed on the UNESCO World

Table 1. Profile of interviewees.

Organisation	Sector	Number of participants
Iran's Cultural Heritage, Handicrafts and Tourism Organization (ICHHTO)*	Public	2
Regional offices of ICHHTO	Public	3
The association of hoteliers	Private	2
Hotel owners/manager	Private	5
Traditional guesthouse owner/manager	Private	4
Tour operator/travel agency manager	Private	6
Tour guide	Private	4
Airline manager/staff	Private	3
Car rental employee	Private	2
Tourism lecturer/professor/consultant	Academia	4

*In 2019, ICHHTO converted into the ministry.

Heritage List, while 61 sites are tentatively listed (United Nations Educational and Scientific and Cultural Organization (UNESCO),), 2022). Based on the World Travel and Tourism Council (WTTC) report, the total contribution of travel and tourism to Iran's GDP was 6.5% of the total GDP and 5.4% of total employment (1.344 million jobs) in 2019 (WTTC, 2020). Although women's participation in tourism-related activities increased with the relaxation of economic sanctions in 2015 (Farahani & Dabbaghi, 2018), there is no reliable data on female labour force participation rate in Iran's tourist and hospitality industry. While, nearly half of Iran's population is female, and women account for a rising proportion of university graduates, they are a minority of the employed population (UN, 2020).

The US withdrawal from the nuclear deal and the reestablishment of sanctions in May 2018 halted Iran's booming tourist industry. Following the return of American sanctions, several major European airlines suspended their flights to Iran with negative consequences for Iranian tourism while substantial international investment in the Iranian accommodation sector was also dramatically halted (Khodadadi, 2018). The return of sanctions not only reduced international visitation but also reinforced a negative image of Iran in some Western markets, which the Iranian tourism industry had been working to overcome (Seyfi & Hall, 2018).

The ramifications of such reimposition of sanctions also prompted renewed discussion on the ethics and justice of economic sanctions and these issues will be examined in the next sections from the perspectives of Iranian tourism actors.

Data collection and analysis

Data collection took place in Iran by the first author during two different periods: first in 2016–2017 at the time of the lifting of sanctions and then in 2018-2019 just after the re-imposition of the sanctions. The first author made use of his knowledge of the country to access participants. Some of the participants were personally known to the first author while additional participants were enlisted by snowball sampling, which involved asking participants to recommend further eligible individuals who would be willing to be interviewed (Handcock & Gile, 2011). This was deemed appropriate in obtaining a purposive sample suited for this study. Using ethnographic and participatory approaches that are both interpretative and interactive have been suggested when there is a need to gain more comprehensive information about the phenomenon under investigation (Wise, 2018). Data collection consisted of semi-structured interviews with key informants in the Iranian tourism and hospitality industry. Participants in this study were carefully chosen to represent different sectors of the industry to allow for enriched views. A total of 35 individual interviews were conducted (20 interviews in the first period of data collection followed by 15 interviews in the second period with same interviewees). Sixteen of the interviewees were female. Table 1 shows the profile of interviewees. The interviews continued until data saturation was reached when respondents were repeating the same information provided by prior interviewees, and no additional issues were being identified (Creswell &

Poth, 2018). Additional data was gathered for data triangulation by ethnographic observations (Wise, 2018) as well as analysing policy documents and other literature such as newspaper articles and online forums (Decrop, 1999).

The interviews were conducted using an interview guide and the questions covered themes related to the impacts of the sanctions on the Iranian tourism and hospitality industry and more particularly, its fairness, justice and ethics, as well as personal impacts in an effort to capture different perspectives. The questions were drawn up in light of the justice literature in tourism (e.g. Jamal, 2019; Rastegar et al., 2021) and adapted to the wider literature on sanctions and crisis (e.g. Seyfi & Hall, 2020c). They were left open-ended in order to elicit more spontaneous opinions and allow the participants to contribute their stories and own understandings of justice and ethics without being limited by the structure of data collection (Rowley, 2012).

The interviews lasted between 30 minutes to 75 minutes. The majority of the interviews were digitally recorded, then transcribed after receiving consent. In order to foster an environment to express their opinions freely, the interviews were conducted in Persian, the native language of the interviewees with transcripts and supporting data then back translated in English. For those respondents who did not want their voices recorded, notes were taken during the interview. Because Iranians, particularly officials, are generally hesitant to speak freely about political issues, interviews were conducted anonymously (Farasatkhah et al., 2008), and all participants were reassured that their identity would remain anonymous. The participants were thus given codes to maintain confidentiality (Given, 2008).

To gather in-depth accounts of respondents' experiences, a semi-structured in-depth interview approach involving a series of open-ended questions was utilized which encourages more spontaneous opinions and minimizes potential bias from confining responses to predetermined categories (Creswell & Poth, 2018). The transcripts were examined along with the field notes and cross-checked by the first and second author. To ensure the trustworthiness of the qualitative methods, a series of procedures recommended by Lincoln and Guba (1985) were adopted. First, during data collection, information was provided with interviewees to ensure accuracy, and data was collected from different sources to allow triangulation. Second, the initial codes and transcripts were also reviewed by two tourism researchers to ensure that codes are applied consistently throughout the analysis (Lincoln & Guba, 1985). The data was manually coded as this form of coding is consistent with the interpretive research paradigm and provides more control and ownership over data analysis and keeps the data within the context (Saldaña, 2021). Nevertheless, such approach to coding is prone to potential personal bias although the authors sought to overcome this by the adoption of a systematic approach to analysis. Qualitative thematic analysis was adopted for data analysis. Thematic analysis is a "method for identifying, analysing and reporting patterns (themes) within data' (Braun & Clarke, 2006, p.6). Thematic analysis is a useful method for researchers working within an interpretivist paradigm and is employed to gain a fuller and more nuanced understanding of empirical data (Braun & Clarke, 2006). The five steps of analysis (familiarization of the researcher with the data, generating initial codes, searching for themes, reviewing themes, defining and naming themes) outlined by Braun and Clarke (2006) were used to guide data analysis. The interviews were initially reviewed, transcribed, and then read and re-read multiple times to narrow down the number of codes, further reviewed and then categorized into identifiable themes. Three main themes were formed that are reflected in the next section.

Findings and discussion

This section examines the restrictions imposed on Iranian tourism as a result of the sanctions from the views of key informants. Three main themes emerged from the thematic analysis (Figure 1) that will be discussed below.

Figure 1. Themes emerged from interviews.

Restricted mobility rights

Conversations with participants revealed that long standing sanctions on Iran have had a negative influence on the country's economic growth, with the tourism sector being particularly hard hit throughout the sanctions period. Sanctions have a variety of notable consequences, including 'restrictions on mobility', which many respondents noted as one of the most significant and harmful consequences of sanctions. 'Travel advisories' were noted as one tool of US sanctions against Iran. One of the respondents mentioned that:

> The tensions between Iran and the USA which has been the main reasons for US sanctions has negatively hit the Iranian tourism industry… Over the years, the US government has issued a range of travel advisories prohibiting its nationals from visiting Iran, this has very negatively impacted on tourist visitation from the western countries (Interview #19).

The safety-related contents of state travel warnings play a significant role in shaping the perception of risk by international travellers (Toubes & Araújo-Vila, 2021) which have the potential to damage the image of a destination and affect tourists' destination choices (e.g. Deep & Johnston, 2017; Chaulagain et al., 2019). While, mobility is often associated with flow and freedom (Baerenholdt, 2013), it is also about power and governance (Hall, 2005). Travel warnings are sometimes highly politicized and as Bianchi (2006) notes, they can be perceived as an extension of a state's geopolitical concerns. The latter highlights that tourism is more intimately tied to global geopolitics and security issues via state travel warnings. This reflects Azcárate et al. (2021) framing of this as 'mobility geopolitics', and they argue that tourism is increasingly used by states as a "geopolitical and political economic weapon of choice that often fuels discriminatory social imaginaries of domestic and international destinations" (p.18).

Others respondents noted that the country's negative image has made it a high-risk destination for international visitors. They all believe that the nuclear agreement of 2015 and the following relaxation of sanctions greatly helped to altering Iran's image, resulting in a significant rise in the number of visitors, particularly European tourists, and an increase in foreign investment in tourism. A tour operator mentioned that:

> …. Because of the lifting of sanctions and positive media coverage of nuclear deal, Iran's image in the world's media was changed and the country became a popular tourism destination for European tourists (Interview #23).

This reflects the previous studies' findings on the role of media in shaping or altering a destination image (Avraham, 2015; Rasoolimanesh et al., 2021). Media coverage of political upheaval has the potential to shape individuals' image of destinations (Khodadadi & O'Donnell, 2017).

The US government also imposed sanctions on individuals visiting Iran. Under US law travelers who visit Iran are subject to enhanced scrutiny when later attempting to enter the

United States. This has as led to a collapse in international visitation which was noted by the interviewees:

> Following this new type of sanctions, numerous tours were cancelled. We had many cancelation by European tourists because they feared that visiting Iran may prohibit them from entering the United States in the future (Interview #11).

> This sanction has had a significant impact on Westerners' willingness to visit Iran. The general perception among Western visitors is that visiting Iran will make their future trips to the United States extremely difficult… I don't think they'll go to Iran any longer (Interview #25).

Apart from restricting the mobility of visitors to Iran, sanctions have also had an impact on Iranians' outbound travel. Respondents also mentioned the difficulty of obtaining a visa for Iranians which undermines the 'right to travel' of Iranian outbound tourists. This was noted by several tour operators:

> Because of sanctions, we've had a lot of issues with visa applications in recent years for our Europe travel packages. Payment delays, as well as unilateral and arbitrary cancellations, have been experienced by visa applicants (interview #3).

> After the return of sanctions, several of our package tours to Europe were cancelled since only a few persons on the tour were able to obtain a visa…. imagine that when you plan a tour, you pay for the tickets, hotels etc, but few days before the trip, your customers get a visa rejection (interview #13).

Additionally, the restrictions caused by sanctions are not only limited to the mobilities of individuals but carriers have been suffered as well. Many major European airlines, for example, do not provide direct flights to Iran because firms doing business with Iran risk being blacklisted from the US market. This has had a significant impact on airline capacity to and from Iran which was repeatedly noted by the interviewees:

> …We only have a few options when we plan to travel overseas now that European airlines have ceased direct flights to Iran, and regional airlines that have direct flights from Iran raise their ticket prices, leaving us with no other alternative… (interview #7).

Furthermore, Iran's civil aviation has long been subject to sanctions imposed by the United States, the United Nations, and Europe. Sanctions have mainly targeted Iran's purchase of aircraft and supply of spare parts and technical services. One of the respondents commented:

> … sanctions ban the supply of fuel to Iranian airplanes at major European airports causing significant operational difficulties for the aviation sector (interview #29).

As illustrated, sanctions have largely restricted international mobilities to and from Iran. This contrasts with SDG16, which highlights the significance of ensuring just and inclusive societies for everyone, and Target 16.10, which emphasizes the importance of 'protecting fundamental freedoms, in accordance with national legislation and international agreements'. This has meant respondent concerns over how their rights to mobility have been affected by sanctions. While, for international tourism bodies such as the UNWTO and the WTTC, the right to international travel should be regarded as a fundamental freedom (Breakey & Breakey, 2013), for others, notions of social justice in international travel and notions of mobility remain elusive (Higgins-Desbiolles, 2018) or contested (Torabian et al., 2022). The respondents perceive that these rights exist even if commentators or governments do not. Notions of justice with respect to international mobility remain contrary and contested (Higgins-Desbiolles, 2018, Bianchi et al., 2020), especially given the limited nature of rights to travel across national borders in international law (Torabian et al., 2022).

While Hall (2005, p.134) argues that "mobility is at the heart of our understanding of tourism", Bianchi and Stephenson (2019, p.13) suggest that when the international mobilities of many individuals is "deemed problematic or illegitimate," the right to tourism cannot be fully realized. In the same vein, Torabian and Miller (2017) observe, restricting international movement perpetuates inequalities across the globe, although the right to travel "is frequently perceived as a market issue and is less frequently situated in a socio-political context" (Torabian et al., 2022, p.3).

In a globalized world the effect of sanctions has both distributive justice (e.g., Rawls's theory) and performative justice dimensions. Distributive issues clearly arise with respect to the flow on economic impacts of mobility restrictions which have repeatedly noted by the interviewees of this study. However, performative justice concerns arguably arise because of the more localised social impacts that occur through restrictions on VFR as well as, in the Iranian case, on pilgrimage.

Restriction of digital and financial rights

One of the issues discussed by participants was that the sanctions resulted in a lack of resources to develop a 'well-equipped transportation' and 'hotel sector', making Iran even 'less appealing to tourists'. In addition, online travel booking platforms have been prohibited of listing Iran on their platforms with respondents feeling that their 'digital rights' or 'communication rights' have been affected. As a result, there is no reservation system or platform for travellers planning to visit Iran. One of the interviewees commented:

> ....an online booking system is an important aspect of any country's tourism sector since it instills trust and confidence particularly in foreign visitors....travelers who decide to plan a trip do not have to do so through travel agency; they can do so on their own time... Because of sanctions, we are prohibited to list our hotels on online booking platforms (interview #17).

In addition to the sectoral sanctions, financial restrictions were also noted by informants. The prohibition and restrictions on international payments networks and the banking industry and the effect on their financial 'rights' were repeatedly mentioned by interviewees as one of the noteworthy financial repercussions of sanctions. An interviewee commented:

> Sanctions make it hard to send money to Iran or send money from Iran to other nations using formal banking systems... For all overseas transactions with our tourism business partners abroad (like booking hotel, flights), we must go via exchange offices... For whatever transaction, they constantly demand a large commission (interview #4).

This issue clearly shows that sanctions have created many challenges for different sectors in the Iranian tourism and hospitality industry. Attaining the SDGs necessitates financial and economic development, multilateral commerce, and international investment, all of which can be harmed by sanctions (Madani, 2021). While Goal 17 emphasizes the integration of developing countries into the global economy and the expansion of the multilateral trading system in order to achieve the SDGs, as the findings of this study indicated, economic sanctions hamper such capacities by prohibiting financial transactions and foreign investment in target countries. In the case of Iran this situation effectively financially isolates it from much of the rest of the world (Pratt & Alizadeh, 2018) and creates significant issues of distributive justice with respect to economic well-being.

Gender justice

Economic sanctions against Iran have resulted in a significant increase in inflation, unemployment, and economic uncertainty in general. Sanctions supposedly gender neutral, have disproportionately affected vulnerable groups in society, and in the case of Iran, women have suffered more socio-economic hardships as a result of the sanctions due to the country's male-dominated and patriarchal society. Respondents believe that the effects of sanctions on Iranian women have been substantial.

> Women are the first to lose their employment as a result of the of sanctions, and they have poorer job security than males in the Iranian tourism sector ... Sanctions put women's livelihoods at jeopardy and take away their employment security ... (interview #12).

> Sanctions create economic harm in a variety of ways, including lower-paying jobs in the hotel industry, reduced job security, and no government backing with no protections ... All of this has a negative impact on Iranian women's economic situation (interview #30).

Conversations with female participants revealed that despite Iran's highly gendered economy and religio-patriarchal society and masculinized political culture (Shahidian, 2002), tourism has been a sector that has advanced gender empowerment and increased levels of economic independence for Iranian women who are disproportionately affected by sanctions.

> I go to several places in Iran as a tour guide, something a single female like me may not be able to do ... However, as a certified tour guide, I have no problems with traveling alone ... My career has altered my vision of what it means to be a woman in Iran and has made me feel strong .... (interview #10).

> I really love my job of being a tour guide. when I am with the tours coming from Europe, I feel very happy as I earn very good money and get to meet with people from different cultures. working in tourism helped me to build my self-confidence that I am a productive members of society ... Now that the number of tourists has decreased, I've lost my employment and my network relationships (interview #14).

These narratives show that economic sanctions have had a direct impact on women employment and entrepreneurship as a result of lower tourist numbers, and that they have also contributed to financial discrimination against them by reinforcing pre-existing structural gender discrimination in Iran. As highlighted by the interviewees, sanctions are not gender blind and that the empowerment of women through tourism has been made more fragile with the return of sanctions (Seyfi et al., 2020b). This is consistent with other studies that show that economic sanctions led to increased economic maldistribution between both rich and poor and between genders, with such situations often being reinforced by government policies that reinforce the economic position of men in an already highly patriarchal economy (Drury & Peksen, 2014; Gutmann et al., 2020). While this situation enables a distributive justice critique of such a situation there is also the need for a more fundamental performative justice assessment of empowerment and gender-justice at national and local scales (Houston & Pulido, 2002; Jamal & Hales, 2016). The latter is extremely significant in the Iranian context given that the notion of what constitutes performative justice at the local level may be highly contested. Different from other countries subjected to sanctions, in Iran, tourism and hospitality employment and entrepreneurial activities serve as an 'acceptable' work environment in the conservative and highly patriarchal Iranian society, making tourism and hospitality employment and entrepreneurial activities arguably more significant than other sectors for women's rights (Shahidian, 2002; Farahani & Dabbaghi, 2018; Seyfi et al., 2020b).

For women continuing to travel, albeit domestically and engaging in volunteering or undertaking other tourism-related work activities in the light of sanctions act as a way to be seen and are a performative vehicle by which claims for recognition can be made legitimate (Brownlow, 2011). Such seemingly everyday practices can provide a form of 'performative citizenship' for the gendered and politically marginalised (Gilbert & Phillips, 2003). Such performances, as Houston and Pulido (2002, p.403) suggest, when "enacted in specific historical and geographical contexts ... expose the dynamics of power and exploitation," by operating "simultaneously as a space of possibility and becoming, and as a mechanism for working through existing social contradictions [and injustice] by making them visible" (Houston & Pulido, 2002, p.406). However, women's performance stands in opposition to the ongoing impositions of sanctions and the subsequent normalisation of gender discrimination and particular local gendered, religious and political interpretations of justice that it reinforces (Biyouki & Marinova, 2020). As a result, the economic sanctions imposed on Iran clearly run directly counter to the SDG goals with respect to women rights and empowerment.

Aligning with previous studies, the findings of the present study show some support for the notion that performative tourism-based empowerment is a response to distributive sanctions-based disempowerment. This is especially important in Iran for women's economic empowerment, which is critical to attaining women's rights and gender equality, as well as achieving the SDGs (Farahani & Dabbaghi, 2018; Seyfi et al., 2020a; UN Women, 2020). Given its acceptable nature with Iran's highly gendered economy, tourism is a sector potentially capable of advancing gender equality and women's empowerment (Alarcón & Cole, 2019; Boluk et al., 2019), particularly because it provides a performative space for such social and economic justice actions. However, it can be concluded that distributive constraints resulting from the sanctions-driven economic disruption in Iran are obviously restricting tourism's ability to contribute to better gender justice.

Conclusion and implications

This study has aimed to gain a better understanding of the implications of a widely used foreign policy tool for justice and sustainability in destinations to which sanctions are applied. Guided by a justice lens and grounded in international relations and political science disciplinary base and building on in-depth, empirical data from key tourism informants in Iran, the findings of this study emphasized sanctions as a significant though overlooked crisis event for tourism destinations and actors. Although the broader ethics of sanctions has been examined in the political science and international relations literature, there is a scarcity of studies exploring this in the context of tourism. Additionally, this study addressed the call of previous researchers (e.g., Jamal, 2019; Higgins-Desbiolles, 2018; Rastegar et al., 2021; Jamal & Higham, 2021) for more 'multi-/inter-/trans-disciplinary' empirical work related to justice and tourism. This study provides new insights into the crisis-justice nexus and argues that both distributive and performative justice dimensions can simultaneously be read into the effects of sanctions and other crises and their perceived effects on rights. However, although such effects are noted at the individual level, their recognition at the macro policy level appears more problematic. Nevertheless, this study brings to the fore that Rawlsian distributive justice provides a clear benchmark by which to assess material justice issues, particularly at the national level. Performative justice provides a means of reading some of the responses to economic change, particularly for women, and the way in which the economic and employment effects of sanctions intersects with the gender, religious and political structures of contemporary Iran. However, while recognising the limitations of Rawlsian and Kantian approaches to justice issues in tourism and crisis, we suggest that, to a great extent, notions of universal justice measures continue to frame performative justice actions and grassroots justice measures, including with respect to responding to crises, as they provide an 'ideal' over what could be. Such in-depth insights add significant value to the existing body of literature.

Here the SDGs, which are substantially rights-based, act as a critical benchmark by which to assess the difference between what is desired and what exists, and for tourism researchers to assess how might that gap be closed. In other words, although a primarily distributive justice document, it provides a vital framing role for performative justice issue raising at the local scale by highlighting the gap between ideal and reality and therefore provides directions for future research, especially the potential significance of the relationality between distributive and performative justice and its application over different scales. This means that that the tension between those approaches, especially in a crisis context, may depend on recognition as justice as an essentially contested concept, the application of which inherently depends on its definition in particular contexts. Arguably, this means that future tourism crisis research needs to better engage in broader debates over the relationship between different notions of rights to mobility, communication, employment and economic well-being and social, economic and environmental justice, at a time of major environmental change, and place them at the forefront of sustainability thinking.

Acknowledgement

The authors would like to thank the special issue editors and four anonymous reviewers for their careful reading of the paper and their thoughtful comments and constructive suggestions, which helped us to improve the quality of this paper.

Disclosure statement

No potential conflict of interest was reported by the authors.

ORCID

Siamak Seyfi http://orcid.org/0000-0002-2427-7958
Colin Michael Hall http://orcid.org/0000-0002-7734-4587
Jarkko Saarinen http://orcid.org/0000-0002-9327-6932
Tan Vo-Thanh http://orcid.org/0000-0001-9964-3724

References

Afshari, L. (2019). *International sanctions and sustainable development goals*. https://www.sdwatch.org/post/international-sanctions-and-sustainable-development-goals.

Alarcón, D. M., & Cole, S. (2019). No sustainability for tourism without gender equality. *Journal of Sustainable Tourism*, *27*(7), 903–919. https://doi.org/10.1080/09669582.2019.1588283

Avraham, E. (2015). Destination image repair during crisis: Attracting tourism during the Arab Spring uprisings. *Tourism Management*, *47*, 224–232. https://doi.org/10.1016/j.tourman.2014.10.003

Azcárate, M. C., Mostafanezhad, M., & Norum, R. (2021). Introduction: Tourism geopolitics. In M. Mostafanezhad, M. C. Azcárate, & R. Norum (Eds.), *Tourism geopolitics: Assemblages of infrastructure, affect, and imagination* (pp. 3–28). University of Arizona Press.

Baerenholdt, J. O. (2013). Governmobility: The powers of mobility. *Mobilities*, *8*(1), 20–34. https://doi.org/10.1080/17450101.2012.747754

Bianchi, R. (2006). Tourism and the globalisation of fear: Analysing the politics of risk and (in) security in global travel. *Tourism and Hospitality Research*, *7*(1), 64–74. https://doi.org/10.1057/palgrave.thr.6050028

Bianchi, R. V., Stephenson, M. L., & Hannam, K. (2020). The contradictory politics of the right to travel: mobilities, borders & tourism. *Mobilities*, *15*(2), 290–306. https://doi.org/10.1080/17450101.2020.1723251

Bianchi, R., & Stephenson, M. L. (2019). Tourism, border politics, and the fault lines of mobility. In A. Paasi, E.-K. Prokkola, J. Saarinen, & K. Zimmerbauer (Eds.), *Borderless worlds for whom? Ethics, moralities and mobilities* (pp. 123–138). Routledge.

Bidegain Ponte, N., & Enríquez, C. R. (2016). Agenda 2030: A bold enough framework towards sustainable, gender-just development? *Gender & Development*, *24*(1), 83–98. https://doi.org/10.1080/13552074.2016.1142227

Biyouki, Z. A., & Marinova, D. (2020). What hides behind the scarf: Iranian women's participation in the economy. In S. Samani & D. Marinova (Eds.), *Muslim women in the economy* (pp. 85–99). Routledge.

Boluk, K. A., Cavaliere, C. T., & Higgins-Desbiolles, F. (2019). A critical framework for interrogating the United Nations Sustainable Development Goals 2030 Agenda in tourism. *Journal of Sustainable Tourism*, *27*(7), 847–864. https://doi.org/10.1080/09669582.2019.1619748

Boomen, M. (2014). *The effectiveness and ethics of economic sanctions*. https://sevenpillarsinstitute.org/effectiveness-and-ethics-of-economic-sanctions/.

Braun, V., & Clarke, V. (2006). Using thematic analysis in psychology. *Qualitative Research in Psychology*, *3*(2), 77–101. [Database] https://doi.org/10.1191/1478088706qp063oa

Breakey, N., & Breakey, H. (2013). Is there a right to tourism? *Tourism Analysis*, *18*(6), 739–748. https://doi.org/10.3727/108354213X13824558470943

Brown, G. (2006). Kantian cosmopolitan law and the idea of a cosmopolitan constitution. *History of Political Thought*, *27*(4), 661–684.

Brownlow, C. A. (2011). Between rights and responsibilities: insurgent performance in an invisible landscape. *Environment and Planning A: Economy and Space*, *43*(6), 1268–1286. https://doi.org/10.1068/a436

Chan, H. M. (2005). Rawls' theory of justice: A naturalistic evaluation. *The Journal of Medicine and Philosophy*, *30*(5), 449–465.

Chaulagain, S., Wiitala, J., & Fu, X. (2019). The impact of country image and destination image on US tourists' travel intention. *Journal of Destination Marketing & Management*, *12*, 1–11. https://doi.org/10.1016/j.jdmm.2019.01.005

Cortright, D., & Lopez, G. A. (2018). *Economic sanctions: Panacea or peacebuilding in a post-cold war world?* Routledge.

Creswell, J. W., & Poth, C. N. (2018). *Qualitative inquiry and research design: Choosing among five approaches* (4th ed.). Sage.

Damrosch, L. F. (1994). The collective enforcement of international norms through economic sanctions. *Ethics & International Affairs*, *8*, 59–75. https://doi.org/10.1111/j.1747-7093.1994.tb00158.x

Daniels, S. (1989). The question of jury competence and the politics of civil justice reform: Symbols, rhetoric, and agenda-building. *Law and Contemporary Problems*, *52*(4), 269–310. https://doi.org/10.2307/1191914

Decrop, A. (1999). Triangulation in qualitative tourism research. *Tourism Management*, *20*(1), 157–161. https://doi.org/10.1016/S0261-5177(98)00102-2

Deep, A., & Johnston, C. S. (2017). Travel advisories–destabilising diplomacy in disguise. *Journal of Policy Research in Tourism, Leisure and Events*, *9*(1), 82–99. https://doi.org/10.1080/19407963.2016.1231195

Dizaji, S. F. (2019). Trade openness, political institutions, and military spending (evidence from lifting Iran's sanctions). *Empirical Economics*, *57*(6), 2013–2041.

Drury, A. C., & Peksen, D. (2014). Women and economic statecraft: The negative impact international economic sanctions visit on women. *European Journal of International Relations*, *20*(2), 463–490. https://doi.org/10.1177/1354066112448200

Duffy, R., & Smith, M. (2004). *The ethics of tourism development*. Routledge.

Ellis, E. (2021). The ethics of economic sanctions: Why just war theory is not the answer. *Res Publica*, *27*(3), 409–426. https://doi.org/10.1007/s11158-020-09483-z

Farahani, B. M., & Dabbaghi, H. (2018). Tourism and the empowerment of women in Iran. In S. Seyfi & C. M. Hall (Eds.), *Tourism in Iran: Challenges, development and issues* (pp. 177–192). Routledge.

Farasatkhah, M., Ghazi, M., & Bazargan, A. (2008). Quality challenge in Iran's Higher Education: A historical review. *Iranian Studies*, *41*(2), 115–138. https://doi.org/10.1080/00210860801913321

Farzin, F. (2017). Localising the impact of techno-entrepreneurship in Eastern Iran: Birjand's Science and Technology Park as a local innovation community. *Local Economy: The Journal of the Local Economy Policy Unit*, *32*(7), 692–710. https://doi.org/10.1177/0269094217734327

Fathollah-Nejad, A. (2014). Why sanctions against Iran are counterproductive: Conflict resolution and state–society relations. *International Journal: Canada's Journal of Global Policy Analysis*, *69*(1), 48–65. https://doi.org/10.1177/0020702014521561

Fennell, D. A. (2006). *Tourism ethics*. Channel View Publications.

Flikschuh, K. (2012). Elusive unity: The general will in Hobbes and Kant. *Hobbes Studies*, *25*(1), 21–42. https://doi.org/10.1163/187502512X639597

Gallie, W. B. (1956). Art as an essentially contested concept. *The Philosophical Quarterly*, *6*(23), 97–114. https://doi.org/10.2307/2217217

Gilbert, L., & Phillips, C. (2003). Practices of urban environmental citizenship: rights to the city and rights to nature in Toronto. *Citizenship Studies*, *7*(3), 313–330. https://doi.org/10.1080/1362102032000098896

Given, L. M. (Ed.). (2008). *The Sage Encyclopedia of qualitative research methods*. Sage publications.

Gowlland-Debbas, V. (2021). *United Nations Sanctions and international law*. Brill.

Gutmann, J., Neuenkirch, M., & Neumeier, F. (2020). Precision-guided or blunt? The effects of US economic sanctions on human rights. *Public Choice*, *185*(1-2), 161–182. https://doi.org/10.1007/s11127-019-00746-9

Hall, C. M. (2005). *Tourism: Rethinking the social science of mobility*. Pearson.

Hall, C. M. (2010). Crisis events in tourism: subjects of crisis in tourism. *Current Issues in Tourism*, *13*(5), 401–417. https://doi.org/10.1080/13683500.2010.491900

Hall, C. M. (2017). Tourism and geopolitics: The political imaginary of territory, tourism and space. In D. Hall (Ed.), *Tourism and geopolitics: Issues from central and Eastern Europe* (pp. 15–24). CABI.

Hall, C. M. (2019). Constructing sustainable tourism development: The 2030 agenda and the managerial ecology of sustainable tourism. *Journal of Sustainable Tourism*, *27*(7), 1044–1060. https://doi.org/10.1080/09669582.2018.1560456

Handcock, M. S., & Gile, K. J. (2011). Comment: On the concept of snowball sampling. *Sociological Methodology*, *41*(1), 367–371.

Hatipoglu, E., & Peksen, D. (2018). Economic sanctions and banking crises in target economies. *Defence and Peace Economics*, *29*(2), 171–189. https://doi.org/10.1080/10242694.2016.1245811

Higgins-Desbiolles, F. (2008). Justice tourism and alternative globalisation. *Journal of Sustainable Tourism*, *16*(3), 345–364. https://doi.org/10.1080/09669580802154132

Higgins-Desbiolles, F. (2018). The potential for justice through tourism. *Via. Tourism Review*, 13. Retrieved from http://journals.openedition.org/viatourism/2469. doi: 10.4000/viatourism.2469

Houston, D., & Pulido, L. (2002). The work of performativity: staging social justice at the University of Southern California. *Environment and Planning D: Society and Space*, *20*(4), 401–424. https://doi.org/10.1068/d344

Hufbauer, G. C., Schott, J. J., Elliott, K. A., & Oeggm, B. (2008). *Economic sanctions reconsidered* (3rd ed.). The Peterson Institute for International Economics.

Ilieva, J., Dashtevski, A., & Kokotovic, F. (2018). Economic sanctions in international law. *UTMS Journal of Economics*, *9*(2), 201–211.

Jamal, T. (2019). *Justice and ethics in tourism*. Routledge.

Jamal, T., & Hales, R. (2016). Performative justice: new directions in environmental and social justice. *Geoforum*, *76*, 176–180. https://doi.org/10.1016/j.geoforum.2016.09.014

Jamal, T., & Higham, J. (2021). Justice and ethics: Towards a new platform for tourism and sustainability. *Journal of Sustainable Tourism*, *29*(2–3), 143–157. https://doi.org/10.1080/09669582.2020.1835933

Jamal, T., Camargo, B., Sandlin, J., & Segrado, R. (2010). Tourism and cultural sustainability: Towards an eco-cultural justice for place and people. *Tourism Recreation Research*, *35*(3), 269–279. https://doi.org/10.1080/02508281.2010.11081643

Kant, I. (1948). *Groundwork of the metaphysic of morals* (Translated by H. J. Paton). Routledge.

Khodadadi, M. (2016). A new dawn? The Iran nuclear deal and the future of the Iranian tourism industry. *Tourism Management Perspectives*, *18*, 6–9. https://doi.org/10.1016/j.tmp.2015.12.019

Khodadadi, M. (2018). Donald Trump, US foreign policy and potential impacts on Iran's tourism industry: Post-nuclear deal. *Tourism Management Perspectives*, *26*, 28–30. https://doi.org/10.1016/j.tmp.2018.01.007

Khodadadi, M., & O'Donnell, H. (2017). UK press and tourist discourses of Iran: A study in multiple realities. *Leisure Studies*, *36*(1), 53–64. https://doi.org/10.1080/02614367.2015.1085591

King, T. (1999). Human rights in European foreign policy: success or failure for post-modern diplomacy? *European Journal of International Law*, *10*(2), 313–337. https://doi.org/10.1093/ejil/10.2.313

Lincoln, Y. S., & Guba, E. G. (1985). *Naturalistic inquiry*. Sage.

Madani, K. (2021). Have international sanctions impacted Iran's environment? *World*, *2*(2), 231–252. https://doi.org/10.3390/world2020015

Marinov, N. (2005). Do economic sanctions destabilize country leaders? *American Journal of Political Science*, *49*(3), 564–576. https://doi.org/10.1111/j.1540-5907.2005.00142.x

Mathis, K. (2009). Future generations in John Rawls' theory of justice. *Archiv Für Rechts-Und Sozialphilosphie*, *95*(1), 49–61. https://www.jstor.org/stable/23680984.

McDowell, D. (2021). Financial sanctions and political risk in the international currency system. *Review of International Political Economy*, *28*(3), 635–661. https://doi.org/10.1080/09692290.2020.1736126

Meuleman, L., & Niestroy, I. (2015). Common but differentiated governance: A metagovernance approach to make the SDGs work. *Sustainability*, *7*(9), 12295–12321. https://doi.org/10.3390/su70912295

Morakabati, Y. (2011). Deterrents to tourism development in Iran. *International Journal of Tourism Research*, *13*(2), 103–123.

Mostafanezhad, M., & Norum, R. (2016). Towards a geopolitics of tourism. *Annals of Tourism Research*, *61*(C), 226–228. https://doi.org/10.1016/j.annals.2016.08.003

Mozafari, M. (2016). Iran and science publishing in the post-sanctions era. *The Lancet, 387*(10029), 1721–1722. https://doi.org/10.1016/S0140-6736(16)30260-4

Neuenkirch, M., & Neumeier, F. (2015). The impact of UN and US economic sanctions on GDP growth. *European Journal of Political Economy, 40*, 110–125.

Özdamar, Ö., & Shahin, E. (2021). Consequences of economic sanctions: The state of the art and paths forward. *International Studies Review, 23*(4), 1646–1671. https://doi.org/10.1093/isr/viab029

Power, M., & Campbell, D. (2010). The state of critical geopolitics. *Political Geography, 29*(5), 243–246. https://doi.org/10.1016/j.polgeo.2010.06.003

Pratt, S., & Alizadeh, V. (2018). The economic impact of the lifting of sanctions on tourism in Iran: A computable general equilibrium analysis. *Current Issues in Tourism, 21*(11), 1221–1238. https://doi.org/10.1080/13683500.2017.1307329

Ramet, S. P. (2019). Kant on ethics and politics. *Eastern Review, 8*, 183–199. https://doi.org/10.18778/1427-9657.08.08

Rarick, C. A. (2006). Destroying a country in order to save it: the folly of economic sanctions against Myanmar. *Economic Affairs, 26*(2), 60–63. https://doi.org/10.1111/j.1468-0270.2006.00632.x

Rarick, C. A. (2007). Economic sanctions: failed foreign policy tool and a cost to American business. *Economic Affairs, 27*(3), 65–70. https://doi.org/10.1111/j.1468-0270.2007.00757.x

Rasoolimanesh, S. M., Seyfi, S., Rastegar, R., & Hall, C. M. (2021). Destination image during the COVID-19 pandemic and future travel behavior: The moderating role of past experience. *Journal of Destination Marketing & Management, 21*, 100620. https://doi.org/10.1016/j.jdmm.2021.100620

Rastegar, R., Higgins-Desbiolles, F., & Ruhanen, L. (2021). COVID-19 and a justice framework to guide tourism recovery. *Annals of Tourism Research*, 91*(103161)*: 1–4.

Rastegar, R., Zarezadeh, Z., & Gretzel, U. (2021). World heritage and social justice: Insights from the inscription of Yazd. Iran. *Journal of Sustainable Tourism, 29*(2–3), 521–540. https://doi.org/10.1080/09669582.2020.1782925

Rawls, J. (1971). *A theory of justice*. Harvard University Press.

Ritchie, B. W., & Jiang, Y. (2019). A review of research on tourism risk, crisis and disaster management: Launching the annals of tourism research curated collection on tourism risk, crisis and disaster management. *Annals of Tourism Research, 79*, 102812. https://doi.org/10.1016/j.annals.2019.102812

Rosenberg, E., Drezner, D. W., Solomon-Strauss, J., & Goldman, Z. K. (2016). *The new tools of economic warfare: Effects and effectiveness of contemporary U.S. financial sanctions*. Center for a New American Security.

Rowley, J. (2012). Conducting research interviews. *Management Research Review, 35*(3/4), 260–271. https://doi.org/10.1108/01409171211210154

Rustler, A. (2019). Are sanctions stopping us from achieving global climate goals? *Journal of International Affairs*. https://jia.sipa.columbia.edu/online-articles/are-sanctions-stopping-us-achieving-global-climate-goals.

Saldaña, J. (2021). *The coding manual for qualitative researchers* (4th ed.). Sage.

Schwendinger, H., & Schwendinger, J. (2014). Defenders of order or guardians of human rights? *Social Justice, 40*(1/2 (131-132), 87–117.

Scott, S. V. (2018). The decline of international law as a normative ideal. *Victoria University of Wellington Law Review, 49*(4), 627–644. https://doi.org/10.26686/vuwlr.v49i4.5344

Sen, G. (2019). Gender equality and women's empowerment: Feminist mobilization for the SDGs. *Global Policy, 10*(S1), 28–38. https://doi.org/10.1111/1758-5899.12593

Seyfi, S., & Hall, C. M. (2020a). *Tourism, sanctions and boycotts*. Routledge.

Seyfi, S., & Hall, C. M. (2020b). Sanctions and tourism: Conceptualisation and implications for destination marketing and management. *Journal of Destination Marketing & Management, 15*, 100381. https://doi.org/10.1016/j.jdmm.2019.100381

Seyfi, S., & Hall, C. M. (2020c). Sanctions and tourism: Effects, complexities and research. *Tourism Geographies, 22*(4–5), 749–767. https://doi.org/10.1080/14616688.2019.1663911

Seyfi, S., & Hall, C. M. (Eds.) (2018). *Tourism in Iran: Challenges, development and issues*. Routledge.

Seyfi, S., Hall, C. M., & Fagnoni, E. (2019). Managing world heritage site stakeholders: A grounded theory paradigm model approach. *Journal of Heritage Tourism, 14*(4), 308–324.

Seyfi, S., Hall, C. M., & Vo-Thanh, T. (2020a). The gendered effects of statecraft on women in tourism: Economic sanctions, women's disempowerment and sustainability? *Journal of Sustainable Tourism*, 1–18. https://doi.org/10.1080/09669582.2020.1850749

Seyfi, S., Hall, C. M., & Vo-Thanh, T. (2020b). Tourism, peace and sustainability in sanctions-ridden destinations. *Journal of Sustainable Tourism, 30*(2–3), 372–391. https://doi.org/10.1080/09669582.2020.1818764

Shahidian, H. (2002). *Women in Iran: gender politics in the Islamic Republic*. Greenwood.

Steinberger, P. J. (2008). Hobbes, Rousseau and the modern conception of the state. *The Journal of Politics, 70*(3), 595–611. https://doi.org/10.1017/S002238160808064X

Tahmasebi, S. (2018). *How US sanctions impede the women's movement in Iran*. https://www.atlanticcouncil.org/blogs/iransource/how-us-sanctions-impede-the-women-s-movement-in-iran/.

Takeyh, R., & Maloney, S. (2011). The self-limiting success of Iran sanctions. *International Affairs, 87*(6), 1297–1312. https://doi.org/10.1111/j.1468-2346.2011.01037.x

Teitel, R. G. (2015). *Globalizing transitional justice: contemporary essays*. Oxford University Press.
Torabian, P., & Mair, H. (2022). Insurgent citizens: mobility (in) justice and international travel. *Journal of Sustainable Tourism*, *30*(2–3), 392–407. https://doi.org/10.1080/09669582.2021.1945069
Torabian, P., & Miller, M. C. (2017). Freedom of movement for all? Unpacking racialized travel experiences. *Current Issues in Tourism*, *20*(9), 931–945. https://doi.org/10.1080/13683500.2016.1273882
Toubes, D. R., & Araújo-Vila, N. (2021). The treatment of language in travel advisories as a covert tool of political sanction. *Tourism Management Perspectives*, *40*, 100866. https://doi.org/10.1016/j.tmp.2021.100866
UN Women. (2020). *SDG 5: Achieve gender equality and empower all women and girls*. https://www.unwomen.org/en/news/in-focus/women-and-the-sdgs/sdg-5-gender-equality.
United Nations Educational, Scientific and Cultural Organization (UNESCO). (2022). *Iran (Islamic Republic of)*. https://whc.unesco.org/en/statesparties/ir.
United Nations Security Council. (2020). *Sanctions*. https://www.un.org/securitycouncil/sanctions/information.
United Nations. (2015). *Transforming our world: The 2030 agenda for sustainable development*. https://sdgs.un.org/2030agenda.
UNWTO. (2019). *UNWTO tourism highlights, 2018 edition*. UNWTO.
Winkler, A. (1999). Just sanctions. *Human Rights Quarterly*, *21*(1), 133–155. https://doi.org/10.1353/hrq.1999.0014
Wise, N. (2018). Using ethnographic and participatory approaches in tourism and hospitality research. In R. Nunkoo (Ed.), *Handbook of research methods for tourism and hospitality management* (pp. 115–122). Edward Elgar.
Wood, A. W. (1999). *Kant's ethical thought*. Cambridge University Press.
WTTC. (2020). *Iran. 2021 annual research: Key highlights*. https://wttc.org/Research/Economic-Impact.
Wut, T. M., Xu, J. B., & Wong, S. M. (2021). Crisis management research (1985–2020) in the hospitality and tourism industry: A review and research agenda. *Tourism Management*, *85*, 104307. https://doi.org/10.1016/j.tourman.2021.104307

Tourism policies and inclusive development: the case of Kenya and Rwanda

Christine N. Buzinde and Tanner Caterina-Knorr

ABSTRACT

Drawing on the *inclusive development* conceptual framework, this study utilizes textual analysis to interrogate the development policies for two African nations, Kenya and Rwanda. Composed of three tenets, *social inclusiveness*, *ecological inclusiveness* and *relational inclusiveness*, this framework is rooted in theories of social justice and sustainable development. *Social inclusiveness* was accounted for in the findings vis-à-vis the participation of all in development, as well as protection and capacity building for the most vulnerable. However, participatory public processes aimed at pluralizing knowledges that inform development were not supported. Evidence for *ecological inclusiveness* included accounts on ecological limits, just sharing of the responsibility and risks of climate change, and participation of all stakeholders. Accounts on steps taken to protect the most vulnerable to climate change and engage all stakeholders in mitigation were absent. *Relational inclusiveness* included discussions on rule of law; however, wealth redistribution, critical assessment of taxation policies and protection of public resources from privatization were notably absent. The omissions are attributed to the histories of colonialism and neocolonial realities. Advocacy for locally defined and equitable approaches to development have become more pronounced globally, particularly in Africa, thus it is imperative for the tourism industry to align with these decolonial ideals.

Introduction

Within the context of the Global South, desires such as liberation, self-determination, empowerment, and sovereignty are continually thwarted and undercut by political and economic actions of the West. Social, ecological, and relational injustices abound when these two worldviews collide, particularly when the common language is economic growth and/or "development" above all other ways of knowing and relating. The othering of nonwestern knowledges within westernized systems and structures continues to be a consequential root cause of several contemporary global crises. This "ethnolinguistic fractionalization" is reflected in different ways within tourism policy documents and runs counter-intuitively to the concept of sustainable tourism (Minkov et al., 2021, p. 801). The interrogation of tourism policy documents illuminates the extent to which local knowledge is included within the policy creation process and how policies contribute to social justice within tourism destinations.

Any interpretive assessment of the contents of tourism policy documents first requires critical awareness of the historical, economic, and political dimensions that inform the aforementioned fractionalization. The implications of uncritically interrogating this tension results in accounts that ignore tangible global issues of poverty, wealth inequality, climate change, access to water and healthcare, and other wicked problems (Dredge, 2019). Many of these global crises, which manifest in the Global South particularly within the African continent, can be attributed to neoliberal economics. Within this model, economic growth is regularly conflated with the concept of development and it is regarded as the "panacea for a multitude of (often changing) socio-economic challenges" (Schmelzer, 2015, p. 264; Sharpley, 2020). In order to account for socio-cultural, economic, environmental, and political challenges, international institutions and organizations like the United Nations (UN), World Bank, and International Monetary Fund (IMF) promote economic growth, through privatization and increased trade, as a remedy (Dredge, 2019; UNWTO, ITC, & EIF, 2017). Many of the aforementioned international organizations deploy terms like economic development, sustainable development, inclusive growth, and sustainable growth as key drivers in alleviating global challenges. However, these terms remain entrenched within Western ideals and are embedded within an economic growth paradigm (Sharpley, 2020) that tends to ignore and/or undervalue social and environmental dimensions. For instance, in the context of East Africa, tourism is often utilized as a tool to ameliorate social, environmental, and economic development challenges (see Okello & Novelli, 2014), but an emphasis on economic growth is often most prominent (UNWTO, ITC, & EIF, 2017).

The World Tourism Organization (UNWTO) et al.,) (2017) undertook a review of 48 tourism policies from least economically developed nations (LEDCs; 35 of which are in Africa) and they suggest that these policies should "reinforc[e] the involvement of tourism-related institutions with trade institutions at the national level" as well as "closer collaboration between different stakeholders, including those in charge of environmental sustainability and poverty reduction" (p. 27). Notably, international organizations, like the UNWTO, are the very institutions often directing and authoring tourism policies for many East African nations and they do so while drawing on neoliberal, colonial, and economic growth ideals (see Bianchi, 2002), which in turn, undervalues local ways of knowing and being. According to Dredge (2019), despite good intentions, these processes advance neoliberal development agendas set by the Global North and imposed upon the Global South, consequently "triggering increased poverty and marginalization" in the Global South (p. 41).

Scholars argue that we cannot address social ills like poverty, ecological destruction, and racism without addressing "the dynamics and fundamental contradictions of capitalism" (Bowles & Veltmeyer, 2022, p. 5), particularly given that Global South nations are often not in position to negotiate arrangements in their favor. Notably, the argument presented here is not to dissuade Global South nations from collaborating with foreign agencies on development policy creation, but rather to highlight that such partnerships are often based on a "one dollar one vote" way of doing things versus "one country one vote" (Gupta & Vegelin, 2016, p. 434). Consequently, these partnerships are less likely to promote e quitable aspects of development because their local co-authors and communities, unlike their Western counterparts, are writing from the periphery and are largely neither involved nor tokenistically included within the policy creation and development process (Moscardo, 2011). This lack of attention to and the removal of local actors within the policy making processes often leads to a mismatch of ideals between policy-makers, the private sector, and residents (see Adu-Ampong & Kimbu, 2019; Stone & Nyaupane, 2020).

Many scholars outside of tourism studies, noting these issues of conflict and exclusion, have advocated for *inclusive development* approaches which entail the creation of growth-related goals that are embedded within equity-based frameworks (de Haan, 2015; Rauniyar & Kanbur, 2010). Drawing on extensive experience in international development (with the United Nations) and international law, scholars Gupta and Vegelin (2016) prolegomenon has advanced the relevance of *inclusive development* to global debates on sustainability and the sustainable development goals. Within extant research, the concept of inclusive development has generally been conceptualized as solely related to economic growth (see Rauniyar & Kanbur, 2010), however Gupta and

Table 1. Inclusive development principles.

Social inclusiveness	Ecological inclusiveness	Relational inclusiveness
Ensure equity principles to share the opportunities for development	Adopt ecocentric limits from local to global level	Ensure that public goods and merit goods do not become privatized or securitized
Include the knowledge of all	Equitably share rights, responsibilities and risks	Address all drivers and actors; and combat offshoring, tax havens, and other instruments that lead to concentration of wealth and ecospace
Build targeted capacity building to enable effective participation	Build resilience and adaptive capacity	Address discourses that concentrate wealth to ensure that markets and growth takes place within inclusive development paradigm
Enhance protection for the poorest	"Green" international cooperative instruments	Test instruments for inclusiveness and ensure downward accountability
Engage all in the politics of development	Involve all stakeholders	Ensure global rule of law and constitutionalism

Source: Adapted from Gupta and Vegelin (2016, p. 437).

Vegelin (2016) seminal work critically queries "the need for continuous economic growth in a business-as-usual paradigm" and rather offers a unique account that incorporates social and environmental dimensions (p. 435). The inclusive development framework is founded on three key tenets, namely, *social inclusiveness*, *ecological inclusiveness*, and *relational inclusiveness* (see Table 1). *Social inclusiveness* accounts for various principles including: all citizens have access to development opportunities; the knowledge of all is incorporated in development initiatives; all citizens can engage in development due to capacity building efforts; and, safeguards are in place for the poor and vulnerable (Gupta & Vegelin, 2016). The key goal is to ensure that local human capital is equitably linked to development opportunities. This concept is relevant to tourism because scholars have long argued that unbridled growth policies adopted by nation states often exacerbate the gap between the wealthy and the poor (Dredge, 2019; Zhang, 2021) and such efforts often result in magnified injustices within tourism destinations (see Büscher & Fletcher, 2017; Dredge, 2019; Higgins-Desbiolles, 2008; 2020).

The second concept, *ecological inclusiveness*, also draws on the foundation of equity, as it relates to human-environmental interactions. That is, the concept of "ecology" in this context encapsulates the connection between the natural and human worlds (Sze & London, 2008). *Ecological inclusiveness* encompasses the adoption of "ecocentric limits" as well as the consideration of equitably sharing "rights, responsibilities and risks" related to climate change, building "resilience and adaptive capacity", creating "green international cooperative instruments," and involving all citizens in ecological initiatives (Gupta & Vegelin, 2016, p. 437). Much like the concept of social inclusiveness, ecological inclusiveness requires an understanding of the nexus between marginalized communities and related ecological matters. Ecological inclusiveness is certainly applicable to tourism because destinations are dependent on healthy natural environments and resilient communities through which they develop an environmental friendly travel industry.

Relational inclusiveness "refers to the difference in income, wealth, opportunity and access between the rich and the poor" and it ensures that privatization and securitization processes do not disproportionately disadvantage the poor (Gupta & Vegelin, 2016, p. 439). Other principles related to this concept include: addressing "offshore tax havens, and other instruments that lead to concentration of wealth"; dealing with "discourses that concentrate wealth"; experimenting with "instruments for inclusiveness ... and downward accountability"; and, creating a foundation upon which "global rule of law and constitutionalism" can be practiced (Gupta & Vegelin, 2016, p. 437). This concept is relevant to tourism because private agencies do not always prioritize public interests, which can result in the omission of key initiatives/policies created to improve livelihoods (see Adu-Ampong, 2019; Mowforth & Munt, 2016; Stumpf & Swanger, 2015). Notably, all three concepts can be used to interrogate tourism development policies in the Global South, particularly within the African continent, to ascertain the level of and acccountability to inclusive development. Additionally, this work has implications for sustainability and social justice because the concept of inclusive development is intricately connected to environmental, social, and political justice (Agyeman et al., 2016). That is, social, ecological, and relational inclusiveness can be linked to Rawls's (1971) tenets of social justice, specifically procedural and distributional justice, which focus on inclusive participatory practices and equitable distributions of resources. It is thus important to interrogate how the inclusive development ideals of *social inclusiveness*, *ecological inclusiveness*, and *relational inclusiveness* are accounted for, if at all, within development policies, so as to assess alignment with sustainability and social justice related parameters. It is also vital to interrogate policy documents by posing the question of answerability, a concept which decolonial scholars describe as a way to shift "praxis of ethics" surrounding "referents of learning, knowledge as ontology, and context" from Western coloniality to decolonization (Patel, 2016, p. 73).

While Gupta and Vegelin (2016) anchor the inclusive development framework within Western ways of knowing and being, in order to contribute to the project of decolonization, it is necessary to balance this inclusivity lens with relevant leading voices from the African continent that are intimately involved in similar socio-political processes. In this manuscript, these voices come by way of Kenyan legal scholar and African thought leader, Dr. Patrick Loch Otieno (PLO) Lumumba, given his seminal work on African empowerment through decolonial desires, and Dr. Bagele Chilisa, a Botswanan post-colonial scholar who describes the need to decolonize the African educational system so as to center and amplify non-Western knowledges within discussions/actions on local development. Lumumba (2019) states that "Africa must define for itself what development is and define its story" (online). Similarly, and from a historical perspective, Chilisa (2012) states that colonialism destroyed the "political, social, and economic systems and lead to external political control and economic dependence on the West… resulting in the captive or colonized mind and the imposition of the colonizers' ways of knowing" (Chilisa, 2012, p. 9). From this standpoint, the decolonial project necessitates critical awareness of how notions of inclusivity, equity, and justice manifest; these concepts are all germaine to tourism. Hence, the interrogation of tourism policy documents does not only require cognizance of the socio-historical past and its lingering impact on the present; it also demands attention to spaces of answerability through which the amplification of discourses advanced by postcolonial thought leaders (i.e. Dr. Lumumba and Dr. Chilisa) can take place.

Extant research has engaged in the laborious work of examining tourism policy texts to highlight opportunities and threats to development for the East African member states (Okello & Novelli, 2014). However, critical interrogation of these documents from an inclusive development lens remains scarce. Accordingly, questions such as the following remain answered: what capacity building mechanisms are in place to empower impoverished communities and to involve them in the development process? What protocols are in place to ensure that stewardship and use of environmental assets are equitably shared? What regulatory measures are in place to ensure that public resources remain publicly owned? Accounting for the aforementioned multiple perspectives in the examination of development policies is vital because, as indicated by Bevir et al. (2003), these documents are important repositories of values and philosophies of nation states, and where relevant, their partnering international organizations. Thus, drawing upon the lens of inclusive development and the work of decolonial scholars and thought leaders from the African continent, the purpose of this study is to interrogate two development policies to understand their conceptualizations of development. The development policies represent the agendas for Kenya and Rwanda. The current study is undertaken in solidarity with tourism researchers like Kimbu and Ngoasong (2013) whose work has amplified the voices of the often silenced African citizenry.

Literature review

In its conceptualization inclusive development amplifies the social and ecological dimensions and situates them as the foundation upon which the economic aspects should be considered (Gupta et al., 2015). In order for inclusive development to yield the three key tenets mentioned earlier (social, ecological, and relational), it requires actions towards the presence of: *"relevant epistemic communities and communities of practice and social movements"*; *"interactive governance to enable empowerment"*; and, *"appropriate governance instruments"* (Gupta et al., 2015, p. 541). The *creation of epistemic communities and communities of practice* ensures that an educated and skilled labor force is available to implement the industry changes needed. Tourism is reliant on the availability of a skilled labor force, particularly within isolated tourism dependent destinations so its success is directly connected to the presence of epistemic communities and communities of practice. Notably, these communities are informed and enriched by knowledge emanating from research. But in the context of the Global South, these knowledge production systems have to adopt a decolonial approach that actively includes the "knowledge systems of formerly colonized, historically marginalized, and oppressed groups" (Chilisa, 2012, p. 1). The goal is not to annihilate Western ways of knowing, as the colonial agenda did in nonwestern contexts, but rather to acknowledge and actively seek out other ways of knowing that capitalize on locals' lived experiences and local thought leaders striving for change. Applying this goal to tourism development research for instance, calls for these processes to not be extractive in nature but rather to actively seek opportunities to co-produce knowledge alongside the local citizenry (Buzinde et al., 2020).

According to Lumumba (2019), the continent needs to look upon itself from a decolonial standpoint in order to bring about the multitude of sustainable development changes needed. He has for decades advanced the idea that the remedy for the African development crises lies with Africans and not foreign nations (see AAU, 2019). He states that "Africa has over the years been the victim [and the target] of globalization," which manifested through the means of slavery, colonization, divided countries, and resource extraction, including knowledge (online). This comment resonates with Higgins-Desbiolles (2008) remarks which urge society to seek out a "more humanistic form of globalization" (p. 347). Furthermore, and speaking to the dehumanizing dimension of globalization, Lumumba (2019) indicates that on the African continent, there is a tendency for local intellectual capital to be ignored as the search for expertise is squarely directed to the West, thus perpetuating the colonial legacy.

The presence of *interactive governance* can counter the abovementioned deficiency lens often directed towards Africa. Interactive governance can bolster civic engagement and spur symbiotic interactions between governmental and non-governmental agencies, all working in harmony and alongside the citizenry to enhance collective well-being. In the context of tourism, it accounts for "continual negotiations" between internal and external stakeholders involved with policy creation (Stevenson et al., 2008, p. 2). Poignantly, tourism governance in many nations worldwide is still largely managed from the top down, primarily guided by economic growth indicators and marketing initiatives (Sharpley, 2020). Within such approaches, local residents have little representation in national tourism development plans, thereby resulting in inequitable, unjust, and incomplete tourism policies that lack local resident input (Fleischer & Rivlin, 2009). Additionally, lack of reciprocity between the industry and communities can lead to unidirectional extraction of local benefits, economic or otherwise, which further supports uniformity advanced by globalization. Actions towards governance however have to account for any lingering strongholds that foreign agencies may have on a Global South nation. Lumumba (2019) uses the term "free" to represent the notion of independence whereby African nations "were able to manage [their] political affairs … design [their] own flags … compose and sing [their] own national anthems … but were [they] really free?" (online). He urges for awareness of how the legacy of colonialism has resulted in the erosion of local ways of knowing which inform governance, amongst other societal structures.

The creation of tourism policies is often a highly politicized process, given the multiple levels of governance associated with the task (Bramwell & Lane, 2011). From this vantage point, the presence of *appropriate governance instruments* that are guided and supported by presence of regulatory bodies are vital, particularly as relates to ensuring the protection of the most vulnerable. Many African national judicial systems, much like their broader governance systems, were imported from the West. Matters are further complicated when one accounts for the fact that neoliberal economics is complexly intertwined with issues of ethics, human rights, and democracy. This is particularly resonant in Sub-Saharan African nations wherein the inability of African nations to be politically and economically independent from the powerful global neoliberal economic agendas has been likened to a new type of imperialism which, in turn, has fueled resistant movements, such as the Pan-African movement (Hahn, 2008). Local judicial systems are often unable to and lack the political capital with which to challenge neoliberal economic agendas. Any one African nation may indeed find it difficult to take on the monolith of neoliberal economics imposed upon it, but Lumumba's (2011) thesis of a Pan-African framework provides collective leverage and a more viable solution by addressing "spheres of human development" (p. 3–4) which is incidentally central to the concept of inclusive development. In fact, although the adoption of Gupta and Vegelin (2016) work in this article focuses on the national level, their work also speaks to interconnectivity with other nations, as would be the case if one were to adopt a Pan-African approach. Applied to tourism, inclusive development allows for a focus on *social*, *ecological* and *relational* dimensions, but these ideals have to occur in tandem with efforts to cultivate *epistemic communities and communities of practice,* foster *interactive governance* and, create *appropriate governance instruments.* Without this critical foundation, a justice oriented and sustainability informed approach to tourism is thwarted. It is from this perspective that the development policies explored in this paper are queried by drawing on the lens of inclusive development. Although Pan-Africanists do not center their argument on the concept of inclusive development, many of its tenets are implied in the thesis that Africans have to be the principal architects responsible for socially engineering an approach to development tailored to the needs of the continent, by capitalizing on and being attentive to African talent and values.

Methods

This study is guided by an interpretive inquiry and it utilizes textual analysis to interrogate two policy documents. Derived from semiotic analysis, textual analysis assists in the exploration of the varying types of information that are present or absent and how they inform meaning by highlighting or annihilating certain elements, resultantly, validating a given vantage point (Entman, 1993). From this perspective, texts represent society by "constituting and constructing the world in meaning" and concurrently "contributing to the construction of systems of knowledge" (Fairclough, 1992, p. 64). This research draws on deductive or concept-driven coding was undertaken. This process commenced with an *a priori* identification of system codes that were assigned to the qualitative data (Schwandt, 1997) (see Table 1). Deductive codes were determined using the categories (i.e. *social inclusiveness*, *ecological inclusiveness*, and *relational inclusiveness*) entailed in the conceptual framework (see Table 2). These categories ensured that the connections with the research questions were maintained (Bingham & Witkowsky, 2021). The goal was not to look for the presence of the phrases, *social inclusiveness*, *ecological inclusiveness*, or *relational inclusiveness*, but rather to explore whether what they denote was captured in the texts.

The subsequent steps entailed familiarization with the data and application of codes. Firstly, this process started with arranging the transcripts so as to facilitate ease of reading and cross

Table 2. A priori codes, descriptions, and themes.

Codes	Descriptions	Themes
Social Inclusiveness	Shared equity, access, and opportunities within development processes	Knowledge; access; capacity building; participation; protection for the poor
Ecological Inclusiveness	Shared rights, responsibilities, and risks associated with human-environmental interactions	Ecocentric limits; rights, responsibilities, and risks; resilience; adaptation; green cooperation
Relational Inclusiveness	Shared public goods associated with wealth creation, concentration, and privatization	Access to wealth; privatization; offshore tax havens; downward accountability; laws and regulations

Source: Adapted from Gupta and Vegelin (2016, p. 437).

comparison. Secondly, thorough reading of the policy documents aimed at comprehending the content and context was undertaken (Hall, 1975). Thirdly, iterative readings of the texts occurred in order to identify emergent patterns related to the codes; this step also allowed for continuous questioning of the data to enhance credibility (Lincoln & Guba, 1986). Two researchers independently coded both documents and the results were collaboratively shared to establish intercoder reliability (Kurasaki, 2000). Different levels of coding took place with each immersion of the data. For example, the data indicated that: *social inclusiveness* was linked to capacity building that was geared towards workforce development, involvement and business incentives; *ecological inclusiveness* was connected to protection of the natural environment; and, *relational inclusiveness* was associated with the creation or enforcement of regulations and agreements related to macro level business operations. Notably, this stage of the analysis predominantly focused on the "reality" depicted within the texts instead of focusing on whether the content was accurate or inaccurate (Crotty, 1998). As scholars operating within Western institutions of higher education, we recognize that we write from a space of privilege, but as underrepresented scholars in our various spheres, we assess structures of power from our lived experiences. Furthermore, our connections to the African continent, via ancestry and work respectively, varyingly inform our outlook of the socio-political and socio-historical context in which tourism planning and policy making emerges.

Contextualizing Kenya's and Rwanda's policy documents

Building on Okello and Novelli (2014) work, two of the five East Africa nations were chosen for this comparative study, namely Kenya and Rwanda. Notably, this research is part of a larger study that included analyses of Uganda and Tanzania, however given limited space constraints and the ability to allow for a detailed presentation of data, this paper only focuses on the two established tourism destinations, Kenya and Rwanda; current industry statistics are lacking due the pandemic. Both nations composed their tourism policies in response to different crises, which can offer opportunities for critical reflection and transformative change. The tourism policy for Rwanda was published in 2009 and Kenya's policy was published in 2020. For Rwanda, the financial crisis of 2008/2009 "affected all major tourist producing countries, and [it was purported to] result in intense competition which [could] lead [to] … a static or at worst a decreasing market (Republic of Rwanda & Ministry of Trade & Industry, 2009, p. 7). For Kenya, "the outbreak of Coronavirus (COVID-19) represents an important and evolving challenge to the tourism sector … [and] … has triggered an unprecedented crisis in the economy given the immediate and immense shock to the sector" (Republic of Kenya et al., 2020, p. 14). Thus, this unique commonality allowed for a closer comparison between these two policies which were dissimilar from the policy-creation context of the other three East African nations.

Kenya is characterized as a "deepening and sustaining" level of tourism development (Christie et al., 2013, p. 5) while Rwanda is "scaling up" (Okello & Novelli, 2014, p. 59). Both policies identify tourism as a key economic driver and an important sector that supports national development policies. Additionally, both nations identify wildlife (e.g. safaris) and cultural tourism (e.g. community-based tours) as key products and like many Global South nations, they have similar "wicked problems" (Adu-Ampong & Kimbu, 2019) which can be ideally countered through inclusive tourism development. Finally, Rwanda's policy was produced by the Ministry of Trade and Industry and they identify the national and local governments, private sector, communities, NGOs, media, financial institutions and development partners, and training institutions as "key players in tourism development" (p. 25). Kenya's policy was composed by the Ministry of Tourism and Wildlife and they similarly identify "national and county governments, state agencies, private sector, civil society organizations, local communities and the general public" as key stakeholders for tourism development and the implementation of the policy (p. 35).

Findings

The results of the textual analysis indicate that some examples of the three pillars of inclusive development (*social*, *ecological*, and *relational inclusiveness*) were present in both of the analyzed policy documents. Generally, evidence pertaining to any one of the pillars was captured in a sentence or more with the infrequent occurrences of all three or two of the pillars captured within the same statement. The section below provides details related to Kenya and Rwanda, respectively.

Social inclusiveness in the context of Kenya

The text analyzed showcased that the tenets of *social inclusiveness* manifested through assertions related to: involvement of the community; investment in local capacities; support of local culture; and skill development for locals. Mention of local involvement appeared early in the document in a section that presented the guiding principles as well as throughout the document. The excerpt below is a quintessential example of the ways in which local involvement was discussed:

> Given that tourism products are found in and hosted by local communities, focus will be on *involving all stakeholders* such as national, county governments, tourism operators and host communities in its management, planning, benefit sharing to achieve sustainability... *Community involvement*, participation and collaboration with stakeholders should be continually encouraged to enhance *local capacity* (p. 18).

Community involvement was often linked to "participation of the private sector, communities, individuals in tourism development through provision of fiscal incentives" (p. 30). There was recognition that "guidelines on partnerships between communities and investors" (p. 27) were needed to ensure that both parties were protected, particularly vulnerable groups. For instance, "a special funding mechanism for host communities and vulnerable groups" (p. 34) was implemented to foster business coalitions between poor communities and investors.

Grounded in the need to "alleviate poverty... promote social inclusion" and to support and protect local communities and culture, "a code of ethics for the tourism industry" (p. 27) but also "guidelines on the use of Indigenous knowledge" (p. 34) were outlined so as to ensure the "protection from commercialization of culture" (p. 27). There was recognition that capacity building necessitates participation and coordination between local communities, the private sector, training institutions, and layers of government. In some cases, it was noted that general technology related courses were needed. For instance, the government vowed to "[e]nhance ICT skills development amongst the tourism workforce" (p. 23). In other instances, it was proposed that a focus on general skills needed to be amplified in order to support the tourism industry. The excerpt below offers some examples:

> Improve and expand existing hospitality and tourism training facilities... Encourage technical knowhow and exchange programs within and outside the region for both students and personnel working in the tourism

> sector ... Increase access to training opportunities by offering scholarships and incentives ... Promote youth skills transfer and mentorship programs in the sector (p. 28).

Capacity building also extended to training programs for vulnerable communities employed in or seeking employment in the tourism sector. Of interest was the mention to "[e]ncourage and support capacity building for small, micro and medium enterprises and emerging entrepreneurs in tourism" (p. 28). The focus on "creat[ing] and prompt[ing] partnerships" (p. 28) between government and training facilities was frequently highlighted.

Ecological inclusiveness in the context of Kenya

Within the Kenyan policy, narratives related to ecological inclusiveness included discussions related to sustainability, sustainable tourism, investments, and the protection of natural resources. The document indicated that sustainability is one of the guiding principles for Kenya, as is showcased in the excerpt below:

> [Tourism in] Kenya is nature based, tourism will be undertaken in a manner that does not compromise the quality and value of the resource, or exceed the carrying capacity of supporting ecosystems. It will be managed in a sustainable way by incorporating long term goals in its planning and operations ... It is therefore imperative that standards are enhanced to ... boost economic, environmental and cultural sustainability (p. 2;18; 27).

Sustainable tourism is defined in the policy document as "tourism development meeting the needs of present tourists and hosts while protecting and enhancing opportunities for the future" (p. 40). The implementation of sustainable tourism by the government involves all stakeholders (i.e. local communities, investors, and national, regional, and international governments) "throughout the tourism supply chain" (p. 27) in order to "provide the necessary direction to support a competitive, resilient and sustainable tourism industry" (p. 3). Attention to the sharing of resources and related rights is also captured in the documents, as indicated in the following quote: "encourage peaceful *coexistence among communities and tourism investors* as regards *shared natural resources*" (p. 23). This statement is premised on a well-documented fact that a host community whose access to natural resources is compromised due to tourism development will inevitably be antagonistic towards tourists and developers (Cousins & Kepe, 2004).Accordingly, and given the scarcity of environmental resources, preemptive action is addressed within the policy by regarding and involving essential local participation to facilitate a landscape favorable for business investors, host communities, and negotiations. Of importance is the mentioning of efforts undertaken by the government to "promote the establishment of transboundary and trans frontier conservation areas as a tool for enhancing regional tourism integration" (p. 26). To "enhance linkages between the tourism sector and other sectors, the government will endeavor to ... develop and implement tourism sector environmental impact assessment guidelines" to protect natural resources (p. 33). The reinforcement of pro-environmental behaviors amongst the mass populace was also mentioned vis-à-vis the government's encouragement of "*recycling of water and appropriate disposal of waste*" (p. 23). Anti-environmental behaviors were also addressed in the policy documents as is indicated in the following excerpt: "where there are credible threats of serious or irreversible damage by tourism, lack of full scientific certainty will not be used as a reason for postponing cost-effective measures to prevent damage ... the polluter and users ... should bear the full environmental and social costs of their activities" (p. 18).

Relational inclusiveness in the context of Kenya

Discussions related to *relational inclusiveness* in the case of Kenya included multilateral regulations and responsibilities as well as institutional, investor, and community responsibilities. There were many accounts related to existing rules and regulations as well as the need to create additional guidelines/agreements that facilitate an environment conducive to sustainable business development. For instance, the excerpt below speaks to the foundational guidelines that govern tourism development in Kenya:

> Tourism Act 2011 was enacted and existing statutes relating to tourism were repealed. The Act brought in new institutions and some of the old ones were renamed. The government has also developed the National Tourism Blueprint 2030 that aims at redefining and redesigning the development and management of Kenya's tourism sector (p. 11).

In speaking of the regulatory work needed, the document offered guiding principles by which the work can be undertaken. A quintessential example of this point is the excerpt below:

> The sector will be managed transparently in accordance with principles of *good governance*... Proper enforcement mechanisms with distinct powers and mandates will be prescribed to responsible agencies to ensure that all necessary laws and mechanisms are enforced accordingly and that *all players fully comply and adhere to the set rules* and laws to ensure sustainability... *Bilateral agreements* as well as regional and multilateral instruments should be domesticated and implemented (p. 18)

Reference to regulatory actions also accounted for nationally bound actions as well as international cooperation. For instance, there was mention of strengthening "the policy and oversight role of the Ministry" in order to "ensure that policy and enabling legislation are in tandem with emerging tourism issues" and secondly, to "establish and strengthen coordination mechanisms... consistent negotiations and implementation of tourism-related bilateral and multilateral agreements" (p. 31–32).

The need for legal rubrics and clear institutional processes was articulated throughout the document and it was often stated in reference to the importance of supporting existing institutions. The excerpt below speaks to a set of identified objectives which highlight the regulatory work needed:

> Provide a framework for integration of tourism sector considerations into the various sectoral policies, national and county development planning as well as decision making processes... Transform and strengthen the legal and institutional framework for effective coordination and management of the tourism sector... Promote and support the provision of incentives and other economic instruments that enhance investment in the sector (p. 17).

The policy document also included reference to certain agencies responsible for enacting identified tasks (e.g. maintenance of public roads) that would facilitate sustainable tourism development in Kenya. For instance:

> Implement *regulations, guidelines and measures* for licensing tourist service vehicles (TSVs) and operators... Encourage county governments to support access roads to tourist sites under their jurisdiction... Kenya Wildlife Service to prioritize upgrading and maintenance of tourism roads... Encourage bilateral and multilateral agreements on air travel with a view of opening up airspace for both scheduled airlines and charters... Provide a *regulatory framework* to ensure safe operations of water vessels used in tourism activities... Develop and implement a Code of Conduct for tourism practitioners (p. 21–22).

One specific sector mentioned was the media, which includes national and international promotion, marketing, and reporting. Nationally, the campaign named "*Buy Kenya Build Kenya*" was encouraged to limit economic leakage, through the support of the government (p. 34). Both locally and internationally, publicity was described as a challenge to tourism development because of "unbalanced reporting of insecurity incidences without highlighting the positive developments happening in Kenya" (p. 16). Regional media cooperation was also linked to this category of institutional responsibilities: "harmonize regional policies dealing with tourism development, promotion and marketing" (p. 26). To ensure availability of quality products/services, the policy states that the information regarding and aiding sectoral related standards needs to be considered. For instance, "[d]evelop and implement a code of practice for the tourism sector that sets out the minimum standards and guidelines for best practices" (p. 28). The subsequent section presents the findings related to the Rwandan tourism policy document.

Social inclusiveness in the context of Rwanda

As was the case of Kenya, the Rwandese policy document encapsulated dimensions related to *social inclusiveness* that focused on: involvement of the community; investment in local capacities; support of local culture; and skill development for locals. The text indicated that "local community participation in the development of the sector for their own benefit" (p. 5) was vital for any meaningful change to occur "in the development of tourism" (p. 27). Locals were also regarded as key actors in the journey towards sustainability, as is illustrated in the following statement: "local communities will also have a responsibility to become actively involved in the tourism industry, to practice sustainable development" (p. 8). It was recognized that substantive change is needed to reverse the "[l]ittle involvement of communities in micro, small and medium enterprises (MSMEs) in the sector" (p. 3). Investment in MSMEs was seen as also having the potential to support sustainable growth of the sector. Some of the objectives identified in the policy document mentioned provision of "support to MSMEs, ensuring that they have the capabilities and capacities to enter the tourism value chain, while also ensuring that communities contribute to and benefit from the tourism industry" (p. 6). From this standpoint and as will be further illustrated, investment in small to medium size businesses was regarded as a vital mechanism through which to alleviate poverty amongst vulnerable communities.

Like the Kenyan case, the Rwandan policy mentioned protection of local cultural assets through local community involvement. For instance, the quote below encapsulates this:

> Tourism should also be developed to value, invest in and develop local cultures and protect them from over-commercialisation and over-exploitation. Local communities will also have a responsibility to become actively involved in the tourism industry, to practice sustainable development and to ensure the safety and security of the visitors. Effectively, a *pro-poor approach* will be taken, to ensure that tourism results in increased net *benefits for poor people* (p. 8).

Local culture is also seen as an important strength of visitor attractions:

> Promotion of products that project and enhance the country's people, culture and rich history. Specifically encouraged will be showcasings of the rich history of the Kingdom of Rwanda, the conflict resolution success following the genocide of 1994, the development and promotion of festivals and cultural events leading to an annual calendar of culture and festivals, handicraft production, displays and sales combined in outlets that project the production skills and crafts to the intending purchaser, and community tourism initiatives (p. 11).

As relates to skill development, there was recognition that one of the key challenges faced by Rwanda is "[l]ow capacity and under-skilled human resources" (p. 3). To this end, there was recognition that "[t]he approach and implementation of capacity building and the development of human resources has so far been fragmented" (p. 8). Intentionality regarding capacity building was discussed vis-à-vis "job creation, strengthening professional competencies" (p. 5) for "[t]he long-term goal of the capacity building and human resource development... to ensure that Rwanda's training system will be responsive to the ever-changing needs of the tourism and hospitality industry" (p. 12).

One primary goal of improving local capacities and training programs was to eliminate "the need for human resource imports" (p. 19). There was a sense that capacity building requires the development of "systematic, high-quality training, to create a skilled workforce in value jobs at every level of the tourism and hospitality industry" (p. 5). Similar to Kenya, some suggestions for actionable goals included the following statements:

> The training system will broaden the options of school leavers and those already in employment. It will attract mature Rwandans back to study and training, and provide them with specialized skills and pathways to new careers in the tourism, hospitality and services sectors (p. 12).

Concrete actions related to capacity building were also mentioned and they included a gamut of initiatives including short workshops to degree granting programs, all aimed at ensuring that employability was enhanced for all graduates. Some specific capacity building efforts entailed the following:

> Updating and improving the curriculum of existing travel and tourism related training courses and improving qualifications of teaching staff through an extensive train-the-trainer program. Setting up a tourism and hospitality training institute in Kigali. Developing flexible pathways from vocational skills training into higher education diploma and degree programs in tourism and hospitality studies (p. 13).

Ecological inclusiveness in the context of Rwanda

The manifestation of ecological inclusiveness was related to dimensions within the Rwanda policy document that included accounts about: environmental sustainability, sustainable tourism, assessments, and the protection of natural and cultural resources through various mechanisms.

The document mentions the need for responsible uses of "local resources ... [to] ... avoid over-consumption" (p. 9) and promote "the protection of wildlife and the environment" (p. 6).

Descriptions of environmental sustainability were strongly connected to the goals and definition of sustainable tourism, given Rwanda's dependency on nature-based tourism. As was the case in the Kenyan policy document, a key value expressed in the document indicated:

> Sustainable tourism as the key guiding principle for tourism development, to ensure environmental, social and economic sustainability ... Sustainable tourism implies tourism industry responsibility to the environment through the promotion of balanced and sustainable tourism and focus on the development of environmentally-based tourism activities (p. 9–10)

There were many pro-environmental actions documented within the policy. It was noted in the document that in the past there had been "[l]ittle focus on the environment and sustainability" (p. 3). Of importance and in contrast to the Kenya content, is the fact that a clear change in focus had been adopted and it was one that was centered on evaluating the impacts on the natural environment, of proposed development projects. For instance, it is stated that "assessments will take place prior to any major infrastructure development, including public [and private] consultation in the destination with relevant stakeholders" (p. 17). The assessments and associated management plans will support the development and promotion of "socially and environmentally responsible tourism" (p. 26) and will "oppose developments that are harmful to the local environment and culture of the community" (p. 27). Notably, a key objective mentioned in the text was to "ensure that the tourism sector is planned and developed to the benefit of future generations of Rwandans, in terms of the sustainability of resource use, the protection of wildlife and the environment" (p. 6).

A framework for monitoring and managing the natural environment included the following proposed actions:

> Awareness programs will be implemented to alert tourists and local populations to the need to respect nature, the physical environment, fauna and flora. The district councils ... will enforce good planning and environmental protection laws and regulations for the protection of the natural and cultural sites. Areas will be identified, environmental and social studies conducted, public consultations made, and then they will be legally gazetted. Environmental management plans for protected areas will be developed, and will incorporate biodiversity management, business development (including sustainable tourism) and income generation, human resources and equipment, monitoring and evaluation (p. 18).

To support the continued and proper management of sustainable tourism, the policy stated that "concessions will be preferentially granted to investors with sound environmental, social and economic proposals, which *enhance biodiversity conservation* and benefit local livelihoods" (p. 18); this qualification criterion was unique to the Rwandan content.

Relational inclusiveness in the context of Rwanda

Finally, discussions on *relational inclusiveness* captured in the Rwanda policy document included accounts related to multilateral regulations and land management. There was recognition that the nation has an "[u]nder-developed regulatory framework for the tourism sector" (p. 3). Accordingly, there was mention of resolving the aforementioned deficiency by "[e]stablish[ing] and maintain[ing] regulation that enables the tourism industry to develop in a fast yet sustainable manner" (p. 6). Recognition of the need for a more regulated and sustainable sector through which the needs of locals and tourists were met was evident in the document. For instance, this is illustrated in the following statement: "[i]t is vital that the tourism industry develops in an orderly manner to ensure that the services provided are regulated for the protection of tourists, the public, the private sector, and the state" (p. 9). Additionally, "[c]ivil society groups, private sector, NGOs, and other agencies will be invited to provide a peer review mechanism to support the state decision-making on particular developments" (p. 17). To support this protection, various regional and local governance collaborations are mentioned within the policy. For example, it is mentioned that Rwanda seeks to lure and cater to "the needs of international, regional, and domestic visitors" (p. 5) as well as put in place "mechanism[s] for achieving equitable sharing of benefits among the partner states" (p. 22). Additionally, "[a]t the local government level, specific provincial functions of policy implementation ... are further supported" (p. 26) by an "*inclusive approach* to planning and design, ensuring community consultation, involvement and partnerships, and putting an end to ad hoc and inappropriate development" (p. 30).

There was an emphasis on "responsible land-use planning [as well as] urban and rural development" regulation (p. 26), as is indicated in the following quote:

> Land is extremely scarce in Rwanda and so understandably there are differences in interests in how it should be distributed. Land in Rwanda is now privately owned by the people, and the majority of Rwanda's tourism products are land-based. A framework to facilitate and guide access to land by tourism investors is needed, for land both within and outside protected areas. Land tenure security and *land rights for all stakeholders* need to be clarified and strengthened to facilitate investment (p. 8).

However, the content emphasized that "opportunity costs of setting aside land for product development and how tourism revenue affects the trade-off must be analyzed" (p. 18). This will "enable tourism investments through ensuring the availability and security of land tenure for development" (p. 6). Further, "collaborating with relevant authorities on issues of land is required to ensure that areas identified for tourism investment and activities are developed in an environmentally and socially sustainable manner" (p. 17); the identification of certain landscapes for tourism was unique to the Rwandan example.

Notably, there was recognition that land ownership, which is inherent to most community businesses, could not be mentioned without dealing with the matter of community-based tourism enterprises (CBTEs). As indicated in the excerpt above, this matter is further compounded by the fact that Rwanda has a scarcity of land resources, which are necessary in tourism development. Community based tourism enterprises, be they small or medium businesses, were favored given that they could be owned and managed by locals who can contribute to poverty alleviation. Of interest was the expressed desire to develop regulations by "[p]roviding guidelines for good governance systems within CBTEs"; such measure were regarded as vital to ensuring "transparency and accountability" (p. 15). Furthermore, the policy encouraged the:

> ... development of a concessions strategy ... to stimulate and provide guidance for the development of public-private partnerships, including guidance on tender processes; contract management and exit strategies; stakeholder rights, responsibilities and accountability; revenue sharing; and planning gain for

sustainable investments that conserve the environment, reduce poverty, and enhance local culture and society (p. 11).

Finally, media outlets are mentioned as a "key player in the tourism industry" in connection to the products and services of Rwanda's tourism industry, marketing efforts, and competitiveness (p. 25). One specific initiative related to the media that is similar to the Kenyan case entails "[d]eveloping a 'Buy Rwandan' campaign" which encourages the purchase of local inputs and local employment through tax incentives" (p. 19). More generally, "[m]arket-led… product development and diversification" (p. 5) will require "effective marketing tools" (p. 14) so that "the brand development policy will create a distinctive image and present a new tourism positioning for Rwanda [and] create a competitive advantage" (p. 12).

Discussion

Both tourism policy documents for Rwanda and Kenya elicited findings that generally align with the three tenets of inclusive development: *social*, *ecological*, and *relational inclusiveness*. Notably, there were aspects under each key tenet for which evidence was not available in the policy documents. For instance, and as will be elaborated on in this section, there were more accounts confirming the presence of the social and ecological tenets, but fewer for the relational tenet. Firstly, *social inclusiveness* appeared in the findings *vis-à-vis* participation of all in development as well as protection and capacity building for the most vulnerable. However, participatory public processes aimed at amplifying the principle of "knowledge of all" and local residents to inform development, captured within the *social inclusiveness* concept (Gupta & Vegelin, 2016, p. 347), was not discussed at length in any of the analyzed policy documents. Suffice it to say that in the context of Kenya, there was mention of including Indigenous Knowledges (IK), whereas in the case of Rwanda, information sharing between communities, private sector, and the state was noted. Knowledge is complexly imbricated in the colonial engine and one cannot ignore the violence by which colonized African peoples' knowledge was annihilated, "disconnecting them from what they knew and how they know it" (Chilisa, 2012, p. 9). This historical violence is often perpetuated in the development arenas to the point that "traditional knowledges have typically been seen as obstacles to development" when in fact attention to "the knowledge of a population who have lived experience of the environments in question" can result in more transformative and sustained outcomes (Briggs & Sharp, 2004, p. 661; Manuel-Navarrete et al., 2021). With time, these processes have resulted in multiple injustices that manifest as failure "to transform the lives of the majority of people in the Global South, and especially so in sub-Saharan Africa" (Briggs, 2013, p 232). It is in reaction to such injustices that Lumumba (2019) argues for the urgency in Africans rising to their own aid, using their knowledge and experience to develop the continent. From an agentic vantage point, an African-informed approach to development would thus adopt what Chilisa (2012) refers to as relational knowledge which accounts for African ontologies (i.e. I/we versus I/you) and African epistemologies (knowledge is relational rather than individual). In the context of tourism, relational knowledge can indeed be capitalized upon to inform the creation of differentiated tourism products and processes.

Secondly, evidence for *ecological inclusiveness* included accounts on ecological limits, just sharing of the responsibility and risks of climate change, and participation of all stakeholders in the aforementioned endeavors. By contrast, both documents generally lacked discussions on "greening existing international instruments on trade, investment and development assistance" (Gupta & Vegelin, 2016, p. 439). This particular omission may be attributable to lack of *procedural justice* which in this case can be explained as the minimal leverage some African nations have in negotiating trade-related terms with investors from the Global North or the over-reliance on investor tailored approaches. Also missing in both documents was mention of steps taken to legally protect and build resilience and adaptive capacity for marginalized communities that are most susceptible to climate change and conflicts related to scarcity of resources. This may be attributed to the fact that few resources may be available to invest in research on resilience and adaptive capacities. The notion of resilience was mentioned in the Kenyan document; however, it was in reference to COVID-19 and necessary mitigations in the wake of tourism decline. It is important to note that the effects of climate change have increased and the frequency of climate associated disasters demands the availability of proactive planning and related resources that many affected nations lack (Below & Wallemacq, 2018), particularly in Africa. Scholars claim that "a healthy community must also perforce to be a resilient community, one that has the systems of governance and social capital needed to rebound from difficult times and sustain that which is most positive about its identity for its current and future inhabitants" (Zautra et al., 2008, p. 132). Within tourism destinations, planning should account for procedures focused on resilience and recovery because disasters, whether natural or otherwise, are imminent globally (Edmeade & Buzinde, 2021).

Thirdly, as it pertains to *relational inclusiveness*, both documents included accounts regarding rule of law; however, wealth redistribution, critical assessment of taxation policies, and protection of public resources from privatization were notably absent within the analyzed documents. Furthermore, the principle, to "address discourses that concentrate wealth" (Gupta & Vegelin, 2016, p. 347) was not supported in the analyzed documents, however, in the case of Rwanda, there was cursory mention of an objective to ensure "sustainable economic growth and equitable distribution of wealth for the benefit of all Rwandans" (p. 5). Additionally, there were three other principles for which support was not detected in the policy documents, namely: testing "instruments for inclusiveness and ensure downward accountability"; ensuring "that public goods and merit goods do not become privatized or securitized;" and, "address[ing] all drivers and actors" in order to "combat offshoring, tax havens, and other instruments that lead to concentration of wealth and ecospace" (Gupta & Vegelin, 2016, p. 347). Notably, the aforementioned absent principles necessitate the presence of stable regulatory platforms to protect public goods and ensure equitable distribution of wealth (Gimba, Seraj & Ozdeser, 2021) so as to reduce economic leakage within tourism destinations (Ankomah & Crompton, 1990). Additionally, matters are complicated when one accounts for the histories of colonialism and political instability endured by both nations. Offering a historical rationale for the current circumstances faced by many African nations, Nketsia (2013) states that

> [w]hen we look at the African independence explosion, we must take into consideration that not one African nation came to power using a conventional African structure of government. Every one of them used an imitation parliamentary procedure taken from Europe. Africa will never succeed using European parliamentary techniques (p. 36).

Nketsia's (2013) thesis is important because African governance structures are intricately related to regulatory outcomes and they influence people's perception of whether their nation is procedurally just or not. Procedural injustices in Africa are often framed as corruption which according to Lumumba (2011) is rooted in "the negative colonial legacy, poor leadership … weak institutions of governance … and constant insecurity and conflicts" (p. 5). It is from this vantage point that Lumumba (2019) argues for regeneration in order to re-liberate the continent and assess itself by drawing on human indices rather than "on metrics that show growth as defined by the West" because many basic needs are still unmet, decades after independence (online). This latter point is vital because it calls attention to who has the right to define development for Africans but it also urges scholars and practitioners alike to not blindly propose approaches to development that are neither human centered nor empower local residents as equal decision makers. It also highlights the fact that we cannot look at development without acknowledging the economic, environmental, political, and psychosocial impact of the colonial past. Cognizance of this socio-historical and socio-political set of circumstances allows for an affirmation of

Africanness (agency) necessary for the decolonial (re)centering of Africans in the discussion regarding African development.

Some strides related to the above direction are worth mentioning. For instance, adopting an asset-based approach to development, the African Union has proposed Agenda 2063 which articulates a Pan African Vision that includes "[a]n integrated, prosperous and peaceful Africa, driven by its own citizens, representing a dynamic force in the international arena" (Agenda 2063, n.d.). It is important to mention that this agenda contains paralls to but goes beyond an inclusive development framework and it accounts for cooperation with all nations on the continent to accomplish the following development goals: higher standards of living for Africans; a knowledge revolution; transformed economies; environmentally sustainable and climate resilient of communities; continental collaboration on infrastructure development; regulation of financial infrastructure; human rights and justice; African cultural renaissance; peace and transformational leadership; gender equity; and youth development. Agenda 2063 is a locally defined approach to development and one that ought to be linked to other similar documents such as tourism policies, as so doing would ensure that the tourism sector, upon which many African nations are dependent (Ankomah & Crompton, 1990; UNWTO, ITC, EIF, 2017), contributes to the accomplishment of these goals for the betterment of the African continent.

Conclusion

This study draws on the concept of inclusive development to explore the development policies of two African nations, Kenya and Rwanda. Inclusive development comprises three tenets, namely: *social inclusiveness* which focuses on equitable distribution of opportunities for development; *ecological inclusiveness* which denotes the espousal of regulations that advance equitable sharing of ownership, stewardship, and risk; and *relational inclusiveness* which describes the protection of public goods from privatization (Gupta & Vegelin, 2016). Inclusive development is derived from theories of social justice and it is based on the premise that sustainable development cannot take place without attention to key social and ecological dimensions, and power structures that characterize Global South/North relations, as well as regional/national dynamics within Global South nations. The findings of the current study indicate that some elements of all three tenets were encapsulated within both policy documents and this is a promising outcome, as it can support efforts in the aforementioned nations to engage in further efforts to create more just and sustainable development, be it related to tourism or otherwise. However, it is important to mention that there were specific dimensions related to use of local knowledges, support for green entrepreneurship, enhancement of resiliency, and enforcement of regulatory parameters that were not fully addressed. These absences have to be placed within the context of colonial histories and the neocolonial realities of many African nations. Recognition of this thus ensures that remedies occur within decolonial frameworks that prioritize local ways of knowing and local definitions of development. The African Union's Agenda 2063 is a quintessential example of a grassroots approach to solving local development challenges by drawing on local knowledges. As African nations continue to find ways to link Agenda 2063 to their various sectoral goals (e.g. tourism), there promises to be more African centered and inclusively co-created solutions to the challenges of which the continent is dealing with and/or has been dealt.

Building on Okello and Novelli (2014) research, the current exploratory work engages the important conversation of social justice in relation to development. In an era wherein communities are protesting unjust realities and advocating for locally informed approaches to development, it is imperative for tourism industry experts and scholars alike to engage in discussions related to social justice and tourism development, particularly in relation to new and locally informed development formulas such as Agenda 2063. Future research can address cross-sectoral development policy agendas to assess how they (dis)engage with the concept of inclusive development. Alternatively, a broader geopolitical assessment can be undertaken to offer cross-comparative opportunities to/for development. Lastly, it is incumbent upon researchers and policy-makers to create necessary space for other ways of knowing in addition to composing legally, socially, economically and environmentally sound approaches to community informed tourism development.

Disclosure statement

No potential conflict of interest was reported by the authors.

References

AAU. (2019). Association of African Universities Talks: Professor P.L.O. Lumumba Interview. Retrieved from https://www.youtube.com/watch?v=Qul6uYfH5-g

Adu-Ampong, E. A. (2019). The tourism-development nexus from a governance perspective: a research agenda. In R. Sharpley and D. Harrison. (Eds.), *A research agenda for tourism and development* (pp. 35–52). Edward Elgar Publishing.

Adu-Ampong, E. A., & Kimbu, A. N. (2019). The past, present and future of sustainability in tourism policy and planning in Sub-Saharan Africa. *Tourism Planning & Development*, *16*(2), 119–123. https://doi.org/10.1080/21568316.2019.1580836

Agenda 2063. (n.d.). *The Africa we want*. Retrieved November 1, from https://au.int/agenda2063/overview

Agyeman, J., Schlosberg, D., Craven, L., & Matthews, C. (2016). Trends and directions in environmental justice: from inequity to everyday life, community, and just sustainabilities. *Annual Review of Environment and Resources*, *41*(1), 321–340. https://doi.org/10.1146/annurev-environ-110615-090052

Ankomah, P. K., & Crompton, J. L. (1990). Unrealized tourism potential: The case of sub- Saharan Africa. *Tourism Management*, *11*(1), 11–28. https://doi.org/10.1016/0261-5177(90)90004-S

Below, R., & Wallemacq, P. (2018). Natural disasters 2017, Centre for Research on the Epidemiology of Disasters (CRED), Brussels.

Bevir, M., Rhodes, R. A., & Weller, P. (2003). Traditions of governance: Interpreting the changing role of the public sector. *Public Administration*, *81*(1), 1–17. https://doi.org/10.1111/1467-9299.00334

Bianchi, R. (2002). Towards a new political economy of global tourism. In R. Sharpley and D. Telfer. (Eds.), *Tourism and development: Concepts and issues* (pp. 265–299). Channel View Publications.

Bingham, A. J., & Witkowsky, P. (2021). Deductive and inductive approaches to qualitative data analysis. In Vanover, C, Mihas, P. & Saldana J. (Eds). *Analyzing and interpreting qualitative data research: After the interview* (pp. 133–147). SAGE Publications.

Bowles, P., & Veltmeyer, H. (2022). Introduction to critical development studies: Four characteristics with illustrations from seven decades. In H. Veltmeyer and P. Bowles (Eds.), *The essential guide to critical development studies*. Routledge.

Bramwell, B., & Lane, B. (2011). Critical research on the governance of tourism and sustainability. *Journal of Sustainable Tourism*, *19*(4-5), 411–421. https://doi.org/10.1080/09669582.2011.580586

Briggs, J. (2013). Indigenous knowledge: A false dawn for development theory and practice? *Progress in Development Studies*, *13*(3), 231–243. https://doi.org/10.1177/1464993413486549

Briggs, J., & Sharp, J. (2004). Indigenous knowledges and development: a postcolonial caution. *Third World Quarterly*, *25*(4), 661–676. https://doi.org/10.1080/01436590410001678915

Büscher, B., & Fletcher, R. (2017). Destructive creation: Capital accumulation and the structural violence of tourism. *Journal of Sustainable Tourism*, *25*(5), 651–667. https://doi.org/10.1080/09669582.2016.1159214

Buzinde, C. N., Manuel-Navarrete, D., & Swanson, T. (2020). Co-producing sustainable solutions in indigenous communities through scientific tourism. *Journal of Sustainable Tourism, 28*(9), 1255–1271. https://doi.org/10.1080/09669582.2020.1732993

Chilisa, B. (2012). *Indigenous research methodologies*. Sage publications.

Christie, I., Fernandes, E., Messerli, H., et al. (2013). *Tourism in Africa: Harnessing tourism for growth and improved livelihood*. World Bank/International Bank for Reconstruction and Development. Retrieved 2021, November 21, from www.worldbank.org/afr/tourism.

Cousins, B., & Kepe, T. (2004). Decentralisation when land and resource rights are deeply contested: A case study of the Mkambati eco-tourism project of South Africa. *The European Journal of Development Research, 16*(1), 41–54. https://doi.org/10.1080/09578810410001688725

Crotty, M. (1998). *The foundation of social research: Meaning and perspective in the research process*. Sage Publications.

de Haan, A. (2015). Inclusive growth: More than safety nets? *The European Journal of Development Research, 27*(4), 606–622. https://doi.org/10.1057/ejdr.2015.47

Dredge, D. (2019). A policy research agenda for tourism and development. In R. Sharpley and D. Harrison. (Eds.) *A research agenda for tourism and development* (pp. 35–52). Edward Elgar Publishing.

Edmeade, J., & Buzinde, C. N. (2021). Educators' personal resilience in the context of disasters triggered by natural hazards: The case of the United States Virgin Islands (USVI). *International Journal of Disaster Risk Reduction, 66*, 102571. https://doi.org/10.1016/j.ijdrr.2021.102571

Entman, R. M. (1993). Framing: Toward clarification of a fractured paradigm. *Journal of Communication, 43*(4), 51–58. https://doi.org/10.1111/j.1460-2466.1993.tb01304.x

Fairclough, N. (1992). *Discourse and social change*. Polity.

Fleischer, A., & Rivlin, J. (2009). More or better? Quantity and quality issues in tourism consumption. *Journal of Travel Research, 47*(3), 285–294. https://doi.org/10.1177/0047287508321204

Gimba, O. J., Seraj, M., & Ozdeser, H. (2021). What drives income inequality in sub-Saharan Africa and its sub-regions? An examination of long-run and short-run effects. *African Development Review, 33*(4), 729–741.

Gupta, J., & Vegelin, C. (2016). Sustainable development goals and inclusive development. *International Environmental Agreements: Politics, Law and Economics, 16*(3), 433–448. https://doi.org/10.1007/s10784-016-9323-z

Gupta, J., Pouw, N. R., & Ros-Tonen, M. A. (2015). Towards an elaborated theory of inclusive development. *The European Journal of Development Research, 27*(4), 541–559. https://doi.org/10.1057/ejdr.2015.30

Hahn, N. S. C. (2008). Neoliberal imperialism and pan-African resistance. *Journal of World-Systems Research, 8*(2), 142–178.

Hall, S. (1975). Introduction. In A. Smith (Ed.), *Paper voices: The Popular press and social change 1935–1965* (pp. 11–24). Chatto and Windus.

Higgins-Desbiolles, F. (2008). Justice tourism and alternative globalisation. *Journal of Sustainable Tourism, 16*(3), 345–364. https://doi.org/10.1080/09669580802154132

Higgins-Desbiolles, F. (2020). Socialising tourism for social and ecological justice after COVID- 19. *Tourism Geographies, 22*(3), 610–623. https://doi.org/10.1080/14616688.2020.1757748

Kimbu, A. N., & Ngoasong, M. Z. (2013). Centralised decentralisation of tourism development: A network perspective. *Annals of Tourism Research, 40*, 235–259. https://doi.org/10.1016/j.annals.2012.09.005

Kurasaki, K. S. (2000). Intercoder reliability for validating conclusions drawn from open-ended interview data. *Field Methods, 12*(3), 179–194. https://doi.org/10.1177/1525822X0001200301

Lincoln, Y. S., & Guba, E. G. (1986). But is it rigorous? Trustworthiness and authenticity in naturalistic evaluation. *New Directions for Program Evaluation, 1986*(30), 73–84. https://doi.org/10.1002/ev.1427

Lumumba, P. L. O. (2011). *Fighting corruption in Africa: The case for an African association of anti-corruption authorities* [Paper presentation]. Presented at the African anti-Corruption Conference, Bujumbura, Burundi.

Lumumba, P. L. O. (2019). Globalization challenges and the impact on the sustainable development of Africa. Convocation Lecture delivered at University of Technology, Nigeria. https://www.youtube.com/watch?v=kaxNK7liRxc

Manuel-Navarrete, D., Buzinde, C., & Swanson, T. (2021). Fostering horizontal knowledge co- production with Indigenous people by leveraging researchers' transdisciplinary intentions. *Ecology and Society, 26*(2), 1255-1271 https://doi.org/10.5751/ES-12265-260222

Minkov, M., Kaasa, A., & Welzel, C. (2021). Economic development and modernization in Africa homogenize national cultures. *Journal of Cross-Cultural Psychology, 52*(8-9), 801-821. https://doi.org/10.1177/00220221211035495Moscardo, G. (2011). Exploring social representations of tourism planning: Issues for governance. *Journal of Sustainable Tourism, 19*(4-5), 423–436. https://doi.org/10.1080/09669582.2011.558625

Mowforth, M., & Munt, I. (2016). *Tourism and sustainability* (4th ed.). Routledge.

Nketsia, N. K. (2013). *African culture in governance and development: The Ghana paradigm*. Ghana Universities Press.

Okello, M., & Novelli, M. (2014). Tourism in the East African Community [EAC]: Challenges, opportunities, and ways forward. *Tourism and Hospitality Research, 14*(1–2), 53–66. https://doi.org/10.1177/1467358414529580

Patel, L. (2016). *Decolonizing educational research: From ownership to answerability*. New York.

Rauniyar, G., & Kanbur, R. (2010). Inclusive growth and inclusive development: A review and synthesis of Asian Development Bank literature. *Journal of the Asia Pacific Economy*, *15*(4), 455–469. https://doi.org/10.1080/13547860.2010.517680

Rawls, J. (1971). *A theory of justice*. Harvard University Press.

Republic of Kenya, Ministry of Tourism & Wildlife, State Department for Tourism. (2020). *Revised national tourism policy 2020 on enhancing resilience and sustainable tourism in Kenya*. Nairobi, Kenya: Republic of Kenya. Retrieved March 3, 2021, from https://www.tourism.go.ke/wp-content/uploads/2020/09/28th-August-National-Tourism-Policy-Review.pdf.

Republic of Rwanda, Ministry of Trade and Industry. (2009). *Rwanda tourism policy*. Kigali, Rwanda: Republic of Rwanda. Retrieved on March 3, 2021, from https://www.minicom.gov.rw/fileadmin/user_upload/Minicom/Publications/Policies/Tourism_Policy_Cleaned_.pdf.

Schmelzer, M. (2015). The growth paradigm: History, hegemony, and the contested making of economic growthmanship. *Ecological Economics*, *118*, 262–271. https://doi.org/10.1016/j.ecolecon.2015.07.029

Schwandt, T. A. (1997). *Qualitative inquiry: A dictionary of terms*. Sage Publications, Inc.

Sharpley, R. (2020). Tourism, sustainable development and the theoretical divide: 20 years on. *Journal of Sustainable Tourism*, *28*(11), 1932–1946. https://doi.org/10.1080/09669582.2020.1779732

Stevenson, N., Airey, D., & Miller, G. (2008). Tourism policy making: The policymakers' perspectives. *Annals of Tourism Research*, *35*(3), 732–750. https://doi.org/10.1016/j.annals.2008.05.002

Stone, L. S., & Nyaupane, G. P. (2020). Local residents' pride, tourists' playground: The misrepresentation and exclusion of local residents in tourism. *Current Issues in Tourism*, *23*(11), 1426–1442. https://doi.org/10.1080/13683500.2019.1615870

Stumpf, T. S., & Swanger, N. (2015). Tourism involvement-conformance theory: A grounded theory concerning the latent consequences of sustainable tourism policy shifts. *Journal of Sustainable Tourism*, *23*(4), 618–637. https://doi.org/10.1080/09669582.2014.959967

Sze, J., & London, J. K. (2008). Environmental justice at the crossroads. *Sociology Compass*, *2*(4), 1331–1354. https://doi.org/10.1111/j.1751-9020.2008.00131.x

World Tourism Organization (UNWTO), the International Trade Centre (ITC), and the Enhanced Integrated Framework (EIF). (2017, July). Tourism for *sustainable development in the least developed countries*: Leveraging *resources for sustainable tourism with the enhanced integrated framework* (2017). UNWTO Elibrary. https://doi.org/10.18111/9789284418848

Zautra, A., Hall, J., & Murray, K. (2008). Community development and community resilience: An integrative approach. *Community Development*, *39*(3), 130–147. https://doi.org/10.1080/15575330809489673

Zhang, J. (2021). The effects of tourism on income inequality: A meta-analysis of econometrics studies. *Journal of Hospitality and Tourism Management*, *48*, 312–321. https://doi.org/10.1016/j.jhtm.2021.07.009

Tourism policy, spatial justice and COVID-19: lessons from a tourist-historic city

Brendan Paddison and Jenny Hall

ABSTRACT

In many historic and post-industrial cities, tourism is often positioned as an important component for urban regeneration. Yet, the promise of sustainability and social transformation is often empty as policymakers focus on sustaining tourism over supporting greater social, economic, and environmental sustainability. Furthermore, the pandemic has drawn attention to the unsustainable nature of the neoliberal model of tourism engagement. Due to the paucity of research exploring spatial injustice in urban tourism, this study examines the impact policymaking and governance structures have on urban destinations and the potential inequalities this creates. Drawing on Edward Soja's approach to Lefebvre's The Right to the City (1968), this research explores how lessons learned during the pandemic in the tourist-historic city of York, UK, could transform tourism in historic urban spaces. Taking an interpretive case study approach, semi-structured interviews were conducted with leading stakeholders to understand the spatial dimensions of the lived experiences of policymaking. The hopeful signs emerging from York's response to the pandemic demonstrate how communities can reclaim voice to build sustainable and purposeful models of engagement. This paper contributes to our understanding by demonstrating the transformative potential that future policymaking could have for reducing the negative impacts of tourism.

Introduction

Tourism has become the panacea for renewal and regeneration in many post-industrial cities as the saviour of business economies and an instrument for poverty reduction. Yet, the promise of sustainability and social transformation is often empty as policymakers have concentrated on sustaining tourism at the expense of greater social, cultural, economic, and environmental sustainability (Aall, 2014; Higgins-Desbiolles, 2020). Consequently, concerns regarding carrying capacity, environmental degradation, and the availability and quality of tourism work have drawn attention to the unsustainable nature of the current industrial, neoliberal models of tourism. In addition, the pandemic has intensified social and economic inequalities (Jamal & Higham, 2021) and heightened issues of urban vulnerability (Sharifi & Khavarian-Garmsir, 2020), particularly for those destinations where tourism and hospitality have a major economic role.

For White (2019, p. 290), policies that support tourism can be harmful socially and ecologically, whereby the mantra of twenty-first-century global capitalism is "consume, be silent and die," for which tourism is front and centre. As Joppe (2018, p. 201) suggests, "the term 'policy' is very fuzzy in tourism because of its social, behavioural and diverse nature." Policies that govern tourism tend to be developed for other purposes such as transport, employment, planning, health and are authored by policymakers with little care or investment in tourism. This can create injustices of all kinds (Joppe, 2018). Public policy in tourism at all geographical levels has seen uneven failures and successes in meaningfully engaging stakeholders in decision-making processes and in addressing sustainability (Jenkins, 2014). Consequently, Joppe (2018) calls for research that evaluates the factors involved in formulating policy processes that engage communities in meaningful decision-making that can lead to positive change (Hall, 2011; Joppe, 2018). Whilst previous research has explored spatial injustice in historic cities (Jover & Dıaz-Parra, 2020), this study offers a novel contribution examining the experiences of policymaking and policymakers.

As destinations emerge from the pandemic, striving to achieve a balance between development and sustainability is at the forefront of tourism debates (Butcher, 2021; Rastegar et al., 2021). If concerns regarding the economic, social, cultural and environmental impacts of tourism are to be addressed, the unlocking of the industry is an opportunity to think radically about how tourism policy can be reimagined for a *just* tourism future (Rastegar et al., 2021). It is within this context that this study aims to critically appreciate how injustice is socially and historically constructed by exploring the effects of globalisation, governance and policymaking in the tourist-historic city of York, United Kingdom (UK). Tourist-Historic Cities (Ashworth & Tunbridge, 2000; Fainstein & Judd, 1999) are characterised by urban structures, natural resources, architecture and culture, where the historic core has become the object of tourist consumption. However, the creation of place-based heritage products can threaten the sustainability of these spaces (Timothy, 2011). Due to York's historic status as an UNESCO World Heritage site and as one of the most visited UK city destinations (Visit Britain, 2022), York was considered to "yield the best data" (Yin, 2009, p. 91) concerning spatial (in)justice. The spatial (in)justices evident in the city manifest in exclusionary decision-making processes managed by a tourism policymaking elite, the corrosive impact of consumer capitalism on economic policymaking, and the impact this has had on public planning and stakeholder engagement. COVID-19 was a moment to reassess and evaluate the emerging themes concerning the governance of tourism through collaborative action and a reclaiming of values.

This study draws upon Edward Soja's (2008, p. 2) geographical and spatial treatment of Lefebvre's The Right to the City. Soja advocates that spatial injustice is not just about outcomes but also "a way of looking at justice from a critical spatial perspective." The emphases here are on the spatiality of justice and injustice, not only in the city, but at all geographical levels, from the local to the global, that impact on the agency of host communities to appropriate urban space (Soja, 2010). Soja's conceptualisation offers an approach to realise solutions to spatial exclusion at the intersection of "race, class, gender and other, often closely associated, forms of human inequality and oppression" (Soja, 1996, p. 22). This research builds on Soja's call to identify and understand "the underlying processes producing unjust geographies" (Soja, 2008, p. 4) within tourism policymaking and governance. Subsequently, our aim is twofold. First, we draw on Soja's spatial approach to justice to examine how public policy and governance processes create inequalities in the tourist-historic city. Second, we identify how lessons learned during the COVID-19 pandemic show promise for addressing spatially unjust policy processes in historic city spaces.

Literature review

Lefebvre's the right to the city

In 2016, at the United Nations Habitat III conference in Quito, Ecuador, a New Urban Agenda was launched with Lefebvre's The Right to The City (1968) at its core. The policy "recalls the need to maximise the participation of city dwellers in local governance to avoid further marginalization" (SDG Knowledge Hub, 2016). Lefebvre's radical vision of city life managed and appropriated by its citizens, beyond the control of the state, is a call to action to free policy and governance processes from capitalism (Purcell, 2014). Lefebvre offers a vision of revolutionary change that imagines a world beyond capitalism, state control and neoliberal consumerism (Lefebvre, 2003). He advocated for a more holistic understanding of the value of city social life that is not reducible to economic imperatives alone by embracing the diversity of human experiences in lived spaces. Theorising The Right to The City, Lefebvre articulates how rights are borne of political struggle performed by mobilised and active citizens reclaiming power, symbolising the "rise to a renewal of political life" (Lefebvre, 1990, p. 32, in Purcell, 2014). Lefebvre (1968) recognised that property rights alienate citizens from urban space that privileges centres of consumption through the process of regeneration (Lefebvre et al., 2000). Turning to consumerism to help regenerate de-industrialised cities has resulted in "the late-capitalist city [existing] in a state of living-death" (Raymen, 2019, p. 87) that is built upon an asocial environment of consumption that supports the flow of capital at the expense of the heterogenous and organic social life of the city (Atkinson, 2017; Cederström & Fleming, 2012). History and sociality have been erased from post-industrial centres by keeping space to its specificity (de Jong & Schuilenburg, 2006), whereby an artificial version of social and cultural activity is supplanted into a city's non-places (Augé, 1995). To counter such loss, city life is re-created through the careful curation of spaces by public authorities through a process of selective appropriation (MacLeod, 2002). This leads to spatial exclusion and marginalisation of social practices (Raymen, 2019).

Such practices are at odds with Lefebvre's vision of city life. He argues that "the right to the city cannot be conceived of as a simple visiting right or as a return to traditional cities. It can only be formulated as a transformed and renewed right to urban life" that works to achieve greater diversity and counter exclusion (Lefebvre et al., 2000, p. 158). This renewal is conceptualised as "The right to the oeuvre [work], to participation and appropriation (clearly distinct from the right to property)," where citizens concretely produce lived city space (Lefebvre et al., 2000, p. 174). Appropriation is conceived in the imaginary of physiolocal and psychosocial city life through customs and enactments that have a use value free of the exchange value rooted in consumption. Yet, hyperregulation, private property rights and luxury private housing built on a culture of consumption have corroded diversity and destroyed what consumer capitalism needs the most—the organic innovative use of urban space (Cederström & Fleming, 2012). Thus, to understand how neoliberal urban policy and governance structures shapes, creates and appropriates city spaces requires a holistic approach for understanding how spatial injustice arises and impacts on the city ecology.

Although Lefebvre's (1974/1998) theory of space in urban society has received significant attention across the social sciences (Fainstein & Fainstein, 1982; Harvey, 2009; Massey, 1984; Soja, 1985, 2000), it is subject to criticism, particularly for privileging space to the exclusion of time, whereby the production of space is perceived as an end point, in contrast to being in constant flux in space-time (Unwin, 2000). In addition, Lefebvre's theory on the production of space was challenged as Universalist, silent on how space transforms into political action and achieves social change (Unwin, 2000). Although similar criticism has been levelled towards scholars such as Harvey and Soja, Unwin (2000) acknowledges that Soja's conceptual approach could offer practical solutions to intersectional forms of human inequality and injustice (Soja, 1996). Soja's spatial treatment of The Right to the City examines how policy and governance structures create spatial injustice through exclusionary processes that operate at material, psychological and social levels, to which we now turn.

Spatial injustice and unjust geographies

The theorisation of The Right to The City in the context of spatial justice has produced a wealth of academic interest and multiple interpretations (Attoh, 2011; Earle, 2017; Marcuse, 2009;

Mitchell, 2003). Notably, and pertinent to this paper, is Soja's (2008) advancement of spatial justice which insisted that achieving The Right to The City involves interrogating the processes and relations that produce injustice in cities to enable them to be re-ordered (Iveson, 2011). Soja (2008) asks how radical and progressive approaches to injustices in cities can be realised and argues that putting "spatial" in front of "justice" is crucial in theory and in practice (Soja, 2010). The emphases here are on the spatiality of justice and injustice, not just in the city, but at all geographical levels (Iveson, 2011; Soja, 2010).

For Soja (2008, p. 2), spatial injustice is "an intentional and focused emphasis on the spatial or geographical aspects of justice and injustice." Therefore, it is not just about outcomes but a lens for critically evaluating justice from a spatial perspective (Soja, 2008). The spatiality of justice is actively causal, with Soja (2008, p. 4) noting that:

> Spatial (in)justice can be seen as both outcome and process, as geographies or distributional patterns that are in themselves just/unjust and as the processes that produce these outcomes. It is relatively easy to discover examples of spatial injustice descriptively, but it is much more difficult to identify and understand the underlying processes producing unjust geographies.

It is to these underlying processes and the unjust geographies that public policymaking creates in the historic city that we will attend. Injustice manifests in powerful social interests of the various governing processes produced by the political organisation of space. Combined with institutionalised segregation of residential and public space, this produces locational spatial discrimination (Soja, 2010). The normal workings of the urban system create uneven (under)development in lasting structures of privilege and advantage, producing unjust geographies in cities (Soja, 2010). Accordingly, Jamal and Camargo (2014, p. 12) argued for just destinations "whose tourism planning, policymaking and practices enable fair treatment of its environmental and social-cultural resources (tangible and intangible), and facilitate the well-being of place, people and pasts." They advocate for a distributive justice that shares benefits amongst inhabitants of the host destination. Although not explicit, this geographical approach highlights the spatial injustice that arises through economically focused tourism policy that perpetuates the colonization and commodification of host-destinations (Barton & Leonard, 2010; Devine & Ojeda, 2017). Poor distribution of economic benefits, exploitation of community cultural identity and environmental degradation mark tourism as spatially unjust and draw attention to the relationship between tourism development and spatial justice (Leśniewska-Napierała et al., 2019). As such, scholars have called for spatial justice to be recognised as a goal of sustainable tourism development (Leśniewska-Napierała et al., 2019) that promotes a balanced approach to cooperation between stakeholders. The purpose is to achieve democratic justice that upholds the rights and power of local people to make decisions concerning ecological, heritage and cultural preservation, along with economic efficiency that balances the negative impacts of tourism (Higgins-Desbiolles, 2018; Ramón-Hidalgo & Harris, 2018).

Despite this, spatial justice approaches have seen little attention in city tourism scholarship aside from Diaz-Parra and Jover (2021) research in Seville, Spain. Diaz-Parra and Jover (2021) draw on Lefebvre's The Right to The City to identify the spatial injustices of over tourism in Southern European cities. Yet, their research does not consider the factors in policymaking that can lead to spatial injustice in historic cities, nor the lived experiences of policymakers and the spatial impact on urban life. The absence of local social life from the city is in part attributable to public policy that favours neoliberal economics over social life. Moreover, the absence of an engaged local public creates an artificial and potentially less satisfying experience for visitors (Diaz-Parra & Jover, 2021).

The re-theorisation of The Right to The City in the context of spatial justice has produced a wealth of academic interest and multiple interpretations (Attoh, 2011; Earle, 2017; Marcuse, 2009; Mitchell, 2003). In the last two decades, numerous public organisations have articulated The Right to The City as part of the human rights agenda. Global policymakers such as UNESCO

(2006) and UN-HABITAT (2010) have interpreted The Right to The City through policy. However, scholars have criticised recent applications of Lefebvre's The Right to the City as a solution for every kind of urban spatial social cause (see Purcell, 2014). Indeed, Purcell (2014) argues that we need distinctive and diverse applications of The Right to The City that are specific and politically transparent. Yet, on one point, academics, activists and policymakers all share the view that The Right to The City is a struggle "to augment the rights of urban inhabitants against the property rights of owners" (Purcell, 2014, p. 142) to achieve spatial justice for city inhabitants.

Spatial justice, governance and stakeholder engagement

For Soja (2010), the political organization of space is a powerful source of spatial injustice. Astleithner and Hamedinger (2003) describe the shift from *government* to *governance*, and the adoption of a New Public Management (NPM) approach, as the political restructuring of cities. Closely aligned with a neoliberal ideology, with a focus on reduced state intervention, marketisation and privatisation (Beaumont & Dredge, 2010; Hall, 2011), NPM saw local government adopt characteristics typically associated with the private sector, with an increased focus on efficiency and market-driven policy and strategy (Paddison & Walmsley, 2018). The reorientation of local government has not necessarily improved democratic practices or transparency in decision-making as initially intended (Beaumont & Dredge, 2010; Dredge, Ford & Whitford, 2011; Moscardo, 2011). Indeed, many argue that it has resulted in a "closing up" of the policy process (Bramwell & Lane, 2011; Jamal & Watt, 2011), which has largely focused on the growth of tourism with success measured in the number of jobs created, the multiplier effect and the level of inward investment received (Paddison & Walmsley, 2018). Whilst there is an acknowledgement of the range of actors involved in urban governance and decision-making (Le Feuvre et al., 2016), there are challenges in engaging with a wide variety of interests. This can potentially further cultivate conflict and power imbalances (Bornhorst et al., 2010; Coles & Church, 2007). These challenges are potentially intensified in the context of heritage and historic urban destinations due to their complexity (Paddison & Biggins, 2017; Waterton & Watson, 2013). In this context, tourism becomes appropriated by corporate interests, with little, if any, attention given to the environmental, ethical or social implications of tourism (Higgins-Desbiolles et al., 2019). The democratisation of tourism, driven by social norms associated with the commercial imperative of the "tourist experience", has resulted in socio-economic spatial injustice. Tourism-focused policymaking is killing the thing we profess to value and has led to a decline in the well-being of cities, with Venice and Barcelona notable examples.

There is a paucity of research that examines how governance hierarchies and their actors' impact host communities (Beaumont & Dredge, 2010). Subsequently, governance structures and processes need to be examined, with a particular focus on who is involved and who is excluded from the decision-making process (Hall, 2000; Kimbu & Ngoasong, 2013) and the injustices that this creates. The structures and approaches to stakeholder engagement are critical in facilitating a representative and balanced perspective of the destination community (Dredge, 2001). Creating opportunities for community stakeholders to actively participate in policymaking and development planning in a way that is sensitive to long term needs and impacts is important (Blackstock, 2005; Paddison & Walmsley, 2018). Dredge (2006) calls for research that explores the application of theory and its development in the context of community engagement in tourism planning. A sustainable recovery from the pandemic requires solutions based on local interests which bring positive social, economic, environmental and ethical change (Higgins-Desbiolles, 2020). For a burgeoning number of scholars (Harvey, 2003, 2009; Jamal & Higham, 2021; Marcuse, 2009; MacLeod, 2002), promoting justice and equity is at the forefront of the change needed, no less in historic tourist cities such as York. Agyeman and McLaren (2017, p. 25) advocate a bridge between socio-spatial and socio-economic divides through the notion of sharing and the right to appropriate or remake city spaces. The concern here is to "achieve inclusion through prioritizing historically neglected neighbourhoods for new high-quality shared infrastructure, and public services, buildings, and spaces, through attending to patterns of design and management to create culturally inclusive spaces" (Agyeman & McLaren, 2017, p. 25).

The tensions which emerged regarding sustainable tourism development and post-COVID recovery provide a rationale for interpreting governance and policymaking within historic urban destinations. The nature of the planning and management of the tourist-historic city has been well articulated in the tourism literature (Ashworth & Tunbridge, 2000; Page, 1995). Yet, there is a lack of research regarding how public policy and governance structures perpetuate spatial injustice in tourist-historic cities through exclusionary and discriminatory processes, where those who live in historic cities become the possession of tourism (White, 2019). We trouble how public policy promulgates hyper-consumerism in historic cities by highlighting the socio-cultural harm this produces. Urban human life has fallen victim to the environmental injustices of tourism policy that promote hyper-consumption and capitalism. It is within this context that this study seeks to critically appreciate how spatial injustice is socially and historically constructed in the tourist-historic city of York, UK.

Following the decline of the railway and chocolate industries, tourism is now York's leading economic sector. Prior to the pandemic, the city received an estimated 8 million visitors a year. The sector supports approximately 24,000 jobs and contributes £608 million to the local economy (Visit York, 2022). As a major heritage visitor destination, York's popularity is afforded to its rich and diverse history, with notable attractions including York Minster, the National Railway Museum, and the Jorvik Viking Centre (Visit York, 2022). The historic centre of York is unique due to its geographically compact nature, its internationally important Minster, well preserved ancient Viking and Roman heritage and archaeology and mediaeval quarter. The city supports a thriving independent business community, with 65% of all businesses occupying space in the city centre (City of York Council, 2022). The development and management of tourism in York have been widely documented (Augustyn & Knowles, 2000; Croft, 2018; Meethan, 1997; Mordue, 2005; Paddison & Biggins, 2017; Paddison & Walmsley, 2018). However, as the city emerges from the pandemic, it is timely to critically assess the spatial injustices embedded in the city's governance and policymaking structures and the implications this has for its resident communities. We respond to Hall's (2007) call for research that examines how the neoliberal discourse has affected policymaking and implementation in tourism.

Methodology

This research adopts an interpretative methodological approach that places reliance on people's interpretations of situations and behaviours (Goodson & Phillimore, 2004). The interpretive paradigm "looks for culturally derived and historically situated interpretations of the social life-world" (Crotty, 1998, p. 67). As such, the researcher has an important role here, seeking to understand the context and then make an interpretation of what is found. The researcher must look to understand the multiple interpretations offered so that they can build a holistic understanding of the phenomena being studied (Veal, 2011). Interpretivism rejects the positive notion that there is one objective truth and acknowledges the existence of multiple realities which can only be understood from the perspective of those involved (Goodson & Phillimore, 2004). Given that this study focuses on exploring social, cultural, and environmental injustice produced through neoliberal tourism policy and governance, an interpretative methodological approach was considered appropriate.

A case study approach offers a method of inquiry to explore, analyse and interpret a single or a range of instances of the same phenomenon (Yin, 2009). The growth of tourist-historic cities and the tensions which have emerged between tourism development and the local community

Table 1. Study participants.

Participant	Sector
Pb1	Public sector
Pb2	Public sector
Pb3	Public sector, charitable
Pb4	Public sector, charitable
Pb5	Public sector
Pr1	Private sector
Pr2	Private sector
Pr3	Private sector
Pr4	Private sector
Pr5	Private sector
Pr6	Private sector
R1	Resident
R2	Resident
R3	Resident

provide a rationale for interpreting the governance of tourism within urban destinations. As a result, cities which demonstrated characteristics of Fainstein and Judd (1999) tourist-historic city were identified as possible case study destinations and thus enabled the narrowing of the selection process. Furthermore, an element of pragmatism also determined the selection of cases. Specifically, access to informants and documentary sources sufficient to provide rich data for the study influenced the choice of case study. Indeed, Yin (2009) advocates this pragmatic approach in case study selection where he argues that the researcher should "choose the case that is likely, all other things being equal, to yield the best data" (Yin, 2009, p. 91). Consequently, York was selected as an appropriate case study.

Following a review of the literature, in-depth, interpretive semi-structured interviews were conducted. A purposive sample of 14 key informants (Table 1) including business leaders, public officials, elected council members and residents were interviewed between April 2021 and March 2022. Potential informants were contacted on a personal basis and identified through either holding a significant role in tourism policy and governance or as a member of the community. The sample size was small and targeted to enable in-depth discussion held through a secure online platform (Microsoft Teams). Both researchers shared a similar background, experience and language of the participants. This enabled acceptance as insiders which helped to build rapport. The identities of the research participants have been anonymised because it is recognised that those holding senior leadership roles (public or private sector) could be identifiable. Respondents have been classified as per their sector, with Pb denoting Public Sector, Pr representing Private Sector and R signifying residents. There was no fixed or predetermined sample size. Interviews were conducted until it was felt the study provided a robust insight and when data saturation was achieved with no new insights emerging.

A semi-structured (Saunders, 2021) interview schedule was developed around key themes that emerged from the literature. This included policy and governance, sustainability, perceptions and experiences of policymaking and the impact this has had on urban life, how policy and governance have changed following COVID-19 and the spatial justice implications of this on building a sustainable tourism future for the city. The topic guide was just that and conversations were facilitated to cover the ground most pertinent to the interviewee. Interviews lasted approximately 60 minutes, were transcribed, and later transferred to the NVivo 12 software package for analysis. Descriptive themed analysis produced 18 codes and sub-codes that were grouped under themes such as justice, injustice, social change, sustainability, and governance. Braun and Clarke (2006) six-phase approach for thematic analysis was adopted (Table 2), which provides rigour and validity in qualitative data analysis enabling crucial links to be made between theory and the data collected. Ethical approval was sought and all interviewees provided written consent to participate in the research.

Table 2. Thematic analysis framework.

Phase	Description of the process
1. Familiarising yourself with your data	Transcribing data, reading and re-reading the data, noting down initial ideas.
2. Generating initial codes	Coding interesting features of the data in a systematic manner across the entire data set, collating data relevant to each code.
3. Searching for themes	Collating codes into potential themes, gathering all data relevant to each potential theme.
4. Reviewing themes	Checking if the themes work in relation to the coded extracts (Phase 1) and the entire data set (Phase 2), generating a thematic "map" of the analysis.
5. Defining and naming themes	On-going analysis to refine the specifics of each theme, and the overall story the analysis tells, generating clear definitions and names for each theme.
6. Producing the report	Selection of vivid, compelling extract examples, final analysis of selected extracts, relating back of the analysis to the research question and literature, producing a scholarly report of the analysis.

Source: Adapted from Braun and Clarke (2006).

Findings and discussion

Several themes emerged from the analysis and provide a structure to the discussion which follows. First, we examine the spatial injustices apparent in economic policy and local development. Second, we review how the underlying governance processes and spatial discrimination impact stakeholder engagement in the decision-making process. Finally, we reflect on the lessons learned from COVID-19 and the hopeful signs emerging for the future of tourism in the city.

Spatial justice: economic policy and local development

Economic inequalities and social division associated with neoliberal globalization create spatial injustice, particularly within a city context (Soja, 2008). Our research found that when tourism is an expression of neoliberal policy, as evident in York, it is subject to the same negative forces of disinvestment that characterise broader capitalism. In the case of York, the decline of the traditional manufacturing industries during the 1980s and 1990s obligated the local authority *"to find suitable economic growth alternatives to help create jobs and stimulate investment"* (Pb2). This primarily occurred through the promotion of York as an international tourist destination (Paddison & Walmsley, 2018). York's historic streets became a centre of consumption, with economic policy embracing consumerism to help regenerate de-industrialised York (Mordue, 2005). However, as concerns regarding the quality of employment in tourism grew, rather than imagine a world beyond capitalism and neoliberal consumerism (Lefebvre, 2003) that might support quality employment opportunities, economic policy instead focused on encouraging *"high value jobs"* (Pr3) in sectors such as bioscience and the digital industries. Although tourism employment in York represents approximately 14% of the city's economy (City of York Council, 2022), as one respondent noted, the sector is now considered *"a subset of economic growth, perceived as those crappy jobs"* (Pb1). Despite recognition that *"there's lots that we [the local authority] need to do to improve pay, quality, skills and career progression in tourism"* (Pb2), the local authority feels *"powerless"* (Pb2) due to a *"lack of funding, resource and political willingness"* (Pb5) to alleviate the socio-economic injustices produced through low pay, precarious working conditions and poor progression opportunities in tourism (Sun et al., 2022). This sense of powerlessness was particularly felt by policymakers in the city, with one respondent commenting that *"tourism in York is politically marginalised"* (Pb1) and therefore subject to discriminatory policymaking where tourism is *"not taken as seriously as other economic development areas"* (R1). This underlying policy process of privileging "high value" sectors over tourism has created uneven and ad-hoc geographical approaches to planning and development in the city leading to spatial injustice associated with reduced social mobility and economic inequality (Lee, 2009). York's approach, whilst not uncommon, is at odds with Lefebvre's vision of city life. Policymaking and governance processes are entangled in capitalism creating socio-economic spatial discrimination that results in the marginalization of local communities.

Tourism is not part of a national public investment agenda, particularly with regards to creating high value jobs (Font et al., 2019). This, therefore, limits the extent to which public authorities can address workforce inequalities at a local level, as evident in York. However, even though recent reductions and pressures on local public finances were well understood, many respondents felt that the local authority should provide the *"strategic leadership"* (R3) in addressing economic inequalities and a lack of investment in the sector. For example, one business leader commented that: *"one of the weaknesses with the local authority is that it's good at talking about strategy, but not actually good at delivering on anything"* (Pr2). This is compounded by the development and delivery of policy being perceived as a *"closed and highly politicised process"* (Pr1), leading to significant spatial injustices resulting from economic policy that fails to recognise and address key social, cultural and environmental needs. The lack of collaboration between policy and governing organisations exacerbates spatial injustice concerning "business growth" (Pr5), decision-making and compounds the precarity of employees working in the sector. The national drive, founded on capitalism, to stimulate high growth businesses and jobs, particularly in the science and digital sectors, marginalises tourism. As one local resident claimed, *"there will not be real change until there's a cash injection into local services which would then allow [local authority] leaders to take tourism seriously"* (R1). Consequently, local people remain in low paid jobs, with poor working conditions and limited career progression opportunities. This perpetuates harmful forms of tourism and community exploitation that impacts on sustainability. Investment in high value sectors creates geographic marginalisation, resulting in locational exclusivity in high value areas and the associated issues of gentrification (see Diaz-Parra & Jover, 2021).

Our research found that respondents felt York lacked a *"cohesive local plan"* (Pr5), including a strategy for the development of the city's physical environment, with an ad-hoc approach to redeveloping city spaces and apathy to actively engage host communities in the decision-making process. This was attributed to the *"short-term lifespan of political office"* (R2), leading to a reluctance by politicians to *"make big decisions"* (Pr5) and invest in the creation of a long-term strategy for the city. Soja's vision of the city is based on realising solutions to spatial exclusion through enabling communities to tackle underlying spatial discriminatory processes that govern how city space is appropriated. Our research has shown that, in the case of York, decision-making in tourism is controlled by a managerial elite, which has limited representation from across the city. As a result, it appears that major building and investment projects *"are happening in isolation"* (Pr6), with *"local people, potential investors and developers not able to see what is going on or the opportunities that exist"* (R1). This has created significant frustration among our respondents, with one interviewee expressing: *"what's the point of government if you're not going to have a plan for your city? Everything is done in a piecemeal way and there's no thought to how it all works together"* (Pr5). A lack of coordinated partnership work perpetuates spatial injustice where local stakeholders are largely powerless in the decision-making process and are subject to locational discrimination produced by ad-hoc planning controls. Indeed, respondents felt that this lack of coordination has led to underinvestment and degradation of York's public spaces, with one respondent commenting how *"the quality of our streetscape is just shocking"* (Pr1), admitting that *"millions were required to transform the main shopping streets into something that meets European standards"* (Pr1). The social, cultural, and environmental fabric and health of York is degraded by a lack of coordination, leadership and power to influence decision-making and investment. This reduces York's attractiveness and effectiveness as a space for social and cultural interaction pertinent to the needs of its residents (Jamal & Camargo, 2014). Investment is not just about fiscal benefits but should also be concerned with investing in public pride, social, cultural, and environmental health by shifting perspectives about how spaces are appropriated. Yet, there is little political incentive to develop collaborative approaches to decision-making when public policymaking is underpinned by a neoliberal free-market philosophy. To address spatial degradation and discrimination, political recognition of tourism's economic place within the city, combined with planning solutions that support stakeholder decision-making, are needed (Jamal & Camargo, 2014).

Spatial justice: governance and stakeholder engagement

The political organization of space is a powerful source of spatial injustice (Soja, 2010, p. 3). In the wake of New Public Management policies (Paddison & Walmsley, 2018) to devolve public sector control and responsibility for tourism and economic policy to the private sector, the City of York Council established a destination management organisation (DMO) "Make it York," with its tourism division branded as "Visit York." Structurally, the organisation is founded on an economic principal to generate its own income. Consequently, as described by one respondent, the local authority's approach to *"devolving responsibilities"* (Pr2) like tourism to external organisations has placed a *"commercial imperative on those organisations"* (Pb4); a situation which is also true of the city's Museums Trust. As a result, even though these organisations are charitable, they are duty bound to make a profit. This *"has an impact on how you work with your communities"* (Pb3) that leads to spatial injustice. As noted, several scholars have argued that the reorientation of local government has had a negative impact on democratic practices (Bramwell & Lane, 2011; Jamal & Watt, 2011; Moscardo, 2011). This was evident in York, in part due to the organisational structure and funding model adopted by the DMO. A proportion of the DMO's income is derived from membership subsidies. Beaumont and Dredge (2010) remind us that the adoption of a membership model can prevent the involvement of all relevant stakeholders in destination management. This can result in the exclusion of key stakeholders in the policy process (Hall & Jenkins, 1995), leading to corporate interests dominating decision-making and policy implementation. Such spatial injustices were noted by one respondent:

> I'm aware that there's a number of individuals and businesses in the city who don't pay for membership but should have a seat at the table. Destination management should include all those who want to be involved, not just those who can afford to pay a fee. (Pr4)

The result is a top-down approach, with highly targeted consultation that excludes several key stakeholder groups. As one responded noted, *"Make it York needs to take its destination management much more seriously and that involves [a] coordinated approach"* (Pr1). This spatial injustice compounds host communities' sense of powerlessness and passivity when it comes to involvement in local decision-making and appropriating city spaces. This has resulted in an acknowledgement that *"York people feel quite excluded from the city centre"* (Pb1), creating an absence of social life and cultural expression in the city. When public policy is founded on neoliberal capitalist consumerism, localism or the social and cultural health of the city are not valued, managed, or nurtured because they are not perceived to achieve a fiscal return (White, 2019). Spatial injustice is manifest through public officials looking beyond the city for creative solutions, inward investment and expertise that exacerbates the problem of how to bring *"localism back into the city"* (Pr1). For example, to ensure financial sustainability and to address the loss of "the social" from city life, public authorities and DMO's re-animate city spaces through the careful curation of licensing buskers, Christmas markets and carol singers. Only social activities sanctioned by public authorities are privileged with spatial legitimacy and permitted to inhabit city spaces. This leads to spatial exclusion and marginalisation of unauthorised social practices. To address this, it was felt that the public authorities *"need to [connect with] stakeholders in a genuine open forum to develop a strategy that allows the city to move forward as a whole"* (Pr2) to achieve *just* appropriative processes that reanimate organic city social life.

The establishment of the Tourism Advisory Board (TAB) during the pandemic was, in part, an attempt to broaden consultation and engagement in tourism decision-making. As one public official noted, *"prior to the pandemic we'd never had such a varied group come together and be more pragmatic"* (Pb2). They described the TAB as *"far more inclusive"* and *"the voice of the sector"* (Pb2). The wider implication of the establishment of the TAB is described by another respondent: *"Make it York always have ideas that don't necessarily speak for the whole city"* (Pb1) and there is a need for *"checks and balances"* (Pb1) that enables the sector to actively lead decision-making. This approach to destination management is one that seeks cooperation between stakeholder groups and goes someway to help achieve equity, justice and enhance democratic practice (Leśniewska-Napierała et al., 2019). Whilst it was apparent that further stakeholder representation could be enhanced, the advisory board offers an opportunity to re-think traditional neoliberal economic approaches to destination management (Higgins-Desbiolles, 2018; Ramón-Hidalgo & Harris, 2018). COVID-19 accelerated these processes of appropriation to which we now turn.

Spatial justice and COVID-19: hopeful signs for appropriating the tourist-historic city

Notwithstanding the impact of COVID-19 on the tourism industry, our research found that in the case of York, it also created a unique moment that galvanised communities in active participation and management of the challenges that the tourism sector faced. As one responded noted, *"because of COVID the city maximised its small size to get the right stakeholders together to have a strategy for what we were communicating about the city"* (Pb4). The local authority acted as a facilitator in managing this process bringing together various stakeholders in a *"unique moment of good coordination"* (Pr2) where *"being involved in these communications meant that we were persuaded to open, even though it wasn't economically viable for us"* (Pr3). The pandemic resulted in policymaking organisations, such as the local authority and the DMO, collaborating with businesses to appropriate spaces through acting on *"feedback from the business community"* (Pr5). As one public official noted:

> We listened and responded to what businesses wanted. For example, at a time when hospitality couldn't have people indoors, a lot of them wanted to create outdoor seating areas. We worked with local businesses and the council to enable this to happen in a timely and effective way. This would never have happened here [in York] without that kind of working relationship. (Pb5)

Through carefully considered processes of appropriation for managing the reopening of the city in August 2020, York outperformed its projected visitor numbers with some attractions, such as the city's art gallery, receiving 60% more visitors than previous years. The local authority facilitated a move towards more collaborative processes and meaningful decision-making. During the pandemic *"partners came together and worked collaboratively in a way that hasn't been done before"* (Pr5) that began to address spatial discrimination and distribution that created socio-economic benefits for local people. If this approach was adopted permanently it would facilitate *"genuine stakeholder engagement in shaping the future of the city"* (Pr2). However, as one responded argued, such an approach would require the *"willingness on the part of the local authority to step away from politics in order to make a process like this happen more permanently"* (Pr2).

Our analysis identified that, during the pandemic, there was a recognition that what makes a city vibrant is its socio-cultural life and its ecological health is predicated on collaboration and the organic appropriation of city spaces. Neoliberal growth economics cannot be the foundation for what tourism *is*. Indeed, policymakers in York understood the importance of this and that *cultural product development needs to be "something that's done as an extension of people's natural behaviour that they like to do instinctively"* (Pr2). Implementing it, however, is another matter, as one respondent commented:

> The pandemic has opened [a] vista of what is possible in terms of engaging with communities and allowing different voices to be heard. That moment might not last long, of course, but it is a moment to act! (Pb1)

Seizing the moment is a key challenge and calls for appropriative approaches that facilitate stakeholder engagement to reanimate the socio-cultural life of the city (Leśniewska-Napierała et al., 2019; Purcell, 2014). Reclaiming a values-based approach for developing and delivering cultural products offers a sustainable and appropriative approach for reshaping tourism (Moscardo, 2018). Public officials described how volunteering and collaborative action demonstrated through the pandemic offered solutions and accelerated collective community efforts to reclaim amenities, such as pubs and libraries, and when *"owned by residents, spaces that people feel they can take charge of and develop cultural products, show great potential"* (Pb1). There is a shift in recognition that *"public spaces and venues need to facilitate more mixed and exciting uses to increase levels of footfall"* (Pb5) to tackle unjust distributional patterns that privilege certain types of cultural product and economies over others to facilitate local economic and socio-cultural appropriation.

If York was to recognise the potential of engaging with its local communities to reanimate the cities social life, tourists would have a greater opportunity and *"sense of connecting with a distinctive community and environment"* (Pb1). The disconnect between a focus on economic growth and what is needed in York is starkly illustrated through the ways in which the marginalisation of tourism as an industry in the city leads to social, cultural and environmental disinvestment and degradation. To reclaim a socially and culturally vibrant ecology, a shift is needed to stem the economic leakage associated with the forces of global commodification that result in the loss of city social and cultural life (Soja, 2010). During the pandemic, the local authority established processes of governing through *"consultations that have become better and more important"* (Pb2). Such an approach tackles spatial discrimination and creates a balanced appropriative ecology that leads to citizens reclaiming city spaces, promoting cultural expression and engagement. Indeed, as one responded noted: *"I think the future's bright and we [have] to shape [tourism] and plan for it rather than letting it just happen"* (Pr5).

Conclusion

Cities are central to the study of contemporary global tourism and are invaluable in examining the dynamics, impacts and broader sustainability issues related to tourism (Aall & Koens, 2019; Ashworth & Page, 2011). As destinations emerge from the pandemic, there is an urgent need to seek a socially just, responsive, and environmentally sensitive future, particularly in our urban spaces (Rastegar et al., 2021). This calls for research examining the impact policymaking and governance structures have on urban destinations and the inequalities this creates. Edward Soja's spatial treatment of Lefebvre's The Right to the City offers a critical approach to evaluate how policymaking in historic cities creates spatial injustice through exclusionary processes that operate at material, psychological and social levels. Understanding city-life through a policymaking lens illustrates how local communities are subject to unjust geographies that are spatially discriminatory. As evident in York, a fragmented approach to public policymaking, founded on neoliberal consumer capitalism, has resulted in distributional injustice and spatial exclusion. If we are to reimagine a *just* tourism future (Rastegar et al., 2021), understanding how spatial injustice is socially and historically constructed is critical to finding new ways to manage tourism. The lessons learned from COVID-19 for addressing spatial injustice reveal how communities can re-appropriate city spaces to tackle distributional discrimination. Carefully designed structures and mechanisms are required that enable local people to influence decision-making. The collaborative working practices between institutions across public and private sectors illustrates what could be achieved through reclaiming social values that instituted a re-animation of city life.

This study offers a novel approach by focusing on the experiences of policymakers to understand the ways in which tourism policy is economically marginalised at a political level and the consequences this has on the social life of the city. The politicised short-termism of economic policymaking illustrated by the absence of an economic masterplan demonstrates how York's local communities are excluded from "bigger picture" decision-making concerning the future of the city. As a result, citizens mobilities are reduced both socially and economically through precarious low-paid working conditions and the rights to appropriate city spaces for local cultural expression. Even elite policymakers are disinvested from decision-making processes because of the ad-hoc way tourism policy is realised and governed. The lack of long-term strategic planning marginalises communities and corrodes diversity in our cities.

This research has shown that spatial justice should be recognised as a central goal of sustainable tourism policymaking. This is realised through enabling local stakeholders to achieve democratic justice whereby decisions concerning ecological, heritage and cultural preservation are made in tandem with achieving economic efficiency that balances the negative impacts of tourism, with greater distributive benefits being afforded to local communities (Higgins-Desbiolles, 2018; Ramón-Hidalgo & Harris, 2018). Whilst we acknowledge that a single case study limits the transferability of the research findings, it does provide an in-depth analysis of spatial injustice produced by policy and governance processes in a European historic city destination. Further studies that adopt a comparative case study method would be both interesting and useful in exploring these issues in different contexts. As our research has demonstrated, there is a need for scholars and policymakers to root policymaking in the local social, cultural, and environmental health of a city. Re-imaging the historic city for a sustainable and participative future needs policy that shifts from treating tourism as a poor cousin and instead invests in its social and cultural health through collaboration and co-creation with the local community. As shown in the case of York, lessons learned during COVID-19 have drawn attention to the positive impact of community engagement. However, more work is needed if we are to create a balanced approach to cooperation between stakeholders to achieve spatial justice. We, therefore, call on scholars to identify "the underlying processes producing unjust geographies" (Soja, 2008, p. 4) to realise a sustainable, collaborative future for our city communities based on creating value that is more than the pursuit of profit.

Acknowledgements

The authors are grateful to all the respondents their time, insightful comments and feedback.

Disclosure statement

No potential conflict of interest was reported by the authors.

ORCID

Brendan Paddison http://orcid.org/0000-0003-2896-2837
Jenny Hall http://orcid.org/0000-0002-5200-4308

References

Aall, C. (2014). Sustainable tourism in practice: Promoting or perverting the quest for a sustainable development? *Sustainability*, *6*(5), 2562–2583. https://doi.org/10.3390/su6052562

Aall, C., & Koens, K. (2019). The discourse on sustainable urban tourism: The need for discussing more than overtourism. *Sustainability*, *11*(15), 4228. https://doi.org/10.3390/su11154228

Agyeman, J., & McLaren, D. (2017). Sharing cities. *Environment: Science and Policy for Sustainable Development*, *59*(3), 22–27. https://doi.org/10.1080/00139157.2017.1301168

Ashworth, G., & Page, S. J. (2011). Urban tourism research: Recent progress and current paradoxes. *Tourism Management*, *32*(1), 1–15. https://doi.org/10.1016/j.tourman.2010.02.002

Ashworth, G. J., & Tunbridge, J. E. (2000). *The tourist-historic city: Retrospect and prospect of managing the heritage city (Advances in tourism research)* (1st ed.). Pergamon.

Astleithner, F., & Hamedinger, A. (2003). Urban sustainability as a new form of governance: Obstacles and potentials in the case of Vienna 1. *Innovation: The European Journal of Social Science Research*, *16*(1), 51–75. https://doi.org/10.1080/13511610304510

Atkinson, R. (2017). 'London, whose city'? Le Monde Diplomatique. https://mondediplo.com/2017/07/06london

Attoh, K. A. (2011). What kind of right is the right to the city? *Progress in Human Geography*, *35*(5), 669–685. https://doi.org/10.1177/0309132510394706

Augé, M. (1995). *Non-Places: An introduction to super-modernity*. Verso.

Augustyn, M. M., & Knowles, T. (2000). Performance of tourism partnerships: A focus on York. *Tourism Management*, *21*(4), 341–351. https://doi.org/10.1016/S0261-5177(99)00068-0

Barton, A. W., & Leonard, S. J. (2010). Incorporating social justice in tourism planning: Racial reconciliation and sustainable community development in the Deep South. *Community Development*, *41*(3), 298–322. https://doi.org/10.1080/15575330903444051

Beaumont, N., & Dredge, D. (2010). Local tourism governance: A comparison of three network approaches. *Journal of Sustainable Tourism*, *18*(1), 7–28. https://doi.org/10.1080/09669580903215139

Blackstock, K. (2005). A critical look at community based tourism. *Community Development Journal*, *40*(1), 39–49. https://doi.org/10.1093/cdj/bsi005

Bornhorst, T., Brent Ritchie, J., & Sheehan, L. (2010). Determinants of tourism success for DMOs & destinations: An empirical examination of stakeholders' perspectives. *Tourism Management*, *31*(5), 572–589. https://doi.org/10.1016/j.tourman.2009.06.008

Bramwell, B., & Lane, B. (2011). Critical research on the governance of tourism and sustainability. *Journal of Sustainable Tourism*, *19*(4–5), 411–421. https://doi.org/10.1080/09669582.2011.580586

Braun, V., & Clarke, V. (2006). Using thematic analysis in psychology. *Qualitative Research in Psychology*, *3*(2), 77–101. https://doi.org/10.1191/1478088706qp063oa

Butcher, J. (2021). Debating tourism degrowth post COVID-19. *Annals of Tourism Research*, *89*, 103250. https://doi.org/10.1016/j.annals.2021.103250

Cederström, C., & Fleming, P. (2012). *Dead-man working*. Zero.

City of York Council. (2022). Economic strategy. City of York Council.

Coles, T., & Church, A. (2007). *Tourism, power & space*. Routledge.

Croft, C. D. (2018). Hearing the authentic voice of stakeholders? Implications for governance of tourism strategy-making. *Current Issues in Tourism*, *21*(14), 1670–1689. https://doi.org/10.1080/13683500.2016.1153051

Crotty, M. (1998). *The foundations of social research*. Sage Publications.

De Jong, A., & Schuilenburg, M. (2006). *Mediapolis*. 010 Publishers.

Devine, J., & Ojeda, D. (2017). Violence and dispossession in tourism development: A critical geographical approach. *Journal of Sustainable Tourism*, *25*(5), 605–617. https://doi.org/10.1080/09669582.2017.1293401

Diaz-Parra, I., & Jover, J. (2021). Overtourism, place alienation and the right to the city: Insights from the historic centre of Seville, Spain. *Journal of Sustainable Tourism*, *29*(2–3), 158–175. https://doi.org/10.1080/09669582.2020.1717504

Dredge, D. (2001). Local government tourism planning and policy-making in New South Wales: Institutional development and historical legacies. *Current Issues in Tourism*, *4*(2–4), 355–380. https://doi.org/10.1080/13683500108667893

Dredge, D. (2006). Policy networks and the local organisation of tourism. *Tourism Management*, *27*(2), 269–280. https://doi.org/10.1016/j.tourman.2004.10.003

Dredge, D., Ford, E. J., & Whitford, M. (2011). Managing local tourism: Building sustainable tourism management practices across local government divides. *Tourism and Hospitality Research*, *11*(2), 101–116. https://doi.org/10.1057/thr.2010.064

Earle, L. (2017). *Transgressive citizenship and the struggle for social justice*. Springer Publishing.

Fainstein, D. R., & Judd, S. S. (1999). *The tourist city*. Yale University.

Fainstein, N., & Fainstein, S. (Eds.). (1982). *Urban policy under capitalism*. Sage.

Font, X., Higham, J., Miller, G., & Pourfakhimi, S. (2019). Research engagement, impact and sustainable tourism. *Journal of Sustainable Tourism*, *27*(1), 1–11. https://doi.org/10.1080/09669582.2019.1560673

Goodson, L., & Phillimore, J. (2004). The inquiry paradigm in qualitative tourism research. In J. Phillimore & L. Goodson (Eds.), *Qualitative research in tourism* (pp. 30–45). Routledge Publications.

Hall, C. M. (2000). Rethinking collaboration and partnership: A public policy perspective. In B. Bramwell & B. Lane (Eds.), *Tourism collaboration and partnerships: Politics, practice and sustainability* (pp. 143–158). Channel View.

Hall, C. M. (2007). Tourism, governance and the (mis-)location of power. In A. Church & T. Coles (Eds.), *Tourism, power and space* (pp. 247–269). Routledge.

Hall, C. M. (2011). Framing governance theory: A typology of governance and its implications for tourism policy analysis. *Journal of Sustainable Tourism*, *19*(4–5), 437–457. https://doi.org/10.1080/09669582.2011.570346

Hall, C. M., & Jenkins, J. M. (1995). *Tourism and public policy*. Routledge.

Harvey, D. (2003). The right to the city. *International Journal of Urban and Regional Research*, *27*(4), 939–941. https://doi.org/10.1111/j.0309-1317.2003.00492.x

Harvey, D. (2009). *Social justice and the city*. University of Georgia Press.

Higgins-Desbiolles, F. (2018). The potential for justice through tourism. *Via Tourism Review*, *13*(13). https://doi.org/10.4000/viatourism.2469

Higgins-Desbiolles, F. (2020). Socialising tourism for social and ecological justice after COVID-19. *Tourism Geographies*, *22*(3), 610–623. https://doi.org/10.1080/14616688.2020.1757748

Higgins-Desbiolles, F., Carnicelli, S., Krolikowski, C., Wijesinghe, G., & Boluk, K. (2019). Degrowing tourism: Rethinking tourism. *Journal of Sustainable Tourism*, *27*(12), 1926–1944. https://doi.org/10.1080/09669582.2019.1601732

Iveson, K. (2011). Social or spatial justice? Marcuse and Soja on the right to the city. *City*, *15*(2), 250–259. https://doi.org/10.1080/13604813.2011.568723

Jamal, T., & Camargo, B. A. (2014). Sustainable tourism, justice and an ethic of care: Toward the Just Destination. *Journal of Sustainable Tourism*, *22*(1), 11–30. https://doi.org/10.1080/09669582.2013.786084

Jamal, T., & Higham, J. (2021). Justice and ethics: Towards a new platform for tourism and sustainability. *Journal of Sustainable Tourism*, *29*(2–3), 143–157. https://doi.org/10.1080/09669582.2020.1835933

Jamal, T., & Watt, E. M. (2011). Climate change pedagogy and performative action: Toward community-based destination governance. *Journal of Sustainable Tourism*, *19*(4–5), 571–588. https://doi.org/10.1080/09669582.2011.575227

Jenkins, J. M. (2014). Tourism and public policy: Contemporary debates and future directions. In C. M. Hall, M. Mkono, A. A. Lew, C. M. Hall, & A. M. Williams (Eds.), *Wiley Blackwell companions to geography* (pp. 542–555). Wiley Blackwell.

Joppe, M. (2018). Tourism policy and governance: Quo vadis? *Tourism Management Perspectives*, *25*, 201–204. https://doi.org/10.1016/j.tmp.2017.11.011

Jover, J., & Díaz-Parra, I. (2020). Gentrification, transnational gentrification and touristification in Seville, Spain. *Urban Studies*, *57*(15), pp. 3044–3059.

Kimbu, A. N., & Ngoasong, M. Z. (2013). Centralised decentralisation of tourism development: A network perspective. *Annals of Tourism Research*, *40*, 235–259. https://doi.org/10.1016/j.annals.2012.09.005

Lee, S. (2009). Income inequality in tourism services-dependent counties. *Current Issues in Tourism*, *12*(1), 33–45. https://doi.org/10.1080/13683500802248001

Le Feuvre, M., Medway, D., Warnaby, G., Ward, K., & Goatman, A. (2016). Understanding stakeholder interactions in urban partnerships. *Cities*, *52*, 55–65. https://doi.org/10.1016/j.cities.2015.10.017

Lefebvre, H. (1968). *Le droit`a la ville*. Anthropos.

Lefebvre, H. (1998). *The production of space* (D. Nicholson-Smith, Trans.). Wiley-Blackwell. (Original work published 1974).

Lefebvre, H. (2003). *The urban revolution*. University of Minnesota Press.

Lefebvre, H., Kofman, E., & Lebas, E. (2000). *Writings on cities* (5th ed.). Wiley-Blackwell.

Leśniewska-Napierała, K., Napierała, T., Tobiasz-Lis, P., & Wójcik, M. (2019). *Sustainable tourism development towards spatial justice: Evidence from Masłomęcz, Poland*. DSpace. Retrieved November 26, 2021, from https://dspace.uni.lodz.pl/handle/11089/30512

MacLeod, G. (2002). From urban entrepreneurialism to a "Revanchist City"? On the spatial injustices of Glasgow's renaissance. In *Antipode* (pp. 602–624). Blackwell. https://doi.org/10.1111/1467-8330.00256

Marcuse, P. (2009). From critical urban theory to the right to the city. *City*, *13*(2–3), 185–197. https://doi.org/10.1080/13604810902982177

Massey, D. (1984). *Spatial divisions of labor: Social structures and the geography of production*. Macmillan.

Meethan, K. (1997). York: Managing the tourist city. *Cities*, *14*(6)., 333–342. https://doi.org/10.1016/S0264-2751(97)00024-3

Mitchell, D. (2003). *The right to the city: Social justice and the fight for public space* (1st ed.). The Guilford Press.

Mordue, T. (2005). Tourism, performance and social exclusion in "Olde York". *Annals of Tourism Research*, *32*(1), 179–198. https://doi.org/10.1016/j.annals.2004.06.002

Moscardo, G. (2011). Exploring social representations of tourism planning: Issues for governance. *Journal of Sustainable Tourism*, *19*(4–5), 423–436. https://doi.org/10.1080/09669582.2011.558625

Moscardo, G. (2018). Rethinking the role and practice of destination community involvement in tourism planning. In *Tourism policy and planning implementation* (pp. 36–52). Routledge.

Paddison, B., & Biggins, R. (2017). Advocating community integrated destination marketing planning in heritage destinations: The case of York. *Journal of Marketing Management*, *33*(9–10), 835–857. https://doi.org/10.1080/0267257X.2017.1329226

Paddison, B., & Walmsley, A. (2018). New Public Management in tourism: A case study of York. *Journal of Sustainable Tourism*, *26*(6), 910–926. https://doi.org/10.1080/09669582.2018.1425696

Page, S. J. (1995). *Urban tourism*. CABI.

Purcell, M. (2014). Possible worlds: Henri Lefebvre and the right to the city. *Journal of Urban Affairs*, *36*(1), 141–154. https://doi.org/10.1111/juaf.12034

Ramón-Hidalgo, A. E., & Harris, L. M. (2018). Social Capital, political empowerment and social difference: A mixed-methods study of an ecotourism project in the rural Volta region of Ghana. *Journal of Sustainable Tourism*, *26*(12), 2153–2172. https://doi.org/10.1080/09669582.2018.1546711

Rastegar, R., Higgins-Desbiolles, F., & Ruhanen, L. (2021). COVID-19 and a justice framework to guide tourism recovery. *Annals of Tourism Research*, *91*, 103161. https://doi.org/10.1016/j.annals.2021.103161

Raymen, T. (2019). *Parkour, deviance and leisure in the late-capitalist city: An ethnography*. Emerald.

Saunders, M. (2021). *Research methods for business students* (6th ed.). Pearson Custom Publishing.

SDG Knowledge Hub. (2016, July 12). *Policy brief: The "right to the city" and the new urban agenda | IISD*. SDG Knowledge Hub. Retrieved November 26, 2021, from https://sdg.iisd.org/commentary/policy-briefs/the-right-to-the-city-and-the-new-urban-agenda/

Sharifi, A., & Khavarian-Garmsir, A. R. (2020). The COVID-19 pandemic: Impacts on cities and major lessons for urban planning, design, and management. *The Science of the Total Environment*, *749*, 142391. https://doi.org/10.1016/j.scitotenv.2020.142391

Soja, E. W. (1985). The spatiality of social life: Towards a transformative retheorization. In D. Gregory & J. Urry (Eds.), *Social relations and spatial structures* (pp. 90–127). Macmillan.

Soja, E. W. (1996). *Thirdspace: Journeys to Los Angeles and other real- and-imagined places*. Blackwell.

Soja, E. W. (2000). *Postmetropolis: Critical studies of cities and regions*. Blackwell.

Soja, E. W. (2008, March 12–14). *The city and spatial justice*. Spatial Justice, Nanterre, Paris, pp. 1–5. https://www.jssj.org/wp-content/uploads/2012/12/JSSJ1-1en4.pdf

Soja, E. W. (2010). *Seeking spatial justice* (Globalization and community series). University of Minnesota Press.

Sun, Y. Y., Li, M., Lenzen, M., Malik, A., & Pomponi, F. (2022). Tourism, job vulnerability and income inequality during the COVID-19 pandemic. *Annals of Tourism Research Empirical Insights*, *3*(1), 100046. https://doi.org/10.1016/j.annale.2022.100046

Timothy, D. J. (2011). *Cultural heritage and tourism: An introduction*. Channel View.

UNESCO. (2006). *International public debates: Urban policies and the right to the city*. UNESCO.

UN-HABITAT. (2010). *The right to the city: Bridging the urban divide*. World Urban Forum, United Nations.

Unwin, T. (2000). A waste of space? Towards a critique of the social production of space .... *Transactions of the Institute of British Geographers*, *25*(1), 11–29. https://doi.org/10.1111/j.0020-2754.2000.00011.x

Veal, A. J. (2011). *Research methods for leisure and tourism: A practical guide* (4th ed.). Prentice Hall.

Visit Britain. (2022). Visit Britain. Retrieved June 17, 2022, from https://www.visitbritain.com/gb/en

Visit York. (2022). *Key facts on tourism in York*. Retrieved June 17, 2022, from https://www.visityork.org/members/research-statistics

Waterton, E., & Watson, S. (2013). *Heritage and community engagement: Collaboration or contestation?* (1st ed.). Routledge.

White, R. (2019). Loving the planet to death: Tourism and ecocide. In *Deviant leisure: Criminological perspectives on leisure and harm* (1st ed., pp. 285–304). Palgrave Macmillan.

Yin, R. K. (2009). *Case study research: Design and methods* (Vol. 5). Sage Publications.

Part IV

Confronting crises, building justice and transitioning tourism

The poor on the road: qiongyou as a collective resistance and justice tourism

Yimeng Yang

ABSTRACT

"Qiongyou" is an alternative tourism that has emerged in China and is characterized by a "low-budget" approach. This article argues that qiongyou can be seen as a form of justice tourism. First, qiongyou is a collective socio-cultural resistance of grassroots travelers to the structural injustice of state-led tourism in the context of the financial crisis. Second, qiongyouers provide low-budget tourism participation options for the economically disadvantaged through alternative travel market practices and discursive practices. The study of qiongyou provides an empirical case from a developing country and complements current research on justice tourism with a tourism participation perspective and related justice issues.

Introduction

Justice tourism, as distinct from other alternative tourism (e.g., responsible tourism, authentic tourism, slow tourism, sustainable tourism), has been increasingly advocated by critical scholars. Unlike alternative tourism, which has been criticized for being limited to moral engagement, lacking in reflection on political structures, and gradually being mainstreamed (Ateljevic, 2000; Bianchi, 2018; O'Reilly, 2006), justice tourism is seen as a form of activism that allows for socio-cultural responses and resistance to dominant injustices (Guia, 2021; Higgins-Desbiolles, 2008, 2018; Isaac et al., 2015).

However, much of the current research on justice tourism is based on a critique of overtourism under neoliberalism, arguing that excessive tourists bring destruction and socio-environmental injustice to destination communities, and calling for de-growth of tourism or "local community-centered tourism" (Higgins-Desbiolles, 2018, 2020; Jamal & Camargo, 2014). But such discussions overlook the fact that, in many developing countries, tourism has not developed as a universal right, but rather a policy tool appropriated by authoritarian governments, which results in controversies that are often triggered by the distribution of tourism resources and participation rather than the overaccumulation of neoliberal tourism (Bianchi, 2018; Bianchi et al., 2020; McCabe & Diekmann, 2015). This is why some studies argue that discussing the relationship between tourism and justice from a tourist's perspective may sometimes be more necessary and effective than focusing only on the host community (McCabe & Qiao, 2020, p. 12).

To complement the discussion of this issue, this study will use an empirical case of "qiongyou" (穷游) (literal translation: tourism for the poor), an alternative tourism in China, to examine how it serves as a grassroots tourist-initiated socio-cultural action to resist the dominant tourism system and develop alternative opportunities for tourism participation. This is the main research question of this article.

The rapid development of mass tourism in China began around 1997. It mainly served as a governance technique for the state to boost domestic demand and relieve financial pressure under the impact of the Asian economic crisis. The government at that time introduced several relevant policies in the short term to promote tourism industry growth and cultivate modern tourism consumers. This is why Nyíri (2010) argues that tourism in China is more of a tool for the state to promote modernization, economic development and ideological transmission. It is clear that middle-income people with spending power, rather than "the poor", were the ones the state wanted to attract into the tourism market at the time. This socio-cultural exclusion in tourism development was exacerbated by another round of economic crisis in 2008, which led to the resistance of the newly emerged economically disadvantaged under the crisis, namely qiongyou. These resisters, who identify themselves as "qiongyouers" (穷游族) (literal translation: poor travelers), use rebellious travel strategies to seek alternative means of tourism participation outside of the state-led tourism system. In other words, the emergence of qiongyou particularly reflects the structural issues of tourism in China as a developing country and the agency of grassroots tourists in tourism participation.

Specifically, this article will answer three sub-research questions: (1) How does qiongyou develop into a collective resistance to the state-led tourism system? (2) How do qiongyouers develop alternative tourism markets and thus create alternative opportunities for tourism participation? (3) How do the discursive practices of qiongyouers serve as a cultural resistance to give cultural legitimacy to such alternative tourism participation? The discussion that follows is divided into four sections. The first section will review the research agenda of justice tourism, and link it to the discussion of state-led tourism in China. The second section will detail the context of the case and the methodology of the study. The third section will present the findings of this article, including alternative travel practices and cultural discourses of qiongyouers. The discussion and conclusion will highlight again the theoretical and social significance of qiongyou, as well as the limitations of this study and potential future research directions.

Literature review

Tourism participation and justice issues: the role of grassroots traveler

Increasingly, research has noted the injustices associated with tourism development, such as human rights violations, ecological damage, and sociocultural exclusion, and has begun to reflect on issues, such as who benefits from tourism, how tourism can be more equitable and just, and how tourism can be used to secure justice (Higgins-Desbiolles, 2018; Jamal, 2019; Jamal & Higham, 2021). Unlike alternative tourism (e.g., responsible tourism, slow tourism, sustainable tourism), which has been criticized for being limited to rhetoric and reproducing dominant structures (Mosedale, 2015, p. 171, Bianchi, 2018, p. 89), justice tourism is considered to have a radical potential as a social force to challenge the neoliberal agenda and promote structural transformation (Higgins-Desbiolles, 2006; 2008). Guia (2021) makes a similar point, arguing that justice tourism can further challenge the systemic sources of injustice and inequality (political and business systems), and lead to new identities and practices.

Isaac et al. (2015) study on tourism in Palestine highlights this point well. They examined how mainstream tourism had been used as a tool of domination by Israel and the West, and how alternative tourism contributed to resistance actions in Palestine that challenged dominant tourism systems and ideologies. Such alternatives are seen as justice tourism, a socio-cultural response and resistance to dominant injustice (Higgins-Desbiolles, 2008; 2018; Kassis et al., 2015). In other words, unlike past studies of alternative tourism, what was presented with justice tourism as an analytical perspective was the tension between structural domination and resistance from below, and the broader political context in which they were embedded.

Notably, current research on justice tourism focuses more on destination justice, such as the right to development of local cultures and minorities (Jamal & Camargo, 2014), solidarity between tourists and locals (Choi et al., 2021), the conservation of the natural environment (Sheller, 2021). In other words, a large part of the research agenda of justice tourism is about "local community-centered tourism", which advocates avoiding the destruction of local societies and environments by overtourism under neoliberalism (Higgins-Desbiolles et al., 2019, Higgins-Desbiolles, 2020). However, the above discussion is based more on the experience of advanced post-industrial countries where tourism has become a taken-for-granted pursuit for many citizens (Bianchi et al., 2020: 7). This overlooks the fact that in many developing countries, tourism has developed not as a universal right from the beginning, but as a tool appropriated by authoritarian governments to promote state governance, economic growth, and the exchange of benefits between countries (Bianchi, 2018; Nyíri, 2010). Thus, in these countries, on the one hand, the number of tourists as emerging consumers is indeed increasing, but on the other hand, the often overlooked issue is that the distribution of tourism resources and tourism participation is likely to be unequal and unjust, such as the exclusion of low-income groups (McCabe & Diekmann, 2015; Mihalič & Fennell, 2015; Telfer & Sharpley, 2015). This is why some scholars argue that it is equally necessary to discuss the relationship between tourism and justice from the perspective of tourists, rather than focusing only on the host community (McCabe & Qiao, 2020, p. 12).

There are two groups of critical studies that have touched on the issue of tourism participation. On the one hand, "social tourism" first noticed the structural inequalities in tourism participation and tried to help disadvantaged people (e.g. low-income groups) to have more opportunities to engage in travel through some top-down tourism policies and programs (McCabe & Diekmann, 2015; Minnaert, 2012). However, research on social tourism has focused more on policy-making and effects, and lacks a critical perspective on the structural distributive justice of tourism participation (McCabe & Qiao, 2020). The struggle for the right to travel in the neoliberal context, while relying on the intervention of authorities (Bianchi & Stephenson, 2013; Rastegar et al., 2021), overlooks the alternative possibilities that independent grassroots travelers can offer beyond the market and the state.

On the other hand, Cresswell (1993) sees the hippie travel that emerged in the 1960s and 1970s as a "mobility as resistance." It was by being "on the road" that young people rebelled against the dominant political economic system and claimed the right to visit alternative cultures. They were also the precursors of the later backpackers (O'Reilly, 2005; 2006). However, many studies have also found that most backpackers actually just see their alternative travel as a fashion and identity, which has led to backpacking gradually becoming part of the neoliberal system as a new niche market (Ateljevic, 2000; Jabłonkowska & Stankiewicz, 2018). In other words, while the rise of backpackers exemplifies the bottom-up practical potential of independent travelers, whether it really brings about an alternative agenda needs to be specifically examined in the context of power relations and political economic processes (Thieme et al., 2020).

In contrast, qiongyou is a grassroots tourist-initiated resistance and a direct challenge to the dominant tourism system with alternative tourism participation strategies. Discussing the concept of justice tourism using qiongyou as an empirical case would be a good way to complement a tourist perspective that many scholars have begun to call for. Before going any further, however, it is necessary to review the existing research on state-led tourism in China to show how it differs from the neoliberal logic of tourism development, in order to see the particular significance of qiongyou emerging in the Chinese system.

Alternative agendas in China's state-led tourism industries

Unlike the tourism industry under neoliberalism, in authoritarian regimes like China, tourism is often a state-led development program used to achieve a specific political agenda (Bianchi,

2018, p. 90, Liu, 2020). Ever since China's economic reforms, tourism has played an important role in generating state revenues and promoting modern development (Oakes, 1998; Airey & Chong, 2010). For example, ticket revenues from scenic spots are often an important part of local government finances. The government regulates ticket prices to serve specific local development goals (Wu, 2019). Also, strategically planning the layout of the tourism industry is a means for governments to promote regional economic growth, for example, by developing tourism to promote population mobility and activate more opportunities for innovation and entrepreneurship (Zeng & Ryan, 2012; Zhou & Chan, 2019).

In addition, mass tourism can also serve to channel dominant ideologies and exert the cultural authority of the state (Nyíri, 2010). When China's tourism industries first began to develop in the 1990s, the government sought to cultivate civilized subjects with a modern consumer consciousness by interpreting mass tourism as a leisure culture (Nyíri, 2010, p. 62). Subsequently, a series of tourist attractions, travel agencies, and itineraries, prepared by the government, began to emerge, such as red tourism and world heritage tourism, which presented specific cultural discourses to the public through landscape recreations and guided tours (Li et al., 2010; Lin, 2015, Nakano & Zhu, 2020). Thus, in Nyíri's (2010, p. 63) argument, Chinese tourism, unlike Western-style tourism, which encourages self-learning and challenge, is almost like pre-modern *literati* travel with the main objective of making tourists venerate historical and cultural classics.

In short, tourism in China is more of a government-led consumption with boundaries, controlled zones, and target audiences, and it is used to serve governance goals such as economic development and ideological construction in specific contexts (Nyíri, 2010).

Of course, there are also independent travelers, such as "donkey friends" (驴友), who are actively engaged in alternative practices, even in China's state-led tourism industry. These donkey friends are seen to share the same characteristics of Western backpackers in their quest for self-fulfillment and alternative culture, as well as reflecting the rise of individualism and the emergence of a public sphere in China's political economic transformation (Chen & Weiler, 2014; Lim, 2008; Zhu, 2018). Notably, Luo et al. (2015) found that some donkey friends reduce their travel costs by evading tickets. However, Luo et al. merely characterize ticket evasion as unethical and ignore that this behavior is associated with a group that has a completely different travel strategy than donkey friends, namely qiongyouers. At the same time, more research has also found that donkey friends do not use low-budget as their primary travel strategy, and sometimes spend more money than mainstream tourists. This is because, unlike Western backpackers as "drifters", donkey friends actually care more about unique travel experiences (Zhang et al., 2015; Cai, 2018, p. 82), or simply see alternative tourism as a symbol of Western "modernity" and emulate it (Nyíri, 2018, 92–93). Thus, critics argue that what donkey friends practice is still a romanticized and depoliticized tourism, and does not have the potential to achieve an alternative agenda (Zhu, 2018, Nyíri, 2018, p. 75).

However, compared to the donkey friends that emerged earlier in China and were studied more often, the qiongyou that also developed to a larger scale after 2008 has not received enough attention from the academic community. Xie Jia and Zhu Xuan are among the few scholars who have conducted special studies on qiongyou. For example, Xie and Zhu (2019, p. 128–129) compare qiongyouers to backpackers and highlight that qiongyouers differ from backpackers in having low incomes and preferring low-budget travel. Qiongyouers think that backpackers are fake poor, while they are really poor. At the same time, qiongyouers also try to pursue a unique lifestyle and self-identity through alternative travel, which is again very similar to backpackers (Xie & Zhu, 2019). Building on Xie and Zhu's research, this study further embeds qiongyou in the context of China's political economy, which will help highlight the structural issues of Chinese tourism behind qiongyou, namely the inequality and distributive injustice of tourism participation, and how qiongyou can serve as an alternative tourism participation mechanism, thereby establishing a link to the current research agenda of justice tourism.

Case study: qiongyou in China

Case context: state-led tourism commodification and ensuing crisis

The Chinese government began to support the industrialization of tourism after economic reforms, by changing the goal of tourism development from "political hospitality" to "market operation" in 1978. However, for a long time afterward, the government focused on developing the inbound tourism market, while still adopting a conservative attitude of "not promoting, not publicizing, and not opposing" domestic mass tourism. This was because the government believed that the domestic public still lacked awareness of tourism consumption, and because the development of inbound tourism helped to supplement the scarce foreign exchange revenue in the central treasury at the time (Guangdong Tourism Association, 2018). In 1986, the "National Tourism Work Conference" highlighted tourism as one of the three pillar industries for foreign exchange generation. At the same time, to attract more foreigners, the government built a large number of new hotels, organised tours for foreigners, and nominated a number of World Heritage sites, including the Great Wall and the Forbidden City.

However, the Asian financial crisis broke out suddenly in 1997 causing Japan, South Korea, and countries in Southeast Asia to restrict their citizens' outbound travel as their currencies depreciated and their purchasing power abroad declined. Asian travellers, who accounted for over 70 percent of the Chinese tourism market, consequently fell sharply (Zhang, 1998, p. 14). Furthermore, China's exchange rate control policy at the time remained strong whilst other Asian countries' currencies depreciated, causing Euro-American travellers to abandon their trips to China and choose to travel to Southeast Asian countries, where they could have greater local purchasing power. These two factors caused the Chinese tourism market, which at the time relied heavily on inbound tourism, to be hit hard by the financial crisis. Importantly, local finances also depended on tourism revenue (Zhang, 2015). The decline in tourist flows also led to financial difficulties for local governments. As a result of the multiple crises, the government had to start boosting domestic demand for tourism. Mass tourism for domestic tourists in China thus began to emerge.

After 1997, the state introduced a series of policies to support the development of the domestic tourism industry. Firstly, tourism was recognised by the government in 1998 as a "new national economic growth point", meaning that the goal of tourism development shifted from earning foreign exchange to contributing to the country's overall economic development. Secondly, as Chinese people were not yet used to travelling, and their financial management was more geared towards saving than spending, the government introduced two seven-day, paid holidays, the "May Day holiday" and the "National Day holiday", in 1999 to give more people time to travel after work. At the same time, the government sought to interpret tourism as a "culture of leisure" to cultivate tourism consumption habits among the population (Nyíri, 2010, p. 90). In other words, in China, mass tourism was more of a tool for the state to respond to the fiscal crisis and boost domestic demand, which led to Chinese tourism being consumption-oriented from the beginning.

Consumption-oriented tourism was further strengthened in two ways. First, the government began to designate more scenic areas, dominate ticket pricing, and push up ticket prices starting in 2004. The year 2005 was even dubbed the "year of the price hike", as the rate of ticket price increases far exceeded the rate of population income increases over the same period (The Beijing News, 2005). In addition, the *Regulations on the Administration of Travel Agencies*, introduced in 1996, further promoted the marketization of travel agencies. Between 1997 and 2009, the total number of travel agencies in China increased by more than four times to over 20,000 (Yao, 2017). Travel agencies further entered into commercial agreements with destinations, retailers and hotels to offer travel products, such as package tours with specific scenic spots, itineraries, and consumption items, which dominated the travel patterns of most Chinese tourists at the time (Trivett & Staff, 2013).

In other words, the rapid development of mass tourism in China after 1997 was based on the establishment of a state-led and policy-driven tourism market, which involved resource enclosure, tourism commodification, and consumer subject construction. Obviously, it is also a process of screening tourists based on their spending power and cultural identity.

In 2008, however, a new round of economic crisis hit the Chinese tourism industry once again. Compared to 1997, when the Chinese market had not yet fully opened, 2008 saw China already highly dependent on the export economy. The global financial crisis had consequently affected the domestic job market in China. The ensuing drop in income and increase in unemployment led to a decline in the consumption-oriented tourism industry. Many people have to abandon their travel plans because of financial constraints. However, after more than a decade of state support for the tourism industry and consumer culture, by the early 21st century, tourism had become an important part of the leisure life of high-income earners and the emerging urban consumer class (Nyíri, 2010). They were reluctant to be stuck at home despite the decline in income due to the economic crisis. They declared, "travel even if you are poor" and set off on their travels as qiongyouers, which includes a range of alternative travel strategies and cultural discourses that differ from the mainstream tourism. It was in this context that qiongyou began to emerge.

Data collection and analysis

Netnography

Netnography is a method of studying online communities, helping to understand the symbolism, meanings, and consumption patterns of online consumer groups (Kozinets, 2002). Moreover, as more and more tourists are now collecting travel information and exchanging travel tips online, web ethnography has become an important method in tourism studies (Mkono & Markwell, 2014). It is worth noting that qiongyou emerged in the early years of the 21st century, when web 2.0 was becoming widespread in China. The travel strategies and community building of qiongyouers have relied heavily on the Internet. Therefore, a study of qiongyou cannot ignore the practices of qiongyouers in cyberspace. Netnography will help in this analysis.

This study adopted a passive, "lurker" approach in carrying out netnography. This had two advantages: (1) the researcher has more direct and rapid access to publicly available information on the web; (2) the researcher does not interfere with the expressions and interactions of the participants (Mkono, 2012； Mkono and Markwell, 2014, p. 290). This was also particularly helpful for the study of qiongyou. Because the practices of qiongyouers often involve actions that are illegal or contrary to mainstream social ethics, such as fare evasion, an anonymous cyberspace, free from the interference of the researcher, will allow for open and frank discussion.

With regard to the choice of online communities and platforms, this article follows Kozinets (2002) recommendation to choose communities that explicitly discuss and act under the banner of qiongyou, given their relevance to the research topic, or communities with more active discussions, such as those with higher traffic and more postings. In the light of the above considerations, this article chose three online communities: *Baidu Tieba* (百度贴吧), *Douban* (豆瓣) and *Zhihu* (知乎). The sub-community "qiongyou ba" (穷游吧) in *Baidu Tieba* currently has two million followers and 16 million posts. The sub-community "qiongyou tianxia" (穷游天下) in *Douban* is the largest tourism group on "Douban.com", with nearly 1.3 million members. *Zhihu* is one of the most popular online Q&A communities in China, with nearly 600,000 followers and over 5,500 posts on the topic of qiongyou. In addition, the official QQ group of "qiongyou tianxia" was chosen to observe the discussions and interactions of qiongyouers. This group has 448 members. Finally, this article also focuses on posts about qiongyou on online platforms such as "qyer.com" (穷游网), "mafengwo.com" (马蜂窝) and "Sina Blog."

The data related to this study were collected from the above-mentioned online platforms, including public postings, images, and videos from 2008 to 2018, between October 2020 and January 2022. The author began the search with the keyword "qiongyou" and took screenshots of relevant information, such as personal experience stories and follow-up comments. The frequently mentioned travel strategies, such as hitchhiking, travel tips, fare evasion, and frugality, will be analyzed by summarising them into two categories: (1) economic strategies of qiongyou based on alternative travel markets; and (2) cultural strategies of qiongyou based on alternative discourses. They will be used to answer the second and third sub-questions of this article, respectively. The collected data could be used directly and publicly after personal information has been withheld.

However, netnography has its drawbacks, such as the fact that the researcher cannot direct the questions and that it is often difficult to confirm the authenticity of information online (Mkono, 2012). Therefore, the netnographic data was also supplemented with rich insights from semi-structured interviews with qiongyouers.

Semi-structured interviews and research limitations

Semi-structured interviews were used to add specific details about the experiences of qiongyouers in practicing various travel strategies. Based on the information retrieved and captured in online platforms, the author screened some of the main travel strategies and contacted the posters accordingly, ensuring that at least one interviewer was available for each travel strategy. Nine men and five women were selected as interviewees. Ten of the interviewees' stories are quoted later in the article. All respondents explicitly identified themselves as qiongyouers and emphasized that they had chosen to participate in qiongyou due to economic constraints. Interviews were conducted via video call and recorded with the permission of the interviewees. Each interview was limited to one hour and conducted in Chinese. Netnography and semi-structured interviews were conducted simultaneously to facilitate the matching of web postings and statements of personal experience.

The interview questions encouraged the interviewees to discuss the context and motivations for their qiongyou. The first part of the interview questions helped to map the political economic context from which the qiongyou emerged, which addresses the first research sub-question. The second and third parts of the interview questions dealt with the practices of the qiongyouers to answer the second sub-question asking about how they developed alternative tourism markets, thus providing alternative tourism participation channels for the economically disadvantaged. The fourth part of the interview questions dealt with the cultural meaning and social values of qiongyou, to answer the third sub-question about how qiongyouers can give cultural legitimacy to alternative tourism participation.

In addition to netnography and semi-structured interviews, this article also complemented the analysis of some secondary source material. Policy texts related to China's tourism industries were collected on government portals, as well as relevant news reports through the Chinese news database "duxiu" (读秀), using qiongyou as a keyword.

The methodology of this article also has some limitations. One major constraint is the limited number of interviews. Qiongyouers are not a homogeneous group. Different qiongyouers have different personal backgrounds and travel motivations. Some of them may also be involved in qiongyou just for the sake of interesting experiences. This article will focus on participants whose main goal is to save money. However, qiongyouers with other motivations have also contributed to the rise of qiongyou, the formation of alternative travel markets, and the promotion of tourism participation. In addition, this study focuses on the period from 2008-2018 when qiongyou was at its peak. The travel experiences recounted by the respondents all occurred during this time period as well. Qiongyou's transformation in recent years and its impact on policy in the long run deserve to be examined in future studies.

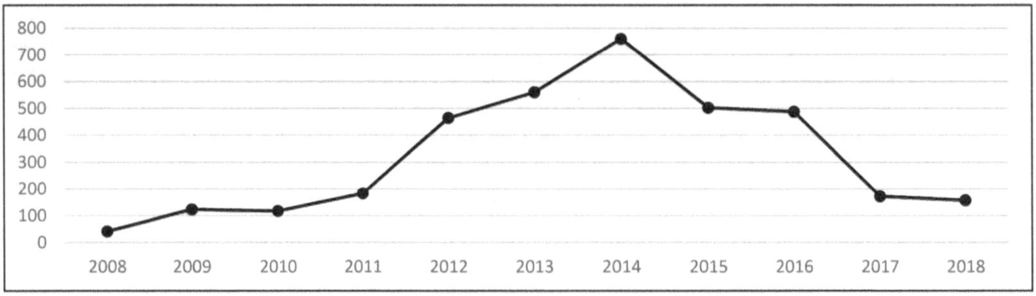

Figure 1. Changes in the number of news stories about qiongyou (source: duxiu).

Findings

Qiongyou as a collective social action of the poor

Emerging out of the sudden economic crisis and relying on the newly developed online media in China, qiongyou quickly gained popularity and grew to a significant scale in a short period of time. According to the Chinese news database "duxiu", the number of reports about qiongyou began to grow significantly in 2008 and multiplied tenfold over the next five years, with many of the stories referring to the financial crisis (Figure 1). At the time, one media outlet launched a series of reports on qiongyou under the theme "2009, the year of qiongyou", where it was mentioned:

> "Save" is perhaps the most mentioned word among travel enthusiasts in 2009, [...] Simply put, "qiongyou" is about maximizing your savings while traveling. So this series of reports shares stories of successful "qiongyou," but also aims to show you more ways to save money." (Chengdu Business Daily, 2009)

There were also qiongyouers who shared their experiences of qiongyou under the financial crisis on online forums:

> Affected by the financial turmoil, she was suspended from work for two months. After staying at home for 27 days, she decided to go to Hainan with her friends to have a break. After returning from Hainan, she posted on a travel forum, "A single girl's trip to Sanya under the financial crisis, with tips on how to save money." (Jiangdong Times, 2018)

Another qiongyouer, though not directly affected by the financial crisis, also got involved in the rise of qiongyou. In the interview, he recalled:

> It was around the same time that I was introduced to qiongyou. [...] I did see some news about it. The economy was bad and people were not well-off, so they had to qiongyou. I was still in college at that time, so I didn't really feel much about the financial crisis. But thinking about it, qiongyou was also quite suitable for me; after all, I was also a "poor student." (Interview #7)

In short, while the 1997 financial crisis gave rise to mass tourism in China to boost domestic consumption, the resurgence of the crisis in 2008 brought about an alternative form of tourism, qiongyou, aimed at saving money and low-budget. The sudden economic crisis and the rise of the Internet contributed to the rapid growth of qiongyou to a larger scale in a short period of time. The participants have also expanded from those directly affected by the financial crisis to a broader group of "poor people." This gives qiongyouers a collectivity and further contributes to the formation of alternative tourism markets and cultural discourses.

From low-budget to alternative tourism markets: the travel practices of qiongyouers

Alternative travel agents: qiongyou designers

Qiongyouers, committed to saving money, initially challenged the tourism market dominated by travel agencies. The package tours arranged by travel agencies not only involve the usual expenses such as sightseeing, hotels and transport, but also include many additional expenses. For example, travel agencies have agreements with many businesses in the tourist attractions. Tour guides take tourists to designated shops to spend money and receive a kickback from tourist purchases. These additional purchases are often compulsory, as they directly affect the travel agency's profit. One qiongyouer complained as a result:

> The group tours go to well-known attractions with high entrance fees, but some of them are actually boring [...] The tour guides always pull us in to "buy, buy, buy" and the guides get angry if we don't buy. So, relying on travel agents, a lot of times that means spending money and not having fun. (Interview #2)

Therefore, qiongyouers often do not rely on mainstream travel agents and have drawn on the travel strategies of backpackers. For example, one student qiongyouer who had cycled into Tibet recalled: "One reason is that I don't want to be too extravagant; after all, the cost of living is limited. Also, backpackers were quite popular at the time. One car, one backpack. I brought everything I could, and it didn't actually cost much money" (Interview #1). At the same time, qiongyouers also try strategies such as hitchhiking, couch-surfing or crowdfunding. One qiongyouer shared on an online platform: "I managed to stop 70 cars on the way and took a qiongyou to Tibet before graduating. The minimum daily cost was only two yuan. [...] This is what travel is really about" (Tieba, 2014). These alternative travel experiences quickly went viral on the Internet. In 2004, a website dedicated to sharing experiences of qiongyou, "qyers.com", was set up by a few qiongyouers. Communities with a common interest in qiongyou also began to form on social media platforms such as Baidu, Douban and QQ. These online communities contributed to the popularisation of the "backpacking" experience, which was unfamiliar to many Chinese tourists at the time.

However, it is not easy for many Chinese tourists, who used to rely on travel agents, to learn and practice these alternative travel strategies on their own. As a result, an alternative travel agent, the "qiongyou designer", has started to appear on the Internet. Most of these designers have extensive experience in qiongyou and know how to buy cheap tickets, where hotels are cheaper, how to get a lift, and what travel equipment to pack. Unlike mainstream travel agents who encourage tourists to spend money, qiongyou designers consider all kinds of money-saving strategies and plan trips for qiongyouers with minimal spending, while charging a small fee for their services. Such alternative travel agencies became popular based on the then emerging online trading platforms, such as Taobao (Newsnight, 2008). One qiongyouer who had sought such services said:

> My husband and I were planning a newlywed trip to Eastern Europe. The travel agency quote was too high. And they couldn't provide the ideal itinerary [...] [The itinerary given by the designer] cost about 30,000 yuan, which was a savings of almost 4,000 to 5,000 yuan compared to the travel agency, while (the designer) only charged 150 yuan. (Beijing, 2009)

Another qiongyouer gave his opinion about alternative agency services in the interview:

> I think it's good. There are more choices. [...] I did qiongyou, but I also did not have so much time to get equipment, to plan the route, so I found people online to help me do this. It may not necessarily be called a qiongyou designer, but there are many such services that help with planning. (Interview #8).

Ticket evasion and related services: local embeddedness of qiongyouers

Another object of resistance by qiongyouers is the government-led ticketing of scenic spots, especially in the context of the general increase in ticket prices in 2004. Ticket evasion thus became another major travel strategy for qiongyouers. Because it is difficult to set strict boundaries in many natural scenic areas or ancient villages and towns, many qiongyouers can find some ways to avoid checkpoints and enter areas without a ticket. At the height of the qiongyou boom, it was easy to find all kinds of ticket-evading experiences and road maps on the Internet

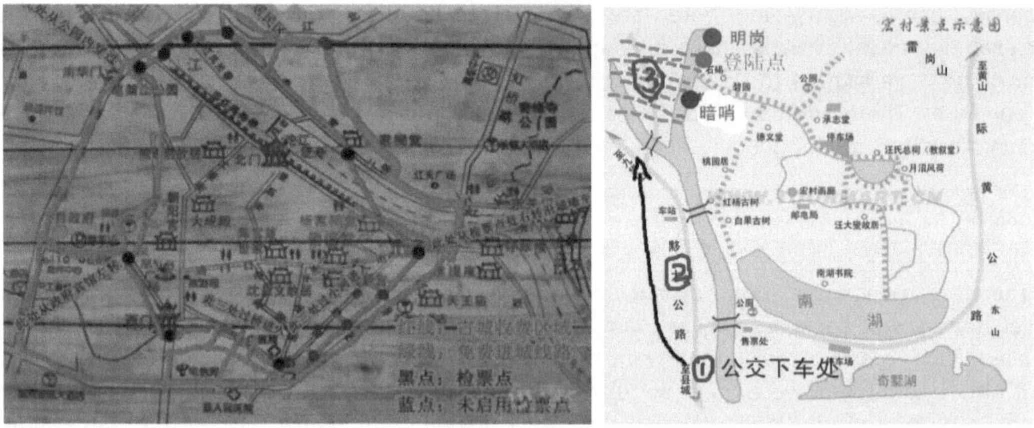

Figure 2. Qiongyouers draw alternative maps for ticket evasion on the official map of the scenic spot.

(Figure 2). For example, one qiongyouer shared his experience of evading paying for a ticket and taking a detour to enter a natural scenic area:

> We got out of the car and followed the online guide and kept following the riverside path. But as it was daytime, you could see down here from that position at the toll booth, so we had to sneak [...] It took us two hours and ten minutes to get to the lake loop and we managed to escape the entrance fee. (Douban, 2014)

A student qiongyouer, who also had a limited budget and had successfully evaded paying for tickets at natural scenic spots, mentioned that it was by evading ticket fees that they were able to visit more scenic spots.

> A local man rowed us to the scenic spot. We would pay him a ferry fee, but it was definitely cheaper than the entrance fee. [...] The tickets for the big scenic spots are not cheap. It's not that we can't afford it at all. If we need to buy a ticket, we will have to hesitate for a while and consider whether we want to go or not. Some we definitely won't go to. (Interview #14)

At the same time, in many ancient villages, the locals, who are more familiar with the spatial planning of the area, often assist qiongyouers in evading tickets. One reason for this is, when a village is designated as a scenic spot by the government, the revenue from the entrance fee is often not given back to the local villagers. Furthermore, the government's increase in ticket prices also results in fewer visitors, which further affects the turnover of local businesses. For these reasons, many villagers are keen to help tourists evade tickets, so that they can attract more visitors and earn some fees for their services. For example, a local businessman in "S town" posted online that he would be willing to provide ferry services to help travelers evade paying for tickets:

> S town has started charging an entrance fee of 50 RMB per person, which is really expensive. [...] Our merchants in S town are actively asking the government to abolish the entrance fee. We would like to say that whenever you come to S town, no matter what place you stay, you can contact us and we will do our best to help everyone. [...] We do make it easy for everyone to avoid the entrance fee points (Mafengwo, 2013).

Compared to the 50 yuan per person entrance fee, the above-mentioned local villager only charges 15-20 yuan for his ticket-evasion service. As there are many qiongyouers seeking such services, many locals offer further packages such as B&B stays and "agritainment" based on the ticket evasion service. One qiongyouer referred to his experience of evading ticket fees in X town:

> The main idea is to save a little if we can. See if there are "local channels" (Interview #13).

> We were staying at a villager's home in a B&B. He would take visitors staying at his house directly into the scenic area [...] I was thinking it was quite good. It saves money on tickets and you get to experience the locals more. Instead, the scenic area is too commercialized inside, and I don't think it's worth spending that much money on tickets (Interview #13).

As can be seen, with the rise of qiongyou, an alternative tourism market is emerging that includes independent travel agents, low-spending itineraries, ticket evasion services, and cooperation between tourists and locals. This further challenges the formerly state-led tour packages, scenic borders, and ticketing systems, offering low-budget tourism participation options for the economically constrained.

Alternative discursive practices of qiongyouers: frugality as justice

From state discourse to popular culture: frugality as a virtue

With the rise of alternative practices of qiongyou, criticism has followed. First, many people consider qiongyou to be a stop-gap measure in the face of financial constraints. "Who wouldn't want to stay in a better hotel, eat better food, go to more attractions and take more comfortable transport if it wasn't for being poor?" (Zhihu, 2020) Second, many people believe that hitchhiking and couch-surfing are simply "getting something for nothing." Ticket evasion is also frowned upon as an illegal practice. Third, female qiongyouers have come in for more criticism. Women's qiongyou is perceived as being more dangerous or as an activity for women trading sex for rewards. Most of these opposing views presuppose "consumption-oriented tourism" and argue that "no money, no travel" (Douban, 2015). Therefore, qiongyou, with the goal of low spending, is seen by many opponents as a selfish even unethical act:

> If travel is a private affair and just for yourself, then why should anyone else pay for your spending? [...] you have to understand yourself that travel is not a necessity, it is a luxury to some extent. So, if it's a luxury, why should (other people) donate to you? (Zhihu, 2018)

As can be seen, mainstream tourism is also accompanied by a dominant discourse, which stems largely from the state-led cultivation of tourism consumer awareness in the late 1990s. In many people's minds, travel is actually a "luxury" that those without money are naturally not entitled to.

At the same time, however, qiongyou has also been interpreted as having the virtue of frugality. Under the influence of Confucianism, China has particularly emphasized the importance of frugality (Kindel, 1985), and this has been reflected in recent national policies. In 2014, the Propaganda Department of the Chinese Communist Party (CCPPD) and National Development and Reform Commission (NDRC) jointly launched the "Frugality for Virtue, Saving for All" initiative, defining frugality as part of the core socialist values and the Chinese Dream (The Central People's Government of China, 2014). In response to the state's call, a variety show called "The Frugal Chinese" was launched in the same year, presenting a variety of frugal living strategies in the form of a reality show. Notably, one of the themes of the program was qiongyou. The celebrities were committed to spending the least amount of money on accommodation, food and transport for two days and one night, showing a number of qiongyou strategies such as rent sharing, hitchhiking and bargaining.

In other words, according to the program's interpretation, it was clear that qiongyou was not just a stop-gap measure, but a positive expression of social virtue. The official media later praised the program and again criticized extravagance.

> The team members experienced different frugal cultures in various places [...] During the qiongyou in Dalian, Liaoning [...] The members developed team spirit and experienced the true meaning of frugality. [...] The trend of high consumption and extravagance is spreading among some people. But frugality is the future of the state, the future of the nation (Guangming Daily, 2015).

As the national discourse became widespread through popular culture, qiongyouers also began to give legitimacy and broader socio-cultural connotations to their alternative travel practices in the name of frugality.

In the name of frugality: the cultural legitimacy of qiongyou

In the view of qiongyouers, the state's support for "frugality" better fuels their resistance. One qiongyouer has often countered their critics in online forums by saying, "If the state is supporting it, there must be a reason for it" (Interview #9). Since then, "frugality" has also been reinterpreted into a variety of socio-cultural meanings, further contributing to the construction of qiongyou's cultural legitimacy. Firstly, for qiongyouers, a high-cost trip does not actually allow for an authentic experience of the site, and it is perhaps better to do it for free. Such a view further justifies the qiongyouers evasion of tickets:

> (Researcher: Do you think that ticket evasion is illegal and immoral?)
>
> Interviewee: "The ticket prices for some attractions are speculated to be high. It's packaged to be overpriced. [...] The travel agencies often take you to attractions, not because they are really interesting, but because they have partnerships with businesses there and force people to shop. So would you say they are moral?" (Interview #2)
>
> Interviewee: "Sometimes the ones that don't cost money are the authentic ones. For example, natural landscapes that have not been developed yet, and also some ancient buildings. Many of the ones that charge admission fees have been renovated by later generations. But like many ancient pagodas and temples over in Shanxi, no one charges tickets at all, but instead they are authentic historical buildings" (Interview #4).

At the same time, some qiongyouers stress that frugality is a prerequisite for environmental sustainability. One qiongyouer commented on travel strategies, such as hitchhiking and couch-surfing, "I don't think it's about taking advantage. Space can actually be shared. When you go out and stay in a hotel, it's a waste to sleep alone in a big room, and it's not environmentally friendly" (Interview #6). There are also environmentalists who show their concern for ecology by qiongyou. One such volunteer named Ye Lan traveled around the world, learning about environmental protection, he also tried to demonstrate an attitude of traveling in poverty and shared his experience of qiongyou. He believes that:

> The first thing you have to do as an environmental advocate is to practice what you preach. Not only should you not waste, but you have to be stricter with yourself. [...] If the money is wasted, it would be against the original intention of promoting environmental protection [...] My mother is usually frugal and never lets meals at home go to waste, so what I do is generally in line with her temperament. (Sina Blog, 2013) (,)

Many female qiongyouers face more questions than their male counterparts. Some of them have to admit that "traveling with 200 yuan is something that male students can try, but not women" (Zhihu, 2019). This is partly because women have more personal safety concerns when traveling, and partly because female qiongyouers are often subject to ethical criticism. However, proponents of women's qiongyou raise the question: "When did the exploration of the world and the experience of the unknown become based on money? Has this even become part of the requirement for women to behave properly and care for their families?" (Zhihu, 2021) Thus, some female qiongyouers argue that qiongyou can reveal a different side of women:

> (Researcher: Many people on the Internet think that qiongyou for women means exchanging your body for a free trip, what do you think?)
>
> Interviewee: "It's still a prejudice against girls. One is that they don't believe that girls can protect themselves, and the other is that it seems like girls can't live without money. [...] The first time I tried qiongyou was in my second year of university. Many people also thought that girls were materialistic. But I think it's not true, girls can also be very simple and frugal. That's why I wanted to try to qiongyou. [...] My parents were supportive and thought it would be a good way to train myself." (Interview #11)

In other words, the qiongyouers see their alternative travel not as expedient or even immoral, but as a manifestation of more progressive socio-cultural connotations such as authenticity, sustainability and female independence. Opposing extravagance and promoting frugality are, in their view, the key to realizing these values. Clearly, the qiongyouers advocacy of frugality further challenges the dominant ideology embedded in the consumption-oriented tourism industry, encouraging more "poor people" to engage in travel despite criticism of "no money, no travel" from mainstream tourists.

Discussion and conclusion

Based on existing definitions of justice tourism (Higgins-Desbiolles, 2008; 2018; Mihalic and Fennell, 2015), this article argues that two main characteristics of qiongyou allow it to be considered as a form of justice tourism. One is that qiongyou is a grassroots tourist's socio-cultural resistance to the dominant injustice under the crisis. Second, qiongyou opens up more inclusive tourism participation outside the mainstream tourism industry, especially for the economically disadvantaged.

Qiongyou can be seen as a collective resistance, and especially reflects Higgins-Desbiolles (2008, 2018) argument of tourism as social activism in response to structural injustice. It emerged in the political economic context of the sudden economic crisis, rising unemployment and socio-cultural exclusion of mainstream tourism in 2008, and coincided with the beginning of the 21st century when the Internet became popular in China. These two factors contributed to the rapid rise in popularity of qiongyou after 2008, so much so that 2009 became known as the Year of Qiongyou in China. Thus, unlike backpacking, qiongyou is characterized by a short-term emergence and rapid scale-up, which gives it a certain collective nature and further economic and socio-cultural impact. The participants have also expanded from those affected by the early crisis to a wider range of actors.

Moreover, with the rise of qiongyou, an alternative tourism participation mechanism featuring low consumption outside the mainstream tourism industry has emerged, which allows for more opportunities and options for the economically disadvantaged to engage in travel. qiongyou can thus be seen as a challenge to state-led enclosure and tourism commodification (Devine, 2017) and as a contribution to a more inclusive tourism participation (Mihalic and Fennell, 2015). This is specifically due to the practices of qiongyouers in two aspects: alternative market practices and alternative discursive practices.

First, the low-budget travel needs of qiongyouers have inspired an alternative tourism market. Travel services such as the qiongyou designer and ticket evasion services have begun to emerge, driving alternative travel itineraries, grassroots travel agencies, and, more interestingly, cooperation between tourists and locals facilitated by inequitable tourism benefit distribution mechanisms. This cooperation in the particular context breaks the dichotomy of empowered tourists and disempowered locals often presupposed in past studies of justice tourism (Bianchi et al., 2020). The emergence of alternative markets also means that the economically constrained can achieve low-spending tourism participation outside of the dominant tourism industry, which aims to stabilize finances and promote consumption.

Second, qiongyouers also attempt to give cultural legitimacy to their rebellion through alternative discursive practices and further challenge the dominant ideology of tourism as consumer culture. Qiongyouers do not see their way of traveling as a stop-gap measure in the face of economic hardship. Rather, they took "poverty" as a positive identity and emphasize that "frugality" is the justified attitude to travel. Like many backpackers, qiongyouers also advocate authentic and ecological tourism, but they believe that frugality is a key step towards realizing these values. The cultural legitimization of qiongyou also challenges the dominant discourse of "no money, no travel", which helps "poor people" who are culturally excluded from mainstream tourism to be more willing to re-engage in travel.

The contribution of studying qiongyou is that it moves beyond the critique of overtourism under the neoliberalism to reveal the inequality and distributive injustice of tourism participation that exists in China's state-led tourism system, which has developed more as a policy tool for crisis governance. This study considers qiongyou as a form of justice tourism that has emerged in China, illustrating its specific travel strategies, cultural discourses, and ensuing set of alternative tourism participation mechanisms, which adds a tourist perspective to current research on justice tourism and highlights the agency of grassroots travelers to initiate alternative agendas in the face of structural injustice.

This study also has some limitations. Although the initial rise of qiongyou was supported by the economically disadvantaged in the economic crisis, there are also some qiongyouers who try to avoid spending in the name of being poor even if they can afford it, especially as the number of participants in qiongyou grows. This study does not seek to identify whether qiongyouers are truly or falsely poor, nor does it see qiongyou as the final solution to a more equitable and just tourism industry. What the case of qiongyou provides is a glimpse into the tensions between the state-led tourism and grassroots tourist participation amidst the global economic crisis and the specific political context of China. This will also present a series of fields of continued debate: how can tourists have the agency to initiate alternative agendas when the dominant tourism system is in systemic crisis? After grassroots action, how can a more equitable tourism participation and a more just tourism system be institutionally established to sustainably respond to potential new crises in the future. These are issues that can be further explored in future studies.

ORCID

Yimeng Yang http://orcid.org/0000-0002-0939-7628

References

Mafengwo (n.d.). Shuhe Ancient Town Tour: Shuhe Ancient Town Ticket Evasion Strategy. Retrieved 30 June 2022, from http://z.mafengwo.cn/wiki/13812-1.html

Ateljevic, I. (2000). Circuits of tourism: Stepping beyond the production/consumption dichotomy. *Tourism Geographies*, *2*(4), 369–388. https://doi.org/10.1080/146166800750035495

Beijing, D. (2009). The qiongyou designer opened an online shop to start a business: Play well but also spend less. Retrieved July 17, 2021 from https://www.chinanews.com/cj/cfgs/news/2009/02-13/1562310.shtml.

Bianchi, R. (2018). The political economy of tourism development: A critical review. *Annals of Tourism Research*, *70*, 88–102. https://doi.org/10.1016/j.annals.2017.08.005

Bianchi, R. V., & Stephenson, M. L. (2013). Deciphering tourism and citizenship in a globalized world. *Tourism Management*, *39*, 10–20. https://doi.org/10.1016/j.tourman.2013.03.006

Bianchi, R. V., Stephenson, M. L., & Hannam, K. (2020). The contradictory politics of the right to travel: Mobilities, borders & tourism. *Mobilities*, *15*(2), 290–306. https://doi.org/10.1080/17450101.2020.1723251

Chen, H., & Weiler, B. (2014). Chinese donkey friends in Tibet–Evidence from the cyberspace community. *Journal of China Tourism Research*, *10*(4), 475–492. https://doi.org/10.1080/19388160.2014.951503

Chengdu Business Daily. (2009). 2009 is the Year of the qiongyouer. Retrieved Jan 27, 2021 from http://m.qyer.com/bbs/thread-169323-1.html.htm.

Choi, J., Lee, S., & Jamal, T. (2021). Smart Korea: Governance for smart justice during a global pandemic. *Journal of Sustainable Tourism*, *29*(2–3), 541–550. https://doi.org/10.1080/09669582.2020.1777143

Cresswell, T. (1993). Mobility as resistance: A geographical reading of Kerouac's 'On the Road. *Transactions of the Institute of British Geographers, N.S*, *18*(2), 249–262. https://doi.org/10.2307/622366

Devine, J. A. (2017). Colonizing space and commodifying place: Tourism's violent geographies. *Journal of Sustainable Tourism*, *25*(5), 634–650. https://doi.org/10.1080/09669582.2016.1226849

Guangdong Tourism Association. (2018). How many of these changes in the tourism industry have you experienced in the 40 years of reform and opening up. Retrieved Jan 27, 2021 from http://www.gdtu.org/newsinfo/718842.html

Guangming Daily. (2015). "Frugal Chinese" Frugality builds the future. Retrieved July 17, 2021 from http://politics.people.com.cn/n/2015/0803/c70731-27399041.html.

Guia, J. (2021). Conceptualizing justice tourism and the promise of posthumanism. *Journal of Sustainable Tourism*, *29*(2–3), 503–519. https://doi.org/10.1080/09669582.2020.1771347

Higgins-Desbiolles, F. (2006). More than an "industry": The forgotten power of tourism as a social force. *Tourism Management*, *27*(6), 1192–1208. https://doi.org/10.1016/j.tourman.2005.05.020

Higgins-Desbiolles, F. (2008). Justice tourism and alternative globalisation. *Journal of Sustainable Tourism*, *16*(3), 345–364. https://doi.org/10.1080/09669580802154132

Higgins-Desbiolles, F. (2018). The potential for justice through tourism. *Via. Tourism Review*, (13)

Higgins-Desbiolles, F. (2020). Socialising tourism for social and ecological justice after COVID-19. *Tourism Geographies*, *22*(3), 610–623. https://doi.org/10.1080/14616688.2020.1757748

Higgins-Desbiolles, F., Carnicelli, S., Krolikowski, C., Wijesinghe, G., & Boluk, K. (2019). Degrowing tourism: Rethinking tourism. *Journal of Sustainable Tourism*, *27*(12), 1926–1944. https://doi.org/10.1080/09669582.2019.1601732

Isaac, R. K., Hall, C. M., & Higgins-Desbiolles, F. (Eds.). (2015). *The politics and power of tourism in Palestine*. Routledge.

Jabłonkowska, J. B., & Stankiewicz, B. (2018). The low–budgeting concept of backpacking. *European Journal of Service Management*, *26*, 123–129. https://doi.org/10.18276/ejsm.2018.26-15

Jamal, T. (2019). *Justice and ethics in tourism*. Routledge.

Jamal, T., & Camargo, B. A. (2014). Sustainable tourism, justice and an ethic of care: Toward the just destination. *Journal of Sustainable Tourism*, *22*(1), 11–30. https://doi.org/10.1080/09669582.2013.786084

Jamal, T., & Higham, J. (2021). Justice and ethics: Towards a new platform for tourism and sustainability. *Journal of Sustainable Tourism*, *29*(2–3), 143–157. https://doi.org/10.1080/09669582.2020.1835933

Jiangdong Times. (2008). Save money and expenses and have fun when you're poor'. Retrieved Jan 27, 2021 from http://www.bzdrs.com/readPaper.do?id=E997097B110A0257E6D9BD96D0D55A86B6A6416864AADD7B0ECF9D774BE1A92A4917C554CE3B8EF0D87C27F8FA152407

Kassis, R., Solomon, R., & Higgins-Desbiolles, F. (2015). Solidarity Tourism in Palestine: The Alternative Tourism Group of Palestine as a catalyzing instrument of resistance. In Rami K. Isaac, Colin Michael Hall, and Freya Higgins-Desbiolles (Eds.), *The politics and power of tourism in Palestine* (pp. 53–68). Routledge.

Kindel, T. (1985). Chinese consumer behavior: Historical perspective plus an update on communication hypotheses. In Agdish N. Sheth and Chin Tiong Tan (Eds.), *Historical perspectives of consumer behavior: National and international perspectives* (pp. 186–190). Association for Consumer Behavior.

Kozinets, R. V. (2002). The field behind the screen: Using netnography for marketing research in online communities. *Journal of Marketing Research*, *39*(1), 61–72. https://doi.org/10.1509/jmkr.39.1.61.18935

Li, Y., Hu, Z. Y., & Zhang, C. Z. (2010). Red tourism: Sustaining communist identity in a rapidly changing. *China. Journal of Tourism and Cultural Change*, *8*(1-2), 101–119. https://doi.org/10.1080/14766825.2010.493939

Lim, F. K. G. (2008). Donkey friends' in China: The Internet, civil society and the emergence of the Chinese backpacking community. In Tim Winter, Peggy Teo, and Tou Chuang Chang (Eds.), *Asia on tour: Exploring the rise of Asian tourism* (pp. 307–317). Routledge.

Lin, C. (2015). Red tourism: Rethinking propaganda as a social space. *Communication and Critical/Cultural Studies*, *12*(3), 328–346. https://doi.org/10.1080/14791420.2015.1037777

Liu, R. (2020). The state-led tourism development in Beijing's ecologically fragile periphery: Peasants' response and challenges. *Habitat International*, *96*, 102119. https://doi.org/10.1016/j.habitatint.2020.102119

Luo, X., Huang, S., & Brown, G. (2015). Backpacking in China: A netnographic analysis of donkey friends' travel behaviour. *Journal of China Tourism Research*, *11*(1), 67–84. https://doi.org/10.1080/19388160.2014.908757

McCabe, S., & Diekmann, A. (2015). The rights to tourism: Reflections on social tourism and human rights. *Tourism Recreation Research*, *40*(2), 194–204. https://doi.org/10.1080/02508281.2015.1049022

McCabe, S., & Qiao, G. (2020). A review of research into social tourism: Launching the Annals of tourism research curated collection on social tourism. *Annals of Tourism Research*, *85*, 103103. https://doi.org/10.1016/j.annals.2020.103103

Mihalič, T., & Fennell, D. (2015). In pursuit of a more just international tourism: The concept of trading tourism rights. *Journal of Sustainable Tourism*, *23*(2), 188–206. https://doi.org/10.1080/09669582.2014.943763

Minnaert, L. (2012). Social tourism as opportunity for unplanned learning and behavior change. *Journal of Travel Research*, *51*(5), 607–616. https://doi.org/10.1177/0047287511431324

Mkono, M. (2012). Netnographic tourist research: The Internet as a virtual fieldwork site. *Tourism Analysis*, *17*(4), 553–555. https://doi.org/10.3727/108354212X13473157390966

Mkono, M., & Markwell, K. (2014). The application of netnography in tourism studies. *Annals of Tourism Research*, *48*, 289–291. https://doi.org/10.1016/j.annals.2014.07.005

Mosedale, J. (2015). *Political economy and tourism: A critical perspective*. Routledge.

Nakano, R., & Zhu, Y. (2020). Heritage as soft power: Japan and China in international politics. *International Journal of Cultural Policy*, *26*(7), 869–881. https://doi.org/10.1080/10286632.2020.1845322

Newsnight. (2008). 'Qiongyou' itinerary tips hit the web. Retrieved July 17, 2021 from http://news.sina.com.cn/o/2008-12-02/124614818385s.shtml.

Nyíri, P. (2010). *Mobility and Cultural Authority in Contemporary China*. University of Washington Press.

O'regan, M. (2010). Backpacker hostels: Place and performance. In Kevin Hannam, and Anya Diekmann (Eds.), *Beyond backpacker tourism: Mobilities and experiences* (pp. 85–101). Channel View Publications.

O'Reilly, C. C. (2006). From drifter to gap year tourist: Mainstreaming backpacker travel. *Annals of Tourism Research*, *33*(4), 998–1017. https://doi.org/10.1016/j.annals.2006.04.002

Oakes, T. (1998). *Tourism and Modernity in China*. Routledge.

O'Reilly, C. C. (2005). Tourist or traveller? Narrating backpacker identity. In Adam Jaworski and Annette Pritchard (Eds.), *Discourse, communication and tourism* (pp. 150–170). Channel View Publications.

Rastegar, R., Higgins-Desbiolles, F., & Ruhanen, L. (2021). COVID-19 and a justice framework to guide tourism recovery. *Annals of Tourism Research*, *91*, 103161. https://doi.org/10.1016/j.annals.2021.103161

Sheller, M. (2021). Reconstructing tourism in the Caribbean: Connecting pandemic recovery, climate resilience and sustainable tourism through mobility justice. *Journal of Sustainable Tourism*, *29*(9), 1436–1449. https://doi.org/10.1080/09669582.2020.1791141

Telfer, D. J., & Sharpley, R. (2015). *Tourism and development in the developing world*. Routledge.

The Central People's Government of China. (2014). The Chinese Ministry of Propaganda and the National Development and Reform Commission have deployed an action to save for all by being frugal and raising morals. Retrieved July 17, 2021 from https://baike.baidu.com/reference/13993870/23e9VymUbKGsYHEVLtNEWu4fwjdqxO_aupgD_4TeOnLL2eRiEs27rX9NFmEZoxt5PZvFG03tdBBT7exAUSP6vtqYbC18oofHalYaGQYBHQh0p3SL

Thieme, J., Hampton, M. P., Stoian, C., & Zigan, K. (2020). The political economy of backpacker tourism: Explorations of tourism actors' embeddedness in Colombia. *Current Issues in Tourism*, 1–26.

Trivett, V., & Staff, S. (2013). Rise of the Chinese independent traveler. *Retrieved August 15*, 2014.

Wu, A. (2019). The development process of the subculture of "qiongyouers" in the Internet era. *Journal of Li Ming Vocational University*, *104*(3), 49–54.

Xie, J., & Zhu, X. (2019). A sociological analysis on the emergence of" Qiongyou": A comparison between the Qiongyouer and the backpacker. *Tourism Tribune*, *34*(1), 124–135.

Yao, Y. (2017). Review of the development of China's travel agency industry and outlook. Retrieved July 17, 2021 from https://www.sohu.com/a/198911755_748790

Zeng, B., & Ryan, C. (2012). Assisting the poor in China through tourism development: A review of research. *Tourism Management*, *33*(2), 239–248. https://doi.org/10.1016/j.tourman.2011.08.014

Zhang, G. (1998). The Impact of the Asian financial crisis on China's international tourism industry and its countermeasures. *Financial and Trade Economics*, *05*, 13–17.

Zhang, Y. (2015). Legislation must be enacted to clarify the use of revenue from tickets in scenic areas. Retrieved Jan 27, 2021 from https://epaper.gmw.cn/gmrb/html/2015-06/29/nw.D110000gmrb_20150629_5-10.htm?div=-1

Zhou, L., & Chan, E. (2019). Motivations of tourism-induced mobility: Tourism development and the pursuit of the Chinese dream. *International Journal of Tourism Research*, *21*(6), 815–824. https://doi.org/10.1002/jtr.2308

Zhu, X. (2018). The comparative study on authentic experience of Western and Chinese backpacker travelers based on narrative analysis: A case study in Yangshuo, China. *Tourism Tribune*, *33*(3), 116–126.

Appendices

Interview #	Gender	Travel experiences: travel strategies and cultural discourses
#1	Male	Student qiongyouer, Once cycled into Tibet as a qiongyouer
#2	Male	Opposed to joining group tours and had experience of ticket evasion
#3	Male	Ticket evasion
#4	Male	Ticket evasion
#5	Male	Hitchhiking and Couch-surfing
#6	Male	Hitchhiking and Couch-surfing
#7	Male	Student qiongyouer, backpacking
#8	Male	Newly working qiongyouer., relies on independent travel agents
#9	Male	Agree with the value of "frugality" proposed by the state
#10	Female	Ticket evasion
#11	Female	Student qiongyouer, advocate female independence
#12	Female	Young stay-at-home mom, independent traveller
#13	Female	Couch-surfing and Ticket evasion
#14	Female	Student qiongyouer, Ticket evasion

Tourism, compounding crises, and struggles for sovereignty

Carter A. Hunt, María José Barragán-Paladines, Juan Carlos Izurieta and Andrés Ordóñez L

ABSTRACT

Applying justice theory in tourism studies has yielded a vibrant flourishing of scholarship in recent decades. Yet, it is still argued that a clear conceptualization of justice tourism is still lacking. Sovereignty theory has seen broad application across many social sciences in recent decades, yet despite a clear connection, the tourism scholarship has engaged minimally with the sovereignty literature. This article aims to assimilate sovereignty theory into the justice tourism scholarship by carrying out a deep historical analysis to demonstrate how destination residents negotiate chronic and acute crises in the Galápagos Islands, a place with no original human population. With global immigration projected to grow and exacerbate environmental conflicts in the coming years, the current research is well-poised to provide urgent and generalizable insights into the sociocultural underpinnings of increasing human mobility, the environmental conflicts that exist between different value systems and worldviews, and the opportunities that exist to promote improved destination management on behalf of human wellbeing in places experiencing intense in-migration. Historical analyses are thus critical to understanding the subjective and temporal nature of struggles associated with justice-centric concepts, including but not limited to sovereignty.

Introduction

Applying justice theory in tourism studies has yielded a vibrant flourishing of scholarship in recent decades, embodied by but not limited to a 2021 special double issue of *Journal of Sustainable Tourism* (reprinted as Jamal & Higham, 2021a). Justice was actively theorized in the broader social sciences for decades, much of it traced to Rawls's work on distributive and procedural justice (Rawls, 1971). Rawlsian applications of theory about distributive and procedural justice eventually trickled into the tourism literature. In the recent "justice turn" in sustainable tourism scholarship (Rastegar, Higgins-Desbiolles & Ruhanen, 2021, p. 2), numerous other forms of justice (e.g. restorative, cultural, Indigenous, social) are now theorized. As scholars have emphasized (Higgins-Desbiolles, 2020; Rastegar, Higgins-Desbiolles & Ruhanen, 2021; Jamal & Higham, 2021a; 2021b), these various justice frameworks can provide a powerful means with

The article should have been included in the Special issue "Tourism, global crises and justice; Guest Editors: Raymond Rastegar, Freya Higgins-Desbiolles and Lisa Ruhanen" (RSUS_31-12)

This article was originally published with errors, which have now been corrected in the online version. Please see Correction (https://doi.org/10.1080/09669582.2023.2288659)

which to assess and promote recovery from both chronic (e.g. climate) and acute (e.g. COVID-19) crises. They also stimulate critical reflection on the tourism sector itself (Higgins-Desbiolles, 2020; Filep et al., 2022). Yet, despite this recent flourishing, it is argued that "a clear conceptualization of justice tourism is still lacking, and its theoretical grounding is still too limited" (Guia, 2021, p.503).

Like justice theory, sovereignty theory has seen broad application across many social sciences in recent decades. Moving far beyond early Westphalian notions of territorial or nation-state sovereignty, recent theoretical contributions in development studies and agrarian studies have expanded to more constructivist analyses of numerous forms of sovereignty, with food sovereignty receiving the most scholarly attention (Patel, 2009; Jarosz, 2014). Each of these forms of sovereignty emphasizes justice-centric ideas of local representation, participation, and influence in local decision-making processes, yet despite a clear connection, the tourism scholarship has engaged minimally with the sovereignty literature (Naylor et al., 2021; Naylor & Hunt, 2021). Integrating sovereignty into justice theory in tourism could thus provide clearer conceptualization and theoretical grounding by establishing that tourism-enabled sovereignty is a necessary precursor to just tourism and destinations.

The Galápagos Islands provide an ideal context to demonstrate the value of a sovereignty analysis. The arrival of people to the archipelago was very late. Yet, the rapid growth of the human population in just the last 70 years has brought claims of sovereignty, legitimacy, and justice that are evolving, adapting, and diversifying as the islands' resident populations grow. Long-standing crises related to the need to prioritize conservation and promote nature-based tourism to support it are now complemented by responses to an acute COVID-19 crisis that temporarily halted elements of those long-standing discussions. Underlying it all is a growing crisis related to the cultural identity of the burgeoning island society. *Is this a place to promote human flourishing, or should resident concerns remain subordinated to protecting the islands' unique ecosystems? Whose interests are best represented in the decisions and policies related to community development? For whom is the Galápagos?*

The purpose of this article is to assimilate sovereignty theory into the justice tourism scholarship by carrying out a deep historical analysis to demonstrate how residents in the Galápagos Islands negotiate chronic and acute crises. Our methodology expands descriptive and interpretive theory-building regarding sovereignty and tourism (Lynham, 2002; Dennis, 2019; Smith et al., 2013; Jamal & Higham, 2021b), and it also responds to the call of Nunkoo et al. (2021, p. 19), who point out that "disciplines such as anthropology and the humanities have yet to be fully integrated into sustainable tourism research." Furthermore, our author team is comprised entirely of current residents of the Galápagos, three of whom are native-born Ecuadorians. Our writing, therefore, addresses a need for greater representation of voices from the global south in scholarly literature (Chambers & Buzinde, 2015). It also ensures the analysis is imbued with a direct emic understanding of the study context. This approach reveals the deeper origins and entangled nature of long-standing struggles for various forms of sovereignty among Galápagos resident sectors.

Literature review

Tourism researchers have pervasively cited justice and related concepts of fairness, equity, and ethics (Jamal, 2019). The connection between justice and tourism is traceable to early writings by tourism anthropologists and development scholars in the 1970s (e.g. Smith, 1977, De Kadt, 1979). However, justice research has evolved significantly since then, influenced by prominent international legislation advocating social and environmental sustainability (e.g. 1987s Brundtland Report and 2015s UN Sustainable Development Goals). Seen as one means of accounting for social sustainability, the justice concept fostered critical views of mass tourism as inherently exploitive of destination residents (Dangi & Petrick, 2021), thereby implying support for alternatives seen as more socially and environmentally "responsible," such as community-based tourism, pro-poor tourism, and ecotourism (Jamal & Higham, 2021b; Jourdan & Wertin, 2020).

Theorizing justice in tourism studies draws heavily upon Eurocentric philosophies such as Aristotelian ideas about the good life or Rawls's ideas about distributive justice and procedural justice (Rawls, 1971; Jamal, 2019; Jamal & Higham, 2021a). Yet as scholars have applied interpretivist, constructivist, Indigenous, feminist, and post-development research methodologies to move beyond Rawlsian perspectives, they have also helped push this work into new theoretical realms and its current state of flourishing (Rastegar, Higgins-Desbiolles & Ruhanen, 2021). As the justice turn has unfolded, tourism scholars' emphases have diversified into additional notions that include but are not limited to the following: organizational, interactional, recognition, performative, destination, cultural, and restorative justice. Issues of human dignity and self-determination remain the main guiding concepts for justice frameworks, which are promoted as prerequisites to environmentally and socially sustainable tourism (Higgins-Desbiolles, 2020; Camargo & Vázquez-Maguirre, 2021).

Although tourism scholars have assessed a "potpourri of justice outcomes" (Guia, 2021, p.504), a consistency within this scholarship is the emphasis, time and again, on tourism's role in fostering wellbeing, particularly wellbeing as understood and recognized within destination communities themselves (Jamal & Camargo, 2014). An interest in justice thus dictates an imperative of understanding whose voices are marginalized, which voices are heard, and how local populations view the fairness of their participation and representation in decision-making processes of most influence over their own social and environmental wellbeing (Jamal & Stronza, 2009; Rastegar, Higgins-Desbiolles & Ruhanen, 2021). Stronger conceptual and theoretical clarity in tourism and justice writing is likely to result from sibling social science or humanities disciplines that embrace historically focused analyses of tourism's role within long-standing struggles to overcome oppression and improve local wellbeing (Nunkoo et al., 2021; Gao et al., 2021). Sovereignty theory may provide such clarity. As further reviews of the tourism-justice scholarship have already been provided elsewhere (e.g. Jamal, 2019; Jamal & Higham, 2021a), in the remainder of this article, we shift our focus to the prolific sovereignty literature that has been largely overlooked to date by tourism scholars.

Introducing sovereignty

The term sovereignty is commonly traced to the 1648 Peace of Westphalia, which characterized nation-states as the supreme authority over geographical boundaries (Stepputat, 2015). Though that conceptualization continues to form the basis of the modern field of international relations, the sovereignty concept has since diversified into various forms of interest both within and outside academia. As in the Rawlsian justice literature, contemporary research on sovereignty has benefitted from constructivist, critical, participatory, and other radical forms of inquiry that have greatly expanded the focus to other realms (McMichael, 2014). For instance, these approaches have resulted in much research on the mobilization of Indigenous communities to achieve nested *intellectual* and *cultural sovereignties*, demarcating them as necessary precursors to overall *Indigenous sovereignty* (Coffey & Tsosie, 2001).

Food sovereignty has stimulated a vibrant area of scholarship (e.g. Patel, 2009; Agarwal, 2014; Edelman et al., 2014; McMichael, 2014). Inspired by social movements (e.g. Ecuadorian food sovereignty movement *Via Campesina*) related to injustices in the control of food systems (Patel, 2009), food sovereignty refers to "the right of peoples to healthy and culturally appropriate food produced through ecologically sound and sustainable methods, and their right to define their own food and agriculture systems" (Nyéléni, 2022, para. 3). Although a deeper review of food sovereignty literature is also beyond our scope here, Jarosz (2014) provides an influential overview of the theory and practice of food sovereignty that highlights several key threads. First, there is a clear distinction between food security (e.g. simple access to food) and food sovereignty, which represents local control and decision-making, thereby recognizing the political dimensions of food systems (Agarwal, 2014). Second, elaborating on this crucial distinction, rural studies scholars are now calling for ideas of food sovereignty to be extended to encompass broader ideas of *livelihood sovereignty* (e.g. MacRae, 2016; Tilzey, 2019), ideas recently invoked in tourism settings (Naylor & Hunt, 2021; Naylor et al., 2021; see below).

A third thread in the sovereignty literature recognizes "formations of sovereignty" as a way of dealing analytically with the political landscapes formed through multiple, coexisting, overlapping, and sometimes competing claims to sovereignty over people, resources, and/or territories." (Stepputat, 2015, p. 129). Such nested formulations represent varying degrees of political agency and autonomy among heterogeneous local communities, thereby situating sovereignty on a relational scale based on locally produced representations of fairness, political participation, and human wellbeing. A sovereignty perspective implies that system reorganization often results from negotiating sovereignty gaps exposed at the global, national, insular, and sub-insular levels during periods of systemic disturbance (Mawyer & Jacka, 2018).

Finally, and broadly consistent with justice literature, sovereignty theory reinforces that understanding what is socially sustainable requires first understanding whose 'reality' dominates and which people are advantaged or disadvantaged in the process. In conceding that communities are often comprised of heterogenous cultures that are simultaneously interwoven and evolving, we can therefore conceive of sovereignty as "not an extraneously existing object" but rather "a living process" that builds and maintains "relationships between people, institutions, technologies, ecosystems, and landscapes across multiple scales," especially temporal scales (Iles & Montenegro de Wit, 2015, p. 483). Sovereignty theory thus highlights the value of historical analyses for understanding these critical community dynamics.

Tourism and sovereignty

Discussions of sovereignty in the context of tourism are very limited to date. What does exist has often invoked traditional Westphalian sovereignty ideas to analyze how tourism redefines diplomatic relations and territorial claims between nation-states (Timothy et al., 2014), such as the transition of Hong Kong around the turn of the century (e.g. Perry Hobson & Ko, 1994). Even more limited writing analyzes tourism's influence on the negotiation of Indigenous sovereignty in Indigenous communities in Hawaii (Williams & González, 2017) and Ecuador (Santafe-Troncoso & Loring, 2021), and in aboriginal communities in South Australia (Nicholls & Higgins-Desbiolles, 2016). Bunten (2014, p.311) examines tourism as a possible threat to cultural sovereignty of indigenous populations, defined as when "tribal peoples have intellectual property and ways of life that are protected and governed under tribal law." Like many references to sovereignty, Higgins-Desbiolles (2020) mentions sovereignty only in passing, not as a particular framing device for analysis.

Responding to the call for broader application of food sovereignty concepts to local livelihood systems, some scholars have recently defined *livelihood sovereignty* in the context of tourism as "the enhanced levels of local resident control and influence over management institutions and decision making regarding the persistence of valued traditional practices, how new production opportunities are integrated into socio-ecological systems, and how local community wellbeing is perpetuated over time" (Naylor & Hunt, 2021, p.6; Naylor et al., 2021). This definition acknowledges that tourism-related livelihood security (i.e. income and employment) *is insufficient to achieve livelihood sovereignty*, as there is a distinct need for local communities to retain local control and decision-making within the tourism sector to ensure it develops in ways that advance community goals (Naylor et al., 2021). Thus, as food security is a necessary but insufficient condition for food sovereignty, tourism-related livelihood security is a necessary but insufficient condition for just tourism.

Historical methods

Humanities-based approaches, including historical and anthropological analyses, remain exceptions rather than the rule in the peer-reviewed tourism scholarship (e.g. Caton, 2016; Camargo & Vázquez-Maguirre, 2021; Jamal, 2019; Nunkoo et al., 2021). Distinct from humanism as a philosophical stance, our humanistic anthropological approach instead straddles anthropology and the humanities disciplines, especially history. Tilly (1983) characterizes historical research as "studies assuming that the time and place in which a structure or process appears makes a difference to its character, that the sequence in which similar events occur has a substantial impact on their outcomes, and that the existing record of past structures and processes is problematic, requiring systematic investigation in its own right instead of lending itself immediately to social-scientific synthesis" (p.79). The historical approach is thus well-suited to understanding the interwoven temporal dynamics of crises, sovereignty, and tourism in the Galápagos.

To carry out this approach in a context where the "native" human population is not Indigenous but rather comprised of a variety of diasporic, immigrant cultural groups, our historical approach can be further characterized as ethnohistory. Ethnohistorical research creates "a more inclusive picture of the histories of native groups through analyses and interpretations that seek to make evident the experiences, organizations, and identities of indigenous, diasporic, and minority peoples that otherwise elude the histories and anthropologies of nations, states, and colonial empires" (American Society for Ethnohistory, 2022). To enable more in the way of descriptive and interpretive theory-building rather than predictive or causal modeling (Smith et al., 2013; Dennis & Indiana University, USA 2019), our ethnohistorical emphasis favors the subjective and interpretive aspects of analysis over more positivist or empirical traditions within the social sciences and tourism studies more specifically (Nunkoo, 2018), it "embraces emic perspectives as tools of analysis" (Hester, 2018, p.258). We, therefore, actively seek to represent biased interests in our study context.

We rely on hallmark archival research methods among primary and secondary sources in libraries, institutions, and private collections rarely circulated beyond Ecuador (McDowell, 2013). Additionally, to account for views of Galápagos residents, we also draw extensively on *our own lived experiences as current residents* of Galápagos. The first author is a university-based researcher who began studying human-environment relations in the Galápagos Islands in 2012. The second and third authors have eight years of experience in tourism research and directing the scientific agenda and knowledge management platforms at the Charles Darwin Foundation. The final author directs the Galápagos Chamber of Tourism, an institution dedicated to conserving the Galápagos through legal and technical support for the tourism sector. We are not observers nor participant observers but full participants in Galápagos society who also happen to be university researchers, scientists, analysts, and directors of institutions directly involved with monitoring, managing, and studying tourism in the Galápagos Islands. Such positionality is rare in tourism research, enabling us to incorporate emic and etic insights into the analysis below.

Compounding crises in the Galápagos islands

Nowhere is understanding the consequences of anthropogenic disturbance more urgent than in UNESCO's first World Heritage Site (Durham, 2021), where Darwin's description of the Galápagos as "a little world within itself" remains apt today. Theories of island biogeography hold that islands provide simplified conditions that put systemic dynamics into sharper relief than would otherwise be the case (MacArthur & Wilson, 1967; González et al., 2008; Baldacchino, 2016). Yet it is not just the simplified systems that characterize the islands but also the non-Indigenous nature of the fledgling human population in Galápagos that make the place a compelling context in which to conduct a justice-centric sovereignty analysis. Unlike contexts for justice research that juxtapose powerful with powerless or Indigenous with colonizer, *the Galápagos had no native population.* Rapid human convergence in recent decades creates a valuable context to analyze, interpret, and explain how diverse sovereignty interests are negotiated in response to chronic and acute crises, including those related to tourism. To the extent that Galápagos represents a set of fragile socio-ecological conditions undergoing accelerating human-induced change, they can yield insights into how local populations around the planet negotiate chronic and acute crises. This section elaborates a historical analysis describing how the acute *COVID crisis* was overlain upon several chronic crises underway in the Galápagos Islands: the long-standing *conservation crisis*, the mounting *tourism crisis*, and an underlying *cultural identity crisis*.

The chronic conservation crisis

The Galápagos remained free of human presence until 1535 (Ospina, 2001), when wayward Europeans made the first recorded visit. Documentation appears in accounts of pirates, buccaneers, and mapping expeditions that visited in subsequent centuries. Numerous nations sought dominion over regional forms of trade, especially lucrative whaling revenues, during an "exploitation" phase of Galápagos history. Yet, for nearly three hundred years since discovery, no nation claimed territorial sovereignty. Finally, prior to Charles Darwin's 5-week visit in 1835, Ecuador annexed the Galápagos Islands in 1832. The earliest recorded protective action in the Galápagos came in 1883 when the National Ecuadorian Constitution endowed unique governance status through special laws (Barragán Paladines & Chuenpagdee, 2017). Freshwater and ease of access led to Floreana experiencing the earliest colonization efforts (Latorre, 1999). After various failed efforts on that island, other freshwater sources on larger Isabela and later on San Cristobal led to "successful" plantation settlements on those islands, often supported by convict labor (Latorre, 1999; Ospina, 2001). Finally, in the 20th century, the most populated island today, Santa Cruz, was permanently settled. Until as recently as 1950, the total permanent presence across all islands was no more than 1400 residents.

After centuries of little growth or demographic change, land reforms on the Ecuadorian mainland rebranded the islands as an agricultural frontier to be conquered, creating new drivers of migration (Ospina, 2001). In the decades leading up to 1950, dozens of scientific expeditions by private universities (e.g. Stanford, Johns Hopkins), research centers (e.g. California Academy of Sciences), wealthy financiers (W.K. Vanderbilt, V. Astor), and assorted Western scientists collected, cataloged, and named the natural history of the islands. As these expeditions brought back descriptions of unique adaptations of the endemic species and otherworldly volcanic landscapes to Western audiences, a powerful imaginary was cultivated that implied a fragile environment full of unique wonders in desperate need of protection from impending impact of land-grabbing farmers. Mobilization of this narrative led to the 1959 creation of the Galápagos National Park (Ecuador's first) and the Charles Darwin Foundation, each headquartered in Puerto Ayora on Santa Cruz island. With the creation of these institutions, a conservation crisis was institutionalized in the islands that deemed it necessary to protect the Galápagos before further human arrival (Barragán Paladines & Chuenpagdee, 2017).

The initiation of organized tours in subsequent decades placed the conservation crisis squarely at the center of the rationale for tourism development (Epler, 1993; 2007). In 1973, the Galápagos was declared an insular province based on its "singularity," and a National Galápagos Institute was established to govern it (Grenier, 2007). State-based infrastructure improved, including roads, hospitals, and schools. Migration to the islands shifted from a frontier-based model focused on subsistence agriculture or fishing livelihoods to an amenity-based model defined by market-based opportunities in tourism (Ospina, 2001). Employment ratcheted up, and living

Table 1. Current population in inhabited islands.

Island	Area (ha)	Inhabitable area %	% of 2015 population*	2015 Population*
Santa Cruz	98,522	44%	62.2%	15,701
San Cristobal	55,697	34%	28.1%	7,088
Isabela	470,344	20%	9.3%	2,344
Floreana	17,229	1.2%	0.4%	111
				25,244

Sources: INEC, 2015.

conditions soon exceeded those in other Ecuadorian provinces, further incentivizing in-migration. Tourism growth made Santa Cruz the population and de facto decision-making center, placing the archipelago on a growth trajectory that continues today (Epler, 2007; Hunt, 2021; Table 1).

By 1984, the Galápagos held UNESCO Biosphere Reserve status, giving it the mandate to account for human activities and "promote solutions reconciling the conservation of biodiversity with its sustainable use... under diverse ecological, social and economic contexts" (UNESCO, 2021). Nevertheless, the stacking of conservation designations in the islands led to conflicts over restrictions on marine resource use. To alleviate rising concerns about impacts of tourism on fragile island environments, the Ecuadorian National Assembly passed the Special Law for Galápagos in 1998. In addition to giving origin to the Galápagos Marine Reserve, the Special Law established a co-management model of governance, new resource management policies, and residency rules to limit immigration. Furthermore, to ensure more tourism benefits flowed to local communities, the Special Law specified that 100% of the revenue received through park entrance fees remain in Galápagos, earmarked for the overall Governing Council of Galápagos, the municipalities in the Galápagos, and the smaller parishes in the islands (CGREG, 2021).

Despite the Special Law's accommodations, bitter conflicts grew during the lobster and sea cucumber booms of the 1990s. In 2000, fishers protesting restrictions burned the national park headquarters, held park staff hostage, and hung giant tortoises to protest fishing restrictions (Lu et al., 2013; Durham, 2008). Uncontrolled growth of tourism, and continued conflict regarding harvest restrictions in the marine sector, exacerbated the conservation crisis and resulted in UNESCO putting the Galápagos on the list of World Heritage in Danger in 2007, whose officials stated, "The principal factor leading to the inscription of the property [as a] World Heritage in Danger arises from the breakdown of its ecological isolation due to the increasing movement of people and goods between the islands and the continent, facilitating the introduction of alien species which threaten species native to the Galápagos" (Strahm & Patry, 2010, p. 6).

In response, restrictions on both fishing and tourism were implemented, and the fishing community gained representation in the Galápagos Governing Council. Consequently, UNESCO removed Galápagos from the list of World Heritage in Danger in 2011. The conservation crisis was nevertheless permanently cemented into narratives related to the islands (Lu et al., 2013). Threats to local environments have grown in scale since the park was created. Even with the careful controls introduced by the Galápagos Biosecurity Agency (ABG) in 2012, the number of introduced species continued to grow and threaten native species. Legal extraction leaves certain fisheries (e.g. Galápagos grouper) teetering on collapse. The logistics of protecting the expansive marine reserve have failed to eliminate illegal fishing or legal overfishing, leading to ongoing media attention to the extraction of shark fins and other protected species by large-scale international fishing fleets (e.g. Alberts, 2020). Such media representation fixates a conservation crisis squarely within the global imaginary of Galápagos.

Evolution of the tourism crisis

The contemporary history of Galápagos is often traced to the installation of a US Army Air Force base on Baltra Island in 1942 (Latorre, 1999). Containing more paved surfaces than the rest of

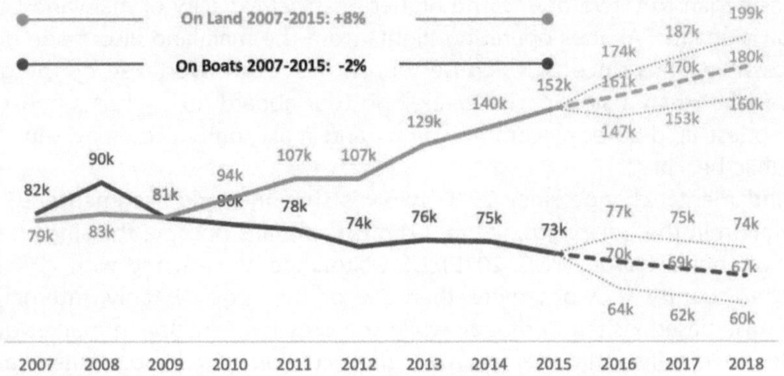

Figure 1. Pre-pandemic distribution of Galápagos tourists on land vs. boats*.
Source: Adapted from Observatorio de Turismo, 2022 (*compound annual growth rate and projection for 2016–2018)

Ecuador at the time, the base made the islands accessible by air for the first time. The US withdrew at the end of World War II, and what remained was transferred to the Ecuadorian government. The facilities were rarely used for years, and only piecemeal tourism efforts occurred. Organized tourism then got underway with Ecuadorian company Metropolitan Touring, which initiated the "floating hotel" model of visitation in the Galápagos in 1969. Metropolitan took passengers onboard luxury cruise ships that embarked on week-long tours. Following carefully controlled itineraries to designated park sites, tourists remained under constant supervision of naturalist guides who were to keep a conservation ethic squarely at the center of the tourism experience. The "floating hotel" tourism model accounted for nearly all tourism between the early 1970s and 2000s. This approach led to the Galápagos being referenced as "the place where ecotourism originated" since it kept environmental impacts minimal while providing visitors with high-quality nature experiences (Honey, 1994). Tourist numbers were to be capped at 12,000 visitors. The National Park carefully managed quotas for visitation to different park sites to control for ecological impacts and preserve the visitors' immersive and uncrowded nature experience.

Yet even in the early days of tourism, the islands were already wrestling with how many visitors were too many. When visitation was still under 20,000 annual visitors, De Groot (1983, p.291) already concluded that "unless decisive action Galápagos, will become another example of man's dangerous habit of preferring short-term economic gains over long term ecological and economic interests." Ten years later, when 42,000 visitors were visiting the islands each year, tourism was acknowledged as "the driving force which, directly and indirectly, dictates the pace and types of changes occurring in the islands" (Epler, 1993, p.1). Despite such dire predictions, tourism growth continued. Yet the lobster and sea cucumber booms in the 1990s raised concerns about overfishing that displaced those regarding tourism's impact (Durham, 2008). Seen as comparably benign, tourism development was further incentivized by 1998s Special Law. The Marine Reserve it created only added to destination marketability (Barragán Paladines & Chuenpagdee, 2017). "Predictions made by De Groot that the "resulting environmental damage will reduce the attractiveness of the islands and tourism will eventually decrease, causing economic damage as well" (1983, p.299–300) did not come to pass. Instead, by 2010, the Galápagos were receiving more tourists per month than were arriving annually in the early 1980s.

While tourism concern abounds in the writings on Galápagos, several distinct changes have emerged since 2010 that are often overlooked (Hunt, 2021). First, starting that year, more visitors stayed in hotels on the islands than aboard cruise boats (Observatorio de Turismo, 2022). The long-established growth of small cruise visitation on a preset itinerary of islands within the Galápagos National Park began to level off as visitor numbers reached the maximum number of overnight berths (Figure 1). By 2008–2009, the Special Law's incentives for locally owned businesses had manifested in more land-based operations. In the years of the global economic crisis, these businesses began to thrive due to the higher financial capacity of mainland Ecuadorians to afford travel to Galápagos. Airlines operating flights from the mainland also made domestic visitation more accessible. Thus since 2010, cruise tourism has been overtaken by the growth in on-island tourism. Rather than a slow-paced week-long tour aboard luxury boats, most visitors now stay in more modest land-based accommodations and make only a few day trips to local park sites, mostly public beaches.

A second fundamental change since 2010 involves shifting market segments. US visitors long dominated visitation in the Galápagos, yet in 2017, Ecuadorians became the single largest market segment (31%) of annual visitors (PNG, 2019). US visitors are now second with 29% of total arrivals, and no other country provides more than 5% of the total. Notably, international visitors dominate the cruise-based visitation model, while the growing Ecuadorian market dominates the on-island model. In the first trimester of 2022, the Ecuadorian market segment accounted for 79,160 visitors to the Galápagos, 88% of whom stayed on land (7% stayed in on-island lodging and 5% stayed with friends/family)(PNG, 2021). In contrast, the second-largest market, the United States, accounted for 33,987 visitors, 65% of whom stayed on boats and 34% stayed on land. This growth of the Ecuadorian market corresponds to a further intensification of the on-island model of tourism in Galápagos that competes with other sand-sun-sea destinations on the coast of the Ecuadorian mainland for this domestic market (Mestanza-Ramón et al., 2021; Hunt, 2021). Such destinations are characterized less by interpretive nature-based experiences with unique and globally significant species and more by beach visits, surf and scuba lessons, upscale cafes selling locally-grown coffee grown, rented electric scooters zipping up and down the streets of Puerto Ayora, bars and discotheques, and an increasing number of craft breweries (Burke, 2021; Carvache-Franco et al., 2021). Moreover, with National Park site quotas to visit specific islands filled by the long-standing agreements with the operators of larger passenger cruises, there is little opportunity for shorter-term, on-island visitors to see the farther reaches of the park. This pivot away from an exclusively nature-based visitor experience and a decoupling from the conservation ethic embodied in small cruise tourism are changing the fundamental character of tourism in the Galápagos (Hunt, 2021).

A third fundamental change since 2010 is intertwined with the last two. It relates to resident demand for recreational access. The national park's management plans were designed around the floating hotel, cruise tourism model. Designed disembarking sites were established in areas throughout the archipelago, and carefully managed schedules for visitation to these sites were created to distribute environmental impact, provide constant supervision by trained guides, and ensure optimal visitor experiences centered on nature interpretation. With the shift to on-island tourism outlined above, visitors to remote park sites are now the minority of overall visitation. The less-regulated park sites nearest to the islands' population centers, which have no visitation quota, have become the most visited sites in the Galápagos National Park (PNG, 2021). Whereas residents can visit these sites with no fee nor guide accompaniment, international and domestic tourists are charged a fee and required to be escorted by certified guides to these same sites. Mainland Ecuadorian tourists now dominate this on-island visitation. As reported in governing council documents (CGREG, 2016), in local media (e.g. Vega, 2020), in peer-reviewed literature (e.g. Cajiao et al., 2020), and as we have experienced on many occasions, increasing congestion elevates tensions with the resident population, who have long considered these local park areas "recreational sites" primarily for their use. Concerns for "recreational sovereignty" have entered the public discourse around park management priorities, including at multi-institutional participatory workshops hosted by the Consejo de Gobierno that two current authors attended.

Despite the concerns over the magnitude of tourism expressed in nearly all writings on Galápagos, cruise-based tourism grew immensely without significant environmental impact. A visitor to a remote park site will likely have a similar experience now as a 1980 visitor. The same cannot be said of the dramatic development in the population centers, where changes have been dramatic. It is, therefore, not the magnitude of tourism alone that is driving challenges in the contemporary tourism crises. Instead, it is a fundamental change in the model of visitation that has come to dominate the islands. This domestic tourism market has become the primary segment, and a qualitatively different tourist experience is now provided and acquired in the islands.

An acute COVID crisis

Annual tourism visitation reached all-time highs of over 270,000 visitors in 2018 and 2019. Then in March 2020, the COVID-19 pandemic caused approximately 25,000 monthly visitors generating $1.5 million in entry fees to plummet to zero (PNG, 2021; Díaz-Sánchez & Obaco, 2021). As occurred in destinations across the globe, the vulnerabilities inherent in a high reliance upon tourism-based economies were made painfully evident. A drop-off in transportation and shipping between the mainland and the islands sparked food security concerns. Individuals who had the means immediately invested in cultivating food products for self-consumption, while others engaged in temporary labor exchanges to ensure access to locally grown foods (Burke, 2021). Food producers (fishers and farmers) organized themselves into new formats to trade their produce. Home delivery services sprouted to help meet food needs and keep restaurants and other establishments afloat. Nevertheless, without a constant flow of tourists, a lack of food and livelihood sovereignty was brought into stark relief for residents, particularly in the two lesser populated islands of Isabela and Floreana, where infrastructure and services are minimal.

By August 2020, residents stranded outside the islands had been allowed to repatriate, and the first researchers and visitors were allowed to enter the islands. As the pandemic continued to unfold, the return of permanent and temporary residents precipitated broader discussions regarding for whom is Galápagos, who should be allowed to obtain temporary residency, and thus who should be allowed to compete for acutely limited resources and livelihood opportunities. While such discussions are encountered on a routine basis in the course of our collective work in Galápagos, these debates were most acutely experienced by two current authors at a series of inter-institutional participatory workshops that focused on redefining the UN Sustainable Development Goals for the Galápagos (Co-Galapagos, 2021). Yet the need to re-activate the economy, of which tourism is the core activity, remained central in the official discourse regarding the prioritization of resources. In the desperation to restart tourism's economic motor, pre-pandemic concerns for excessive visitors and questions regarding how high to raise the entrance fee were immediately set aside in favor of discussions of how to re-activate and get the masses back as soon as possible (Higgins-Desbiolles, 2020).

Nevertheless, as of April 2021, before vaccines were administered anywhere else in Ecuador, nearly the entire Galápagos population received a vaccine, leading to the tourism-friendly assertation that the islands were the first fully vaccinated province in the world. The tourism sector was directly involved in this accomplishment, donating approximately $740,000 in food kits, medical supplies, respirators, PCR tests, and logistical efforts. In the process, it was reconfirmed to the local population how vital tourism was to the health sovereignty of the Galápagos (Werkheiser, 2014). Meanwhile, others capitalized on the moment to reopen old narratives about opening direct international flights to the islands. Direct flights had been discussed in October 2019, five months before the pandemic, when widespread protests erupted in mainland Ecuador, resulting in the temporary closing of mainland airports and highways (Altmann, 2020; El Comercio, 2019). Those closures provided a cautionary preamble regarding the islands' non-sovereign dependence on the mainland. Discussions regarding the viability of international flights directly to the Galápagos Islands that avoided mainland Ecuador altogether made their way to the National Assembly for analysis (Quito Informa, 2020). The disturbance brought about by COVID nearly led to a reorganization of dependencies on mainland Ecuador in the interest of greater tourism sovereignty for certain groups in the islands.

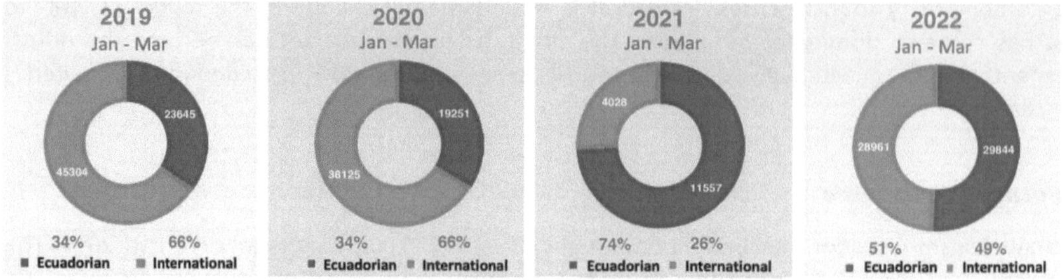

Figure 2. Recent Ecuadorian vs. International visitor arrivals in Galápagos.
Source: Adapted from Observatorio de Turismo, 2022

The pandemic has also sped up changes in the composition of visitors. The slight advantage acquired by the Ecuadorian market (34% of all visitors in 2019) grew to 74% of all visitors in 2021. It remains 51% of total visitation in the first trimester of 2022 (Figure 2). Whereas international visitation has yet to return to pre-pandemic levels, domestic visitation already exceeds them (Figure 3). This gap has widened considerably as mainland residents seek opportunities for leisure via extensive discounts that make tourism to the islands historically affordable. Although these domestic visitors have been critical to re-activating the tourism economy in the islands, they pay very different fee structures than international visitors. Displacing international markets with increased domestic markets thus inhibits recovery of the budgets at the Galápagos Governing Council, the municipalities and parishes, and the National Park. With most domestic visitors staying in the towns, the potential to provide profound nature-based experiences is substantially reduced since time spent inside the national park is limited. The growing domestic market does not yield comparable levels of financial support for conservation (Díaz-Sánchez & Obaco, 2021). The extent to which the extreme dominance of the Ecuadorian market is a temporary condition or an enduring characteristic of the tourism crisis underway pre-pandemic remains to be seen.

The pandemic also heightened another aspect of the tourism crisis – the demand for local recreational access for Galápagos residents. Although there had been pushes for increased domestic (i.e. democratic) park usage to promote greater awareness of the value of the protected area, the pandemic provoked extensive local use of park sites like Tortuga Bay and Garrapatero beaches as outdoor recreation became one of the few sanctioned public activities. The latter site even saw an increase in visitation over 2019 levels (PNG, 2021). As was true around the world, time spent in nature was a powerful coping strategy and an essential means of ensuring one's psychological wellbeing during the pandemic, especially in places that were otherwise under extreme lockdown (Taff et al., 2021). In Galápagos, such shifting park visitation dynamics will likely require careful restructuring of visitation fees and management strategies. Now established, the recreational tensions between residents and domestic tourists are likely to persist beyond the pandemic.

Underlying cultural identity crises

Migration to the Galápagos increased over the 20th century, often stimulated by changing conditions on the Ecuadorian mainland (Table 2). Agricultural reforms and droughts created both push and pull factors for those arriving from the province of *Loja,* who are associated with farming in the islands (Ospina, 2001). The "tuna wars" first instigated those from provinces of *Guayas* and *Manabí,* both located on the Ecuadorian coast, to relocate to the archipelago in favor of improved fishing conditions (Barragán Paladines & Chuenpagdee, 2017). *Tungurahua* province is the traditional land of the *Salasaca* peoples, who continue to arrive in large numbers to work in construction, taxi driving, conservation, and tourism (Ospina, 2001). Individuals from *Pichincha*

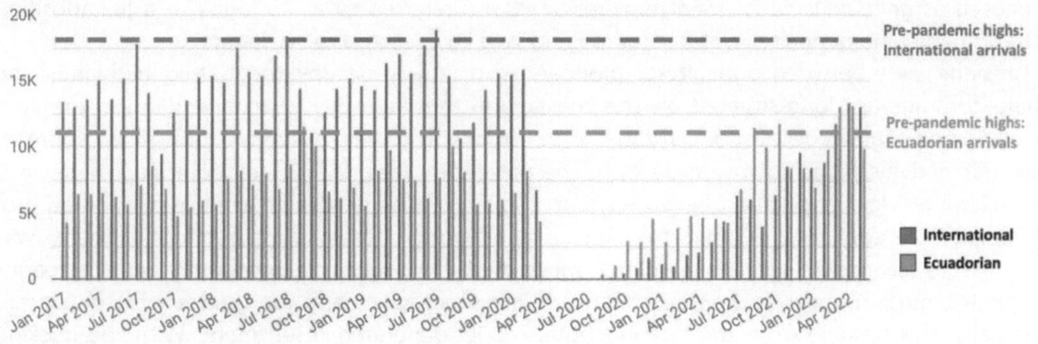

Figure 3. Monthly arrivals to Galápagos by market segment.
Source: Adapted from Observatorio de Turismo, 2022

Table 2. Galápagos residents places of birth.

Place of birth	Galápagos	Guayas	Tungurahua	Manabí	Pichincha	Loja	Chimborazo	Others Provinces	Outside Ecuador	TOTAL
TOTAL	9,125	4,798	3,043	1,551	1,460	1,130	557	3,250	330	**25,244**
%	**36.1%**	19.0%	12.1%	6.1%	5.8%	4.5%	2.2%	12.9%	1.3%	**100%**

Source: INEC, 2015.

(province of Ecuadorian capital Quito) tend to be associated with work in conservation in the park, associated conservation NGOs, or tourism. Norwegian farmers from *Hardangervidda* and utopian visionaries from the United States arrived in earlier periods of the 20th century. Many Germans also migrated to the islands in the inter-war period. This convergence of multiple ethnic groups, cultural worldviews, livelihood strategies, and diverse ways of being within a limited historical timeframe has created a unique and multi-layered society linked by the shared negotiation of the islands' ecological conditions, socio-political arrangements, and special laws.

Today there is thus broad recognition in the islands that the Galápagos do not yet have a cohesive culture but rather diverse cultural influences that lie at the heart of what have been, at times, bitter conservation and development conflicts in the islands. Given the absence of Indigenous presence, the convergence of numerous cultural influences and worldviews in the Galápagos extend into forms of power that bond authority with territory and determine what is considered "just" outcomes for the growing resident population. Typical settler colonial theory is thus inverted, as the *colonos* and *pioneros* (i.e. colonists and pioneers) in the Galápagos have the longest-standing claims to Galapagueño cultural heritage and sovereignty. Additional complexity stems from 1998s Special Law, which enacted residency statuses for the islands. Long-term foreign residents were instantly granted residency, while native-born Galapagueños living on the mainland fell outside residency restrictions (Hoyman & McCall, 2013). Today, residency can only be inherited from permanent resident parents or acquired by marriage with a permanent resident. Others working in Galápagos do so on temporary residency permits. The Special Law thus entrenched "us" and "them" conflicts within the islands and served as a breaking point that enabled distinctions between an Ecuadorian identity and an emerging Galapagueño identity.

The migration, residency, and identity dynamics are made more complex by the underreported in-migration and informalities in the residence-granting and work permitting processes (Villacis & Carrillo, 2012). The population influx can lead to practices not in harmony with the fragile environments encountered in Galápagos (e.g. Grenier, 2007; Barragán Paladines & Chuenpagdee, 2017; Burbano & Meredith, 2021; Burke, 2021). Yet as frontier livelihood strategies (e.g. agriculture and fishing) and the associated identities give way to amenity-based, tourism-related livelihood strategies, recent migrants were able to fold themselves more quickly into the political, economic, and environmental governance structures associated with the acceleration of tourism. As a result, new tensions emerged regarding legitimacy, heritage, and the right to refer to oneself as *galapagueño*. As long-term residents are referred to as "colonos" – a juxtaposition with (nomadic) indigenous peoples – there is no way to "decolonize" Galápagos.

Tensions exist between agricultural pioneers who largely occupy the humid highlands, the fishing communities long-situated on the coasts, and the latter-day arrivals working in the conservation and tourism sector concentrated in the growing towns. Although tourism is a primary economic activity in the towns, aside from the park entry fees, the economic impact is not evident to the resident population because the massive revenues generated are not distributed proportionally or equitably. Under the historical floating hotel tourism model, there was considerable leakage of tourism earnings. Most of the cruises were owned by non-Galápagos companies, most supplies and equipment were imported, and many top guides were not locals. Eventually, this leakage from the floating hotel model generated resentment. With the backing of the 1998 Special Law, local community leaders and NGOs began to advocate for "a new model of ecotourism" that prioritized the maximization of economic benefits for local communities, not just for the external tour operators in charge of the floating hotels.

While these broad strokes oversimplify what are culturally interwoven, temporally dynamic, and highly hybridized worldviews about the Galápagos environment, both natural and human, incompatible cultural sovereignties remain widely seen as being at the heart of conflicts over use of natural resources, loss of endemic species, disruption of the islands' fragile ecosystems, and long-term visions of what Galápagos should be (Grenier, 2007; Durham, 2008; Lu et al., 2013). Moreover, these differences have precipitated a revolving door of leadership at key institutions like the National Park, the Charles Darwin Foundation, and local government offices. Taken together, the divisions hinder broader collective action built upon shared conservation values.

Discussion

It is argued that the struggle to overcome crisis is quite inherent in the idea of sovereignty, as there is an implied need to eradicate unequal or unjust arrangements in favor of more representative, fair, and ethical outcomes, access to resources, or governance systems (Hansen & Stepputat, 2006). Likewise, sustainable tourism "inevitably relates to crisis in some way" (Hopkins, 2021, p. 1430), either by its responses to crises, its contributions to crises, or its use as a form of recovery from crisis. This account of the Galápagos links sovereignty and justice theory via a discussion of crises. This analysis also reflects varying degrees of convergence and divergence of livelihood security and sovereignty across these crises. In all instances, diverse and often conflicting cultural worldviews of the immigrant-based population often inhibited efforts to reach a consensus regarding appropriate policies for environmental protection and sustainable community development (Grenier, 2007).

With global immigration projected to grow and exacerbate environmental conflicts in the coming years (Reuveny, 2007), the current research is well-poised to provide urgent and generalizable insights into the sociocultural underpinnings of this increasing mobility, the environmental conflicts that exist between different value systems and worldviews, and the opportunities that exist to promote improved cooperation on behalf of social and environmental wellbeing in places experiencing intense in-migration. Since a critical step to promoting the collective action and inclusive governance necessary for confronting accelerating change is understanding the sociocultural dynamics within and between groups that dictate thoughts, values, conflicts, and behaviors (Ostrom, 2009), tourism discourse has much to gain from further study of how nested formations of sovereignty manifest in destination contexts (Stepputat, 2015; Naylor & Hunt, 2021).

As seen during the COVID pandemic, there were certain "virtues of insularity" (Baldacchino & Starc, 2021) and a certain degree of "splendid isolation" provided in the Galápagos Islands during the COVID pandemic (Agius et al., 2022), yet these "advantages" did not come without

Table 3. Struggles for sovereignty in Galápagos.

CRISES	Conservation	Tourism	COVID-19	Cultural identity
Inherent Struggle	***Protect nature*** *from past and present human disturbance*	*Ensure tourism is providing benefits for* ***local residents***	*Prioritize* ***permanent*** *resident wellbeing and quality of life*	*Recognize and elevate the heritage of* ***early pioneers***
Nested sovereignties	• *Territorial sovereignty* (Benton, 2010) • *Conservation sovereignty* (Mawyer & Jacka, 2018)	• *Livelihood sovereignty* (Naylor & Hunt, 2021) • *Just tourism* (Jamal, 2019) • *Socialized tourism* (Higgins-Desbiolles, 2020)	• *Health sovereignty* (Werkheiser, 2014) • *Insular, "splendid isolation"* (Agius et al., 2022)	• *Non-sovereign autonomy* (Baldacchino & Milne, 2009) • *Islandian sovereignty* (Prinsen & Blaise, 2017)

heightened concerns for food and livelihood security. As has been noted, "the intersections of crisis are likely to pay reference to the particular time-spaces of crises, but seeks to look not at them as clashing priorities, but instead as highly relational crises which are likely to have been differentially experienced by portions of the population and economy" (Hopkins, 2021, p. 1426). Sovereignty conceptualizes such clashing priorities as "malleable" and "negotiable," thus articulating opportunities for marginalized populations to confront and negotiate established bases of power for their favored outcomes (Iles & Montenegro de Wit, 2015). For these reasons, work that seeks to capture multiple subjectivities of heterogeneous community interests will be timely in many destinations where local concerns are often marginalized by more powerful non-local actors or efforts to restart the tourism industry post-pandemic (Higgins-Desbiolles, 2020).

Sovereignty outcomes are not a function of a given recipe of conditions but are place- and time-specific (Jarosz, 2014; Naylor & Hunt, 2021). Deeper historical analyses like the present work can reveal how numerous nested and entangled sovereignties are simultaneously sought across multiple timeframes, geographic spaces, livelihood sectors, and cultural landscapes (Table 3). Here, the chronic crises in the Galápagos have involved struggles for territorial and diplomatic sovereignty, livelihood sovereignties, and conservation sovereignty. A lack of native population means these various sovereignties are not based on Indigenous claims or heritage but on residence time (e.g. Latorre, 1999; Ospina, 2001). Diverse European, mestizo, and Indigenous colonists create competing claims to territorial sovereignty (Benton, 2010) and sub-insular "Islandian sovereignty" (Prinsen & Blaise, 2017). Now that the work of island sovereignty scholars has spilled over into tourism studies (e.g. Baldacchino, 2016), historical analyses of other destinations are likely to provide valuable insights into the value systems at play as communities respond to social and environmental crises.

Attention to historical dynamics of arrival and settlement also establishes a critical baseline for how advantage and power shift over time. Integrating the humanities and historical analyses into tourism research is therefore vital to improving understanding of the subjective and temporal nature of struggles associated with justice-centric concepts, including but not limited to sovereignty. Recognizing cultural variation and divergence of associated worldviews over time is essential if we hope to manage for more just outcomes of tourism and conservation in and beyond the Galápagos (Jamal & Higham, 2021; Rastegar, Higgins-Desbiolles & Ruhanen, 2021).

Finally, a historical analysis conducted by current residents of the destination is timely given the justice and the decolonial turns underway in the social sciences. Reflective work in and beyond the Galápagos "alerts practitioners to economic and social trends which may impact on global tourism post-pandemic" (Filep et al., 2022, p.1). Our practical contribution to sustainable tourism studies thus demonstrates how the sovereignty framing facilitates the sociocultural and historical understanding needed to effectively analyze, design, promote, and manage just tourism and just destinations (Jamal & Camargo, 2014).

Conclusion

The purpose of this article was to assimilate sovereignty theory into the justice tourism scholarship. The value of sovereignty theory was demonstrated in our ethnohistorical analysis of how Galapagos residents negotiate chronic and acute crises in the Galápagos Islands, a place with no original human population. Such historical analyses are critical for improving our understanding of the subjective and temporal nature of struggles associated with justice-centric concepts, including but not limited to sovereignty. With global immigration projected to grow and exacerbate environmental conflicts in the coming years, the current research is well-poised to provide urgent and generalizable insights into the sociocultural underpinnings of increasing human mobility, the environmental conflicts that exist between different value systems and worldviews, and the opportunities that exist to promote improved destination management on behalf of human wellbeing in places experiencing intense in-migration. It will just require improved recognition of the often-overlooked human history of tourism destinations, the sovereignty-related struggles present in these histories, and how those struggles dictate destination community responses to chronic and acute changes and crises over time.

Disclosure statement

No conflict of interest has been declared by the authors.

Funding

Funding for research upon which a portion of this article draws came from the Fulbright Scholar Program (Award #9437-EC Research), the US National Science Foundation's Cultural Anthropology Program (Award #2020555), and the Penn State University Office of Global Programs.

ORCID

Carter A. Hunt http://orcid.org/0000-0001-6319-8171

References

Agius, K., Sindico, F., Sajeva, G., & Baldacchino, G. (2022). 'Splendid isolation': Embracing islandness in a global pandemic. *Island Studies Journal, 17*(1), 44–22. https://doi.org/10.24043/isj.163

Agarwal, B. (2014). Food sovereignty, food security and democratic choice: Critical contradictions, difficult conciliations. *Journal of Peasant Studies, 41*(6), 1247–1268. https://doi.org/10.1080/03066150.2013.876996

Alberts, E. C. (2020). New evidence suggests China's 'dark' vessels poached in Galápagos waters. Mongabay. Retrieved from https://news.mongabay.com/2020/10/new-evidence-suggests-chinas-dark-vessels-poached-in-Galápagos-waters/.

Altmann, P. (2020). Eleven days in October 2019–The indigenous movement in the recent mobilizations in Ecuador. *International Journal of Sociology, 50*(3), 220–226. https://doi.org/10.1080/00207659.2020.1752498

American Society for Ethnohistory. (2022). ASE mission statement. Retrieved from https://ethnohistory.org/ase-mission-statement/.

Baldacchino, G. (Ed.). (2016). *Archipelago tourism: Policies and practices*. Routledge.

Baldacchino, G., & Milne, D. (2009). *The case for non-sovereignty*. Routledge.

Baldacchino, G., & Starc, N. (2021). The virtues of insularity: Pondering a new chapter in the historical geography of islands. *Geography Compass, 15*(12), e12596.

Barragán Paladines, M. J., & Chuenpagdee, R. (2017). A step zero analysis of the Galápagos Marine Reserve. *Coastal Management, 45*(5), 339–359. https://doi.org/10.1080/08920753.2017.1345606

Benton, L. (2010). Island chains. In *A search for sovereignty: Law and geography in European Empires: 1400–1900* (pp. 162–222). Cambridge University Press.

Bunten, A. (2014). More like ourselves: Indigenous capitalism through tourism. *American Indian Quarterly, 34*(3), 285.

Burbano, D. V., & Meredith, T. C. (2021). Effects of tourism growth in a UNESCO World Heritage Site: Resource-based livelihood diversification in the Galápagos Islands, Ecuador. *Journal of Sustainable Tourism*, *29*(8), 1270–1289. https://doi.org/10.1080/09669582.2020.1832101

Burke, A. (2021). The crossroads of ecotourism dependency, food security and a global pandemic in Galápagos, Ecuador. *Sustainability*, *13*(23), 13094. https://doi.org/10.3390/su132313094

Cajiao, D., Izurieta, J. C., Casafont, M., Reck, G., Castro, K., Santamaría, V., Cárdenas, S., & Leung, Y. F. (2020). Tourist use and impact monitoring in the Galapagos: An evolving programme with lessons learned. *PARKS*, *26*(26.2), 89–102. https://doi.org/10.2305/IUCN.CH.2020.PARKS-26-2DC.en

Camargo, B. A., & Vázquez-Maguirre, M. (2021). Humanism, dignity and indigenous justice: The Mayan train mega-project, Mexico. *Journal of Sustainable Tourism*, *29*(2–3), 372–391. https://doi.org/10.1080/09669582.2020.1758707

Carvache-Franco, W., Carvache-Franco, M., & Hernández-Lara, A. B. (2021). From motivation to segmentation in coastal and marine destinations: A study from the Galápagos Islands, Ecuador. *Current Issues in Tourism*, *24*(16), 2325–2341. https://doi.org/10.1080/13683500.2020.1811651

Caton, K. (2016). A humanist paradigm for tourism studies? In A.M. Munar & T. Jamal (Eds.), *Tourism research paradigms: Critical and emergent knowledges*. Emerald Group Publishing Limited.

CGREG. (2016). Plan Galápagos: Plan de Desarrollo Sustentable y Ordenamiento Territorial del Régimen Especial de Galápagos. Consejo de Gobierno del Régimen Especial de Galápagos (CGREG). https://www.gobiernogalapagos.gob.ec/wp-content/uploads/downloads/2017/04/Plan-Galapagos-2015-2020_12.pdf

CGREG. (2021). Tributo de ingreso. Consejo de Gobierno del Régimen Especial de Galápagos (CGREG).

Chambers, D., & Buzinde, C. (2015). Tourism and decolonisation: Locating research and self. *Annals of Tourism Research*, *51*, 1–16. https://doi.org/10.1016/j.annals.2014.12.002

Coffey, W., & Tsosie, R. (2001). Rethinking the tribal sovereignty doctrine: Cultural sovereignty and the collective future of Indian nations. *Stanford Law & Policy Review*, *12*(2), 191–222.

Co-Galapagos. (2021). Sustainable Development Goal (SDG) Prioritization. *Co-Galápagos*.

Dangi, T. B., & Petrick, J. F. (2021). Augmenting the role of tourism governance in addressing destination justice, ethics, and equity for sustainable community-based tourism. *Tourism and Hospitality*, *2*(1), 15–42. https://doi.org/10.3390/tourhosp2010002

De Groot, R. S. (1983). Tourism and conservation in the Galapagos Islands. *Biological Conservation*, *26*(4), 291–300. https://doi.org/10.1016/0006-3207(83)90093-9

De Kadt, E. (1979). *Tourism: Passport to development. Perspectives on the social and cultural effects of tourism in developing countries*. Oxford University Press.

Dennis, A, & Indiana University, USA. (2019). An unhealthy obsession with theory. *Journal of the Association for Information Systems*, *20*(9), 1404–1409. https://doi.org/10.17705/1jais.00572

Díaz-Sánchez, J. P., & Obaco, M. (2021). The effects of Coronavirus (COVID-19) on expected tourism revenues for natural preservation: The case of the Galápagos Islands. *Journal of Policy Research in Tourism, Leisure and Events*, *13*(2), 285–289. https://doi.org/10.1080/19407963.2020.1813149

Durham, W. H. (2008). Fishing for solutions: ecotourism and conservation in Galápagos national park. In A. Stronza & W.H. Durham (Eds.), *Ecotourism and conservation in the Americas* (pp. 66–89). CABI.

Durham, W. H. (2021). *Exuberant life: An evolutionary approach to conservation in Galápagos*. Oxford University Press.

Edelman, M., Weis, T., Baviskar, A., Borras Jr, S. M., Holt-Giménez, E., Kandiyoti, D., & Wolford, W. (2014). Introduction: critical perspectives on food sovereignty. *Journal of Peasant Studies*, *41*(6), 911–931. https://doi.org/10.1080/03066150.2014.963568

El Comercio. (2019). *El paro genera pérdidas diarias de USD 5 millones en Quito, según la Cámara de Turismo de Pichincha*. El Comercio. Retrieved from https://www.elcomercio.com/tendencias/perdidas-diarias-quito-paro-turismo.html.

Epler, B. (2007). *Tourism, the economy, population growth, and conservation in Galápagos*. Puerto Ayora.

Epler, B. (1993). *An economic and social analysis of tourism in the Galápagos Islands*. Coastal Resources Center, University of Rhode Island.

Filep, S., King, B., & McKercher, B. (2022). Reflecting on tourism and COVID-19 research. *Tourism Recreation Research*, 1–5. https://doi.org/10.1080/02508281.2021.2023839

Gao, J., Lin, H., & Zhang, C. (2021). Locally situated rights and the 'doing' of responsibility for heritage conservation and tourism development at the cultural landscape of Honghe Hani Rice Terraces, China. *Journal of Sustainable Tourism*, *29*(2–3), 193–213. https://doi.org/10.1080/09669582.2020.1727912

González, J., Montes, C., Rodríguez, J., & Tapia, W. (2008). Rethinking the Galápagos Islands as a complex social-ecological system: implications for conservation and management. *Ecology and Society*, *13*(2), 13. https://doi.org/10.5751/ES-02557-130213

Grenier, C. (2007). *Conservación contra natura. Las islas Galápagos*. Editorial Abya Yala.

Guia, J. (2021). Conceptualizing justice tourism and the promise of posthumanism. *Journal of Sustainable Tourism, 29*(2–3), 503–520. https://doi.org/10.1080/09669582.2020.1771347

Hansen, T. B., & Stepputat, F. (2006). Sovereignty revisited. *Annual Review of Anthropology, 35*(1), 295–315. https://doi.org/10.1146/annurev.anthro.35.081705.123317

Hester, R. (2018). *Historical research: Theory and methods*. Ed-Tech Press.

Higgins-Desbiolles, F. (2020). Socialising tourism for social and ecological justice after COVID-19. *Tourism Geographies, 22*(3), 610–623. https://doi.org/10.1080/14616688.2020.1757748

Honey, M. (1994). Paying the price of ecotourism: two pioneer biological reserves face the challenges brought by a recent boom in tourism. *Americas, 46*, 40–48.

Hopkins, D. (2021). Crises and tourism mobilities. *Journal of Sustainable Tourism, 29*(9), 1423–1435. https://doi.org/10.1080/09669582.2021.1905969

Hoyman, M. M., & McCall, J. R. (2013). Is there trouble in paradise? The perspectives of Galápagos community leaders on managing economic development and environmental conservation through ecotourism policies and the Special Law of 1998. *Journal of Ecotourism, 12*(1), 33–48. https://doi.org/10.1080/14724049.2012.749882

Hunt, C. A. (2021). The Galápagos Islands, Ecuador. In Honey, M. & Frenkiel, K. (Ed.), *Overtourism: Lessons for a better future*. Island Press.

Iles, A., & Montenegro de Wit, M. (2015). Sovereignty at what scale? An inquiry into multiple dimensions of food sovereignty. *Globalizations, 12*(4), 481–497. https://doi.org/10.1080/14747731.2014.957587

INEC. (2015). *Análisis de resultados definitivos Censo de Población y Vivienda: Galápagos 2015*. Instituto Nacional de Estadística y Censos (INEC).

Jamal, T. (2019). *Justice and ethics in tourism*. Routledge.

Jamal, T., & Camargo, B. A. (2014). Sustainable tourism, justice and an ethic of care: Toward the just destination. *Journal of Sustainable Tourism, 22*(1), 11–30. https://doi.org/10.1080/09669582.2013.786084

Jamal, T., & Higham, J. (2021a). *Justice and tourism: Principles and approaches for local-global sustainability and well-being*. Routledge.

Jamal, T., & Higham, J. (2021b). Justice and tourism: Towards a new platform for tourism and sustainability. *Journal of Sustainable Tourism, 29*(2–3), 143–157. https://doi.org/10.1080/09669582.2020.1835933

Jamal, T., & Stronza, A. (2009). Collaboration theory and tourism practice in protected areas: Stakeholders, structuring and sustainability. *Journal of Sustainable Tourism, 17*(2), 169–189. https://doi.org/10.1080/09669580802495741

Jarosz, L. (2014). Comparing food security and food sovereignty discourses. *Dialogues in Human Geography, 4*(2), 168–181. https://doi.org/10.1177/2043820614537161

Jourdan, D., & Wertin, J. (2020). Intergenerational rights to a sustainable future: insights for climate justice and tourism. *Journal of Sustainable Tourism, 28*(8), 1245–1254. https://doi.org/10.1080/09669582.2020.1732992

Latorre, O. (1999). *El hombre en las Islas Encantadas: La historia humana en Galápagos*. FUNDACYT.

Lu, F., Valdivia, G., & Wolford, W. (2013). Social dimensions of 'Nature at Risk' in the Galápagos Islands, Ecuador. *Conservation and Society, 11*(1), 83. https://doi.org/10.4103/0972-4923.110945

Lynham, S. A. (2002). The general method of theory-building research in applied disciplines. *Advances in Developing Human Resources, 4*(3), 221–241. https://doi.org/10.1177/1523422302043002

MacArthur, R. H., & Wilson, E. O. (1967). *The theory of island biogeography*. Princeton university press.

MacRae, G. (2016). Food sovereignty and the anthropology of food: Ethnographic approaches to policy and practice. *Anthropological Forum, 26*(3), 227–232. https://doi.org/10.1080/00664677.2016.1201456

Mawyer, A., & Jacka, J. K. (2018). Sovereignty, conservation and island ecological futures. *Environmental Conservation, 45*(3), 238–251. https://doi.org/10.1017/S037689291800019X

McDowell, B. (2013). *Historical research: A guide*. Routledge.

McMichael, P. (2014). Historicizing food sovereignty. *Journal of Peasant Studies, 41*(6), 933–957. https://doi.org/10.1080/03066150.2013.876999

Mestanza-Ramón, C., Pranzini, E., Anfuso, G., Botero, C. M., Chica-Ruiz, J. A., & Mooser, A. (2020). An attempt to characterize the "3S" (Sea, Sun, and Sand) parameters: Application to the Galápagos Islands and continental Ecuadorian beaches. *Sustainability, 12*(8), 3468. https://doi.org/10.3390/su12083468

Naylor, R. S., & Hunt, C. A. (2021). Tourism & livelihood sovereignty: A theoretical introduction and research agenda for arctic contexts. *Societies, 11*(3), 105. https://doi.org/10.3390/soc11030105

Naylor, R. S., Hunt, C. A., & Zimmerer, K. S. (2021). Bridging the troubled waters of tourism studies: from social capital and rural community development to livelihood sovereignty. *International Journal of Tourism Anthropology, 8*(4), 370–390. https://doi.org/10.1504/IJTA.2021.123199

Nicholls, R., & Higgins-Desbiolles, F. (2016). Ngarrindjeri authority: A sovereignty approach to tourism. In Mostafenazhad, M., Norum, R., Shelton, E.J., & Thompson-Carr, A. (Eds.), *Political Ecology of Tourism* (pp. 68–125). Routledge.

Nunkoo, R. (2018). The state of research methods in tourism and hospitality. In Nunkoo, R., (Ed.) *Handbook of research methods for tourism and hospitality management* (pp.3–23). Edward Elgar Publishing.

Nunkoo, R., Sharma, A., Rana, N. P., Dwivedi, Y. K., & Sunnassee, V. A. (2021). Advancing sustainable development goals through interdisciplinarity in sustainable tourism research. *Journal of Sustainable Tourism*, 1–25. https://doi.org/10.1080/09669582.2021.2004416

Nyéléni. (2022). Declaration of Nyéléni. Available online: https://nyeleni.org/IMG/pdf/DeclNyeleni-es.pdf (accessed on 24 May 2022).

Observatorio de Turismo. (2022). Estadísticas en línea. Observatorio de Turismo Galápagos.

Ospina, P. (2001). *Migraciones, actores e identidades en Galápagos*. Consejo Latinoamericano de Ciencias Sociales.

Ostrom, E. (2009). *Understanding institutional diversity*. Princeton University Press.

Parque Nacional Galápagos (PNG). (2019). *Informe Annual: Visitantes a las Areas Protegidas de Galápagos 2019*. Parque Nacional Galápagos.

Parque Nacional Galápagos (PNG). (2021). *Informe Semestral: Ingreso de visitantes a las Areas Protegidas de Galápagos 2021*. Parque Nacional Galápagos.

Patel, R. (2009). Food sovereignty. *Journal of Peasant Studies*, *36*(3), 663–706. https://doi.org/10.1080/03066150903143079

Perry Hobson, J. S., & Ko, G. (1994). Tourism and politics: The implications of the change in sovereignty on the future development of Hong Kong's tourism industry. *Journal of Travel Research*, *32*(4), 2–8. https://doi.org/10.1177/004728759403200401

Prinsen, G., & Blaise, S. (2017). An emerging "Islandian" sovereignty of non-self-governing islands. *International Journal*, *72*(1), 56–78.

Quito Informa. (2020). Concejo exhorta a la Asamblea objetar los vuelos directos internacionales a Galápagos. Quito Informa – Agencia de Noticias del Municipio del Distrito del Quito. Retrieved from http://www.quitoinforma.gob.ec/2020/06/16/concejo-exhorta-a-la-asamblea-objetar-los-vuelos-directos-internacionales-a-galapagos/.

Rastegar, R., Higgins-Desbiolles, F., & Ruhanen, L. (2021). COVID-19 and a justice framework to guide tourism recovery. *Annals of Tourism Research*, *91*, 103161. https://doi.org/10.1016/j.annals.2021.103161

Rawls, J. (1971). *A theory of justice*. Harvard University Press.

Reuveny, R. (2007). Climate change-induced migration and violent conflict. *Political Geography*, *26*(6), 656–673. https://doi.org/10.1016/j.polgeo.2007.05.001

Santafe-Troncoso, V., & Loring, P. A. (2021). Indigenous food sovereignty and tourism: The Chakra Route in the Amazon region of Ecuador. *Journal of Sustainable Tourism*, *29*(2–3), 392–410. https://doi.org/10.1080/09669582.2020.1770769

Smith, V. L. (Ed.). (1977). *Hosts and guests: The anthropology of tourism*. University of Pennsylvania Press.

Smith, S. L., Xiao, H., Nunkoo, R., & Tukamushaba, E. K. (2013). Theory in hospitality, tourism, and leisure studies. *Journal of Hospitality Marketing & Management*, *22*(8), 875–894. https://doi.org/10.1080/19368623.2013.771114

Stepputat, F. (2015). Formations of sovereignty at the frontier of the modern state. *Conflict and Society*, *1*(1), 129–143. https://doi.org/10.3167/arcs.2015.010111

Strahm, W., & Patry, M. (2010). *Reactive monitoring mission report, Galápagos Islands 27 April–6 May 2010*. World Heritage Center UNESCO.

Taff, B. D., Rice, W. L., Lawhon, B., & Newman, P. (2021). Who started, stopped, and continued participating in outdoor recreation during the COVID-19 pandemic in the United States? Results from a National Panel Study. *Land*, *10*(12), 1396. https://doi.org/10.3390/land10121396

Tilly, C. (1983). *Big structure, large processes, huge comparisons*. Russell Sage Foundation.

Tilzey, M. (2019). Authoritarian populism and neo-extractivism in Bolivia and Ecuador: The unresolved agrarian question and the prospects for food sovereignty as counter-hegemony. *Journal of Peasant Studies*, *46*(3), 626–652. https://doi.org/10.1080/03066150.2019.1584191

Timothy, D. J., Guia, J., & Berthet, N. (2014). Tourism as a catalyst for changing boundaries and territorial sovereignty at an international border. *Current Issues in Tourism*, *17*(1), 21–27. https://doi.org/10.1080/13683500.2012.712098

UNESCO. (2021). Biosphere Reserves. United Nations Education, Scientific, and Cultural Organization (UNESCO). Retrieved from https://en.unesco.org/biosphere

Vega, F. (2020). Así se masificaba el turismo en Galápagos antes de la pandemia. *PlanV*. Retreived from: https://www.planv.com.ec/historias/sociedad/asi-se-masificaba-el-turismo-galapagos-antes-la-pandemia.

Villacis, B., & Carrillo, D. (2012). The socioeconomic paradox of Galápagos. In S.J. Walsh & C.F. Mena (Eds.) *Science and conservation in the Galápagos Islands: Frameworks and perspectives* (pp. 69–85). Springer.

Werkheiser, I. (2014). Food sovereignty, health sovereignty, and self-organized community viability. *Interdisciplinary Environmental Review*, *15*(2/3), 134–146. https://doi.org/10.1504/IER.2014.063649

Williams, L. K., & González, V. V. (2017). Indigeneity, sovereignty, sustainability and cultural tourism: Hosts and hostages at 'Iolani Palace, Hawai'i. *Journal of Sustainable Tourism*, *25*(5), 668–683. https://doi.org/10.1080/09669582.2016.1226850

OPEN ACCESS

Decolonising tourism and development: from orphanage tourism to community empowerment in Cambodia

Freya Higgins-Desbiolles, Regina A. Scheyvens and Bhanu Bhatia

ABSTRACT
Tourism has been viewed as a development pathway, with alternative tourisms such as volunteer tourism perceived as promising. However, critics have highlighted how white saviourism and Western ideologies of superiority may underpin both development agendas and activities like volunteer tourism. The COVID crisis has impacted both tourism and international development and calls for rethinking. This case study is situated at the intersections of tourism, development and humanitarianism. It charts the evolution of the Cambodian Children's Trust which emerged in 2007 from the co-founding of an orphanage by an Australian volunteer tourist and a local Khmer leader. Through a process of conscientisation, the orphanage has given way to a community development approach under the leadership of a 100 percent Khmer team in country, leaving footprints of empowering spaces rather than dependency structures. This article addresses the research question of how might we transform the paternalistic desire to "do good" found in both voluntourism and development into a practice of mutual solidarity? Illuminating issues of power and inequality in Western-led models, this article offers a framework for more just partnerships based on Freirian praxis: dialogue building critical consciousness, co-development of transformative praxis, capacity sharing and trust in the capabilities of the people.

Introduction

> *The important thing to remember is that children are not tourist attractions* - Tara Winkler, Cambodian Children's Trust Co-Founder (Winkler, n.d.).

Tourism has been considered a pathway to development for countries of the Global South (Scheyvens, 2002; Sharpley & Telfer, 2015). The COVID-19 pandemic has brought the worst development crisis of the 21st century (Steiner, 2020), highlighting additionally the vulnerabilities that tourism dependency may bring to developing countries. From the Seychelles to Bali to Croatia, communities are grappling with how to address the sudden loss of international tourists as countries have experienced lockdowns and border closures to protect public health during the global pandemic (Hall et al., 2020). What is a moment of important learning on tourism dependency may be over-shadowed by a myopic focus on a rapid return to business as usual after the pandemic crisis (Higgins-Desbiolles, 2020).

This article argues that the larger crisis exposed by the pandemic calamity of 2020–21 is a crisis in development that results from operationalising tourism as a development strategy in the neoliberal framework that perpetuates dependency and inequalities. Development agencies and non-government organisations (NGOs) have fostered strategies built particularly on forms of special interest tourism that carry ideological agendas that must be addressed (Mowforth & Munt, 1998). In this study, voluntourism and orphanage tourism are the focus.

Orphanage tourism has proliferated in the Global South as travellers from the Global North are eager for voluntourism opportunities that provide "feel-good" experiences. Voluntourism and orphanage tourism have been criticised for reinforcing inequitable relationships between the Global North and South; particularly, they often emphasise the "development" of the tourists over host community benefit (Banks & Scheyvens, 2014). These forms of travel are premised on the traveller's ability to "make a difference" and on "white saviourism" rooted in neo-colonialism. Bott has suggested "rescue ideologies" underpin them and they are best understood in terms of the "… broader structures of inequality and exploitation underpinning global tourism markets" (2021, p. 11). A Lumos report argued that "orphanage trafficking is more prevalent in countries where there is a significant tourism industry, with orphanages generally being established in key tourist areas" (2021, p. 7).

With the passage of Australia's Modern Slavery Act 2018, Flanagan (2020) described promising practices for NGOs seeking to strengthen families and reduce institutional care of children. In this context, the story of how Australian Tara Winkler co-founded the Cambodian Children's Trust (CCT) in 2007 is well documented in a book (Winkler & Delacey, 2016) and television documentaries (e.g., ABC, 2010, 2014). These sources help record the evolution in thinking that resulted in CCT transforming from a non-profit child orphanage that hosted orphanage tourism (particularly by CCT donors) to a Cambodian-led organisation that facilitates empowerment of Cambodian communities, works to build exit strategies in its programmes and co-creates projects for community development.

This article offers a case study analysis of CCT in order to better understand development dynamics, filling a knowledge gap identified by Delvetere and De Bruyn (2009) on private initiatives' contributions to development. It also critiques the humanitarian impulse found in niches like voluntourism that too often fail to address the problem of development dependency in tourism and may act to sustain poverty. It addresses the research question of how might we transform the paternalistic desire to "do good" found in both voluntourism and development into a practice of mutual solidarity? Here, we chart the evolution of CCT, illuminating the change in consciousness and practice that has seen CCT become an effective, recognised agent and innovative leader of community development in Cambodia and beyond (Nee, 2020). Turning to Freirian thinking (Freire, 1970), this case study supports conscientisation for those with an inclination to "do good" to instead become partners working in mutual solidarity. This transition is essential in the face of the crisis of development that we now confront (Steiner, 2020). Such efforts are timely and in line with the United Nations Sustainable Development Goals (SDGs), particularly SDG 17 focused on "global partnerships for sustainable development".

Literature review

Tourism and development

Until the recent COVID-19 pandemic crisis, the conventional wisdom held that tourism supports development in countries such as Cambodia (Sharpley & Telfer, 2015). With the almost total shutting off of the tourism supply chain in 2020, it is possible to now see how dependency on tourism can be perceived as a crisis of development. But critical analysts have identified crucial problems for decades. For instance, Mowforth and Munt (1998) delineated in detail "the uneven and unequal nature of tourism development" (p. 141), demonstrating how alternative forms of tourism that have attracted the new middle classes (e.g., voluntourism) do not necessarily provide sustainable development outcomes but rather may materialise as pathways to sustaining poverty. One chapter specifically addressing the role of external NGOs is subtitled "where shall we save next?" suggested these may represent forms of "neocolonialism" and be "devoid of notions of social justice and concern for local people's perceptions" (Mowforth & Munt, 1998, p. 160).

Revisiting development studies' dependency paradigm, "dependency theory positions tourism as a new form of imperialism" (Scheyvens, 2002, p. 28). Western development models have been based on ideologies that conceive of developed countries as mature, "developing" countries as like "children" and developmentalism as the pathway to guiding the latter to evolve into modernity and maturity (Manzo, 1991). Dependency theory and the "development of underdevelopment" are theories explaining how Western approaches to development work in the interests of these wealthy, powerful countries and perpetuate underdevelopment in the countries of the Global South (Mowforth & Munt, 1998). Britton's analysis of Fiji explained how use of international mass tourism as a pathway to enter the global economy required the "subordination of national economic autonomy" (1982, p. 334). However, Mowforth and Munt's (1998) work showed that even alternative tourisms that are thought to be more benign and even beneficial, harbour power differentials and are implicated in ongoing unequal development.

Debate continues about tourism's contribution to development and its poverty-alleviating capabilities – particularly for the poorest (see Harrison, 2012; Meyer, 2012; Scheyvens, 2012). Equity and diversity feature strongly in more recent interventions. Rogerson and Saarinen suggested: "the poor are mostly invisible and without agency in the existing debates on tourism on tourism in the Global South" and that we should also focus on the poor as potential tourists themselves (2018, p. 32). From an Indigenous perspective, Kuokkanen outlined a war against subsistence economies as a strategy to make Indigenous communities and others dependent on capitalist economies; as a strategy of resistance and resurgence, she recommends revival of Indigenous subsistence through social economy frameworks as a pathway to Indigenous governance and livelihoods (2011, p. 232).

However, dependency arguments are being revived in COVID's wake. Despite some analysts dismissing the relevancy of dependency theory in a globalising world, Ghosh (2021) argued it remains relevant to understanding development, underdevelopment and economic relations between the Global North and South. Recent work describing "the drain from the global South" argued:

> "Advanced economies" rely on unequal exchange to facilitate their economic growth and to sustain high levels of income and material consumption. In recent years, the drain has amounted to around [US]$2.2 trillion per year (constant 2011 dollars) in Northern prices... These patterns of appropriation through North–South trade are a major driver of global inequality and uneven development (Hickel et al., 2021, p. 13).

The insights from such analyses alert us to be critically attentive to interventions for development in communities of the Global South.

In recent years, development aid has become more complex, with the entry of new actors undertaking private initiatives, such as the one under study here, that have been labelled a "fourth pillar" (governments, multilateral institutions and established NGOs being the original 3 pillars) (Delvetere & De Bruyn, 2009). Delvetere and De Bruyn (2009) argued that these private initiatives for development: deserve greater research focus; have mixed potential as they offer

"... involvement and enthusiasm, but often lack... essential knowledge and experience"; and potentially offer the Global South partnerships "... which... will be more horizontal, accommodating, generous and durable in nature than with the traditional development organisations" (pp. 920-1).

Voluntourism and orphanage tourism

Voluntourism (volunteer tourism) is one of the niches of tourism generally thought to be more beneficial. Voluntourism involves: "tourists who, for various reasons, volunteer in an organized way to undertake holidays that might involve aiding or alleviating the material poverty of some groups in society... " (Wearing, 2001, p. 1). However, there are critics of this impulse to "do good". Illich (1990) criticised American voluntary service missions to Mexico as paternalistic, neo-colonial interventions that distort local lifeways, tying people into a developmentalist, modernisation model[1] that only benefits a small elite.

More recently, Wearing and McGehee cautioned that voluntourism "... can lead host communities to become dependent upon volunteer tourism sending organizations, undermining the dignity of local residents, exceeding the carrying capacity of the community if not properly managed... " (2013, p. 125). With a critical lens, serious issues of power discrepancies, racism and exploitation become evident. Sin noted "... unfortunately, volunteer tourism, similar to geographies of responsibility and care, has more often than not been framed in the position of the 'First world' being responsible for the wellbeing for a 'Third world', poor, and marginalized subject" (2010, p. 988). Research by Freidus on voluntourism found "... westerners are christened more knowledgeable than locals in whatever project they endeavor to complete... This can reify the hegemony of western cultural authority that undergirds the types of projects volunteer tourists undertake" (2017, p. 1308). This can have contradictory, material impact as volunteers take up valuable job opportunities from more qualified, unemployed locals.

This issue of power and inequality at this interface is even more pronounced for orphanage tourism, a popular component of alternative tourism. A report on the global volunteering market released in June 2018 identified Cambodia among the top 10 global orphanage voluntourism destinations and the USA, UK and Australia as the top countries sending volunteers to orphanages overseas (ReThink Orphanages, n.d.). Pelling reported:

> Between 2005 and 2015 it's estimated that the number of orphanages in Cambodia increased by more than 60 per cent. Most... of that growth has been attributed to a boom in so-called "orphanage tourism", where tourists turn up, spend some time with the children, then leave behind a nice wad of cash (2019).

ReThink Orphanages argued:

> Children in orphanages are often kept in slavery like conditions, fully owned by orphanage directors and exploited for profit through forced "cultural" performances for tourists, forced begging, and forced interaction and play with visitors. Children are often kept in poor health, poor conditions and are malnourished in order to elicit more support in the form of donations and gifts (cited in Joint Standing Committee on Foreign Affairs, Defence and Trade (JSCFADT. ,), 2017, n.p.).

Bott observed that orphanage tourism was added to the "rescue ideologies" of anti-slavery policy:

> Anti-slavery discourse implies that the eradication of orphanage tourism will automatically improve the lives and uphold the rights of children. It fails to acknowledge the painful fact that poverty remains the principal cause of the violation of children's rights because lack of resources seriously impedes, or entirely prevents, their access to basic human rights such as healthcare, water, food and education (2021, p. 3).

Bott (2021, p. 4) noted that rescue ideologies can actually "drive the production" of orphans for an "orphan-industrial complex", where the "misidentification of 'orphans' as a category for development and humanitarian intervention has subsequently been misappropriated by many

Western individuals and charitable organizations" (Cheney & Rotabi, 2014, p. 1). ReThink Orphanages, a coalition of NGOs, described "paper orphaning" and "orphanage trafficking" as when children are "constructed on paper through documentation as orphans"; van Doore explained it further as "… the active recruitment of children into orphanages or residential care institutions in developing nations for the purpose of ongoing exploitation, particularly through orphanage tourism" (cited in JSCFADT. , 2017, n.p.).

It is difficult to accept that orphanage tourism, voluntourism and donor benevolence actually drive child exploitation and trafficking through orphanages in developing countries (JSCFADT. , 2017). Unfortunately, monetising this humanitarian desire can in fact lead to devastating consequences and a vicious circle of exploitation. This circle includes: NGOs who need funds and a reason to exist; voluntourists and donors who naively act on their humanitarian desire; families in the Global South with insufficient resources to care for all of their children; and an aid system that is built on ongoing inequality and dependency.

Initiatives to address child welfare and vulnerabilities in tourism are underway. The first International Summit on Child Protection in Travel and Tourism in 2018 issued a call to action asking: "all key stakeholders to adopt a comprehensive, child rights-centered and multi-stakeholder framework where all actors actively work together to end impunity of the travelling child sex offenders" (Global Study of Sexual Exploitation of Children in Travel & Tourism, 2018). Guidelines entitled "Child Welfare and the Travel Industry" offered comprehensive advice to guide the tourism sector to responsible practices (ChildSafe Movement & G Adventures, 2018). It specifically addressed tourist visits to orphanages and other children's spaces, advising tourism companies:

> Children centres are meant to be a safe place for children. They are not tourist attractions. Visiting school classrooms and orphanages not only disrupts child development but can also put children at risk of emotional or physical harm through direct or indirect contact with travellers with inadequate supervision (ChildSafe Movement & G Adventures, 2018, p. 13).

Based on extensive research in Cambodia, Guiney and Mostafanezhad turned to Polanyi's "double movement" describing the neoliberal commodification of orphans and orphanages in Cambodian development processes and the simultaneous "protective countermovement by anti-orphanage tourism campaigns that challenge the industry's morality and legitimacy" (2015, p. 132). The CCT case presented here offers in-depth insights from an NGO that navigated this transition.

White saviourism

White saviourism has recently received attention from tourism analysts (e.g., Bandyopadhyay & Patil, 2017; Wearing et al., 2018) and this provides the missing piece to the puzzle of motivating ideologies that underpins the development dynamics described here. White saviourism describes a situation when a "… white person guides people of colour from the margins to the mainstream with his or her own initiative and benevolence … [this] has the tendency to render people of color incapable of helping themselves — infantile or hapless/helpless victims who survive by instinct" (Cammarota, 2011, pp. 243-4). As well as individuals, this syndrome is also operative among some development NGOs (Mowforth & Munt, 1998, p. 160). White saviour analyses represent critical interrogation of the humanitarian impulse and the way that it is enacted in a neoliberalised context.

In his assessment of the "white savior industrial complex", Cole (2012) gave a scathing assessment of young Westerners' impetus to do good while lacking wider political insight:

> Africa has provided a space onto which white egos can conveniently be projected. It is a liberated space in which the usual rules do not apply: a nobody from America or Europe can go to Africa and become a godlike savior or, at the very least, have his or her emotional needs satisfied. Many have done it under the banner of "making a difference" (p. 2).

Accompanying this phenomenon of white saviourism in Western development approaches is what Bhabha termed the ideology of "hegemonic modernity" (1994, p. 171) that is predicated on pressing a belief of inferiority and backwardness on those on whom development is to be enacted. Bandyopadhyay and Patil explained:

> The historic roots of racial inferiority/superiority are further perpetuated by contemporary neo-colonial aid relationships in volunteer tourism ... The colonial narratives are retold through modern discourses of volunteer tourism and development, where those in the global South come to believe that they have lower capacity for development augmented by inferior science, technology, and resources... (2017, p. 652).

Cole (2012) argued ideological views of the Global South's inferiority has supported a "White Savior Industrial Complex".

Freire's "pedagogy of the oppressed" and working in mutuality

Certain voluntourism analyses have considered how the relations between communities, mediating organisations and volunteers can be reconfigured for justice (e.g., Guttentag, 2011). Simpson insisted that a "pedagogy of social justice" (PSJ) (2004, p. 690) must underpin voluntourism. Henry (2019) advanced these discussions by outlining three steps to support a PSJ: researcher reflection, action research and continued engagement with returned volunteers to support their ongoing radical consciousness and action. Henry's proposed agenda was intended to foster an approach based on social justice "embedded within and continuing beyond voluntourism..." (2019, p. 563).

This article employs Freire's (1970) analysis of the "pedagogy of the oppressed" and his articulation of mutuality as a way of answering the research question of how we might move beyond "doing good" in voluntourism and development. Here, Freire addresses how those in positions of power and privilege may want to support those who suffer oppression. He outlines the nature of oppression, explaining that the humanity of both the oppressed and the oppressor are undermined. Some in the oppressor category may develop a guilty awareness of their role in oppression and seek to mitigate it; but if these efforts fail to address the structural roots of oppression, they only offer "false generosity", thus maintaining the oppressed in binds of dependency (1970, p. 49). For Freire, "oppression is domesticating" (p. 51) and so members of both the oppressor and oppressed classes must continually struggle for their humanisation and freedom. Critical to this is conscientisation (*conscientização*), which refers to mutual, critical awareness work that the oppressed do in understanding their conditions leading to a determination that "the concrete situation that begets oppression must be transformed" (p. 50). The fundamental basis of this work is the recognition of the oppressed's own knowing (p. 68) and abilities to transform their situation.

The aim of this mutual work is to become, both oppressed and oppressor, "human in the process of achieving freedom" (p. 49). This work is achieved through "praxis: reflection and action upon the world in order to transform it" (p. 51) in an effort to gain full humanity and carry out "revolutionary" change (p. 66). Allies work in solidarity with the oppressed but they are cautioned to not fall into the trap of trying to lead and thereby return to oppressing. The method for this work is dialogical action where allies (revolutionary leadership in Freire's terms) support the oppressed in conscientisation and their praxis towards liberation (p. 67). This transformative work is not something the oppressed can do alone nor something others can do for them; it is derived from working *with* and not *for* (p. 67). "Founding itself upon love, humility, and faith, dialogue becomes a horizontal relationship of which mutual trust between the dialoguers is the logical consequence" (Freire, 1970, p. 91).

This thinking from Freire offers a framework for transforming both tourism and development for working in mutuality rather than in the "false generosity" of white saviourism. It is built on critical consciousness developed through dialogue, co-development of a praxis for changing the material conditions of the community, what we would call "capacity sharing" and trust in the capacities of the people. It is an ongoing process in which the people and their allies come to understand their context and how it might be transformed for the better and co-create a shared humanity in this mutual praxis.

Linking together the concepts set out in the literature review, a framework is revealed pertaining to conventional international development which is based on an ideological worldview that recipients are inferior, supporting practices that may be best described as paternalistic. Integrated relationships feature between voluntourism, development, development aid NGOs and countries of the Global South which perpetuate a cycle of dependency. Yet, as Lediard (2016) demonstrated if justice is placed to the fore, volunteer tourism relations can be transformed "from power-over to power-with" (p. 100). This is the promise of the Freirian conceptualisation of mutuality in praxis. Following a discussion of research methods, we present a case study of an NGO and its co-founder that illustrates the development of a mutual praxis towards liberation from dependency.

Methodology

This research has employed qualitative case study methods to consider the case of the Cambodian Children's Trust and the role of its co-founder Tara Winkler. A case study is "the in-depth investigation of a phenomenon in its context" and has been well utilised in qualitative organisational research (Piekkari & Welch, 2018, p. 357). Case study approaches are effective when research seeks to answer "how" and "why" questions and when the research questions "... require an extensive and 'in-depth' description of some social phenomenon" (Yin, 2014, p. 4). Qualitative case studies offer "an intensive, holistic description and analysis of a bounded phenomenon such as a program, an institution, a person, a process or a social unit" (Merriam, 1998, p. xiii).

This research follows a critical, social constructionist approach recognising that people construct their own understanding of the world through their experiences and reflections on these experiences (Schwandt, 1998). This holds true for both the researcher and the researched. However, as Sauer and Lyle explained, Ricoeurian ethical philosophy can help ground narrative constructionism and overcome its "postmodern predicament" through "an ontology of action, ethics, and the 'responsible self'" (1997, p. 203). Thus, in this perspective of constructionism, the narrative of the self is always in relation to others; "ethical identity concerns not just the being of the self but being with and for others" (p. 221). This combined with critical research praxis, prepares research with tools to name, critique and overcome forces of oppression and support emancipatory collaborations (Giroux & Giroux, 2006).

The first author has been a donor to CCT after learning of Winkler's work through a television documentary (ABC, 2010). This research project was developed following a donor update zoom meeting with Winkler in 2021. In that discussion, a shared interest in decolonising development became clear and we agreed to undertake research that studied CCT's transition from child rescue to community empowerment. The second author contributed vital development studies expertise to the project, and she has no prior or ongoing association with CCT. The third author has worked in development (often with a gender focus), including in Bangladesh. Here, she first experienced the dependency of local communities on NGOs and the limiting effect of this on the transformational capacity of individuals (becoming evident after the rollback of long-standing programs). Inspired by CCT's village hive model to bring power back to communities, she is currently involved in designing impact assessment strategies with CCT.

The data included primary and secondary resources as well as a small number of focused interviews. Primary data included CCT annual reports, program evaluations, Winkler's book *How (not) to start an orphanage... by a woman who did* (Winkler & Delacey, 2016), audio-visual materials (e.g., TED Talks and television documentaries), testimony to the 2018 Australian Parliamentary Inquiry into establishing a Modern Slavery Act and numerous reports about child protection and orphanages in Cambodia (deriving from the Cambodian government and NGOs such as UNICEF and Save the Children). The book assisted in identifying preliminary themes as Winkler was courageously honest in addressing the issue of white saviourism, following a driving ethic of "act to do better when we know better" (additional themes included "CCT as empowering youth of Cambodia", "a rising consciousness leading to action" and a "transition of approach from doing good for, to working with" (Winkler & Delacey, 2016). Descriptive coding was applied to this content setting out a categorised inventory of Winkler's work and CCT (Saldaña, 2015).

Interviews occurred by video-link between May and September 2021 and included both Australian and Cambodian leadership of CCT and its sister organisation CCT Australia (CCTA) (n = 5). Each interview, conducted in English, lasted up to one and a half hours in duration following a semi-structured interview protocol (Appendix A); two of the five interviewees with the most holistic insights were interviewed three times for securing greater detail and clarification. The project was approved through the ethics approval process of the University of South Australia's Business School, which informed the protocols for obtaining informed consent and securing participant rights. Anonymity was waived by all interviewees after an informed consent process. Participants were given access to the interview transcripts for clarification and correction and the chance to review a draft of the themes and findings in support of stakeholder validity checks. Analysis of the interview data followed an inductive approach being very attentive to the voices of the informants. A second round of analysis then followed using abduction as a flexible approach to refine themes with attention to the literature and its gaps (Pierce, 1978).

As critical, social constructivist research, this work presents the evolution of Winkler's and CCT's work from child welfare to community development in Cambodia. This leads to considerations of dependency in both tourism and development. While an independently conducted external program evaluation validating CCT claims is available (Nee, 2020), this aspect was not the focus of our study, which instead prioritised the seldom accessed personal narratives of white saviourism and how conscientisation may dismantle it.

Background to the Cambodian children's trust

> *"To create systemic, lasting change, we focus our efforts on an 'upstream', preventative approach that addresses the root causes of complex social problems that make children unsafe and unable to access their basic needs"* (CCT, 2020, p. 7).

Tara Winkler's involvement in Cambodia began when she travelled to the country on an Intrepid Tour of Southeast Asia in 2005. She planned to stay on after the tour concluded but, because she felt disconcerted holidaying in a place struggling with poverty, she decided to volunteer with the Akira Landmine Museum. Following this, she was invited on some orphanage visits and experienced the compulsion to buy materials to support the children she encountered there.

A visit to one of these orphanages resulted in the founding of CCT in 2007 by Tara Winkler and Jedtha Pon, both of whom rescued 14 children from a corrupt and abusive orphanage in the Cambodian province of Battambang.[2] Following this rescue, CCT initially operated as a well-resourced orphanage. Winkler sought donor support from her native Australia and some of these donors and others visited and volunteered at CCT as orphanage tourists.

While operating the orphanage, Winkler became aware that such institutional care wrought unintended harm on children's development and after becoming fluent in Khmer learnt that all the children had family. Their families had only given them up to institutional care due to poverty, thinking that the orphanage would give their children access to education and better futures. As a result of this awareness, CCT undertook a coordinated effort at reunification and reintegration of the children with their families and communities of origin. CCT then became the

Figure 1. CCT holistic approach to services enabled within the Village Hive model.

first orphanage in Cambodia to transition from orphanage to a family-based care approach which evolved into the "Village Hive" model (more below).

CCT is now a Khmer-led organisation focused on community-run child protection to strengthen families and keep children safe through promoting local agency, community empowerment and sovereignty. CCT has followed a reflective and transformative praxis to working with communities on identifying and meeting their development needs all the while partnering in capacity sharing. CCT's efficacy is demonstrated by the way it has evolved to implement constant improvement and the long-term sustainability of its programs. As an NGO, it particularly developed a commitment to embedding exit strategies in its programmes as a result of this learning process, consciously rejecting the fostering of dependency while planting seeds for community empowerment. CCT's operations in Cambodia are supported by the fundraising, promotions, donor liaison and impact evidence-gathering of its sister organisation CCT Australia (CCTA), made up of a team of four Australians.

CCT established a social enterprise restaurant called *Jaan Bai* ['rice bowl'] in 2013. The restaurant is now independently operated and managed by Cambodian youth in a profit-share agreement. In 2015, CCT identified the need for strong Information and Communication Technology (ICT) training for youth, commencing a partnership with all five public high schools and the teacher training college in Battambang City to develop an ICT programme. This programme is now integrated into the national high school curriculum and is fully funded and operated by the public schools, with to date over 27,000 students having completed the course and more than 1,100 teachers trained to deliver the ICT curriculum. Further evidence of success is found in the number of students who have advanced to enrol in higher studies in computer science (n = 827) and who have won or placed in national and international STEM competitions (CCT, 2019).

Learning from the success of the Jaan Bai and the ICT programmes, CCT launched the first formal pilot of its Village Hive model in 2017 (Hive standing for Holistic Interventions Vulnerability to Empowerment). The approach is co-created with local communities from ideation to implementation through CCT-led workshops (see Appendix B "Co-creating child protection networks within villages"). Its purpose is to combat and prevent multidimensional poverty by bringing multisectoral services into villages and mobilising community-wide action to protect children from maltreatment, prevent child-family separation, reunite families and foster long-term resilience and self-sufficiency within families. This addresses the sustainability of the village community and recognises the intersectionality of the social space. Today, CCT's Village Hive reaches 104 villages in Battambang province and has prevented thousands of children from

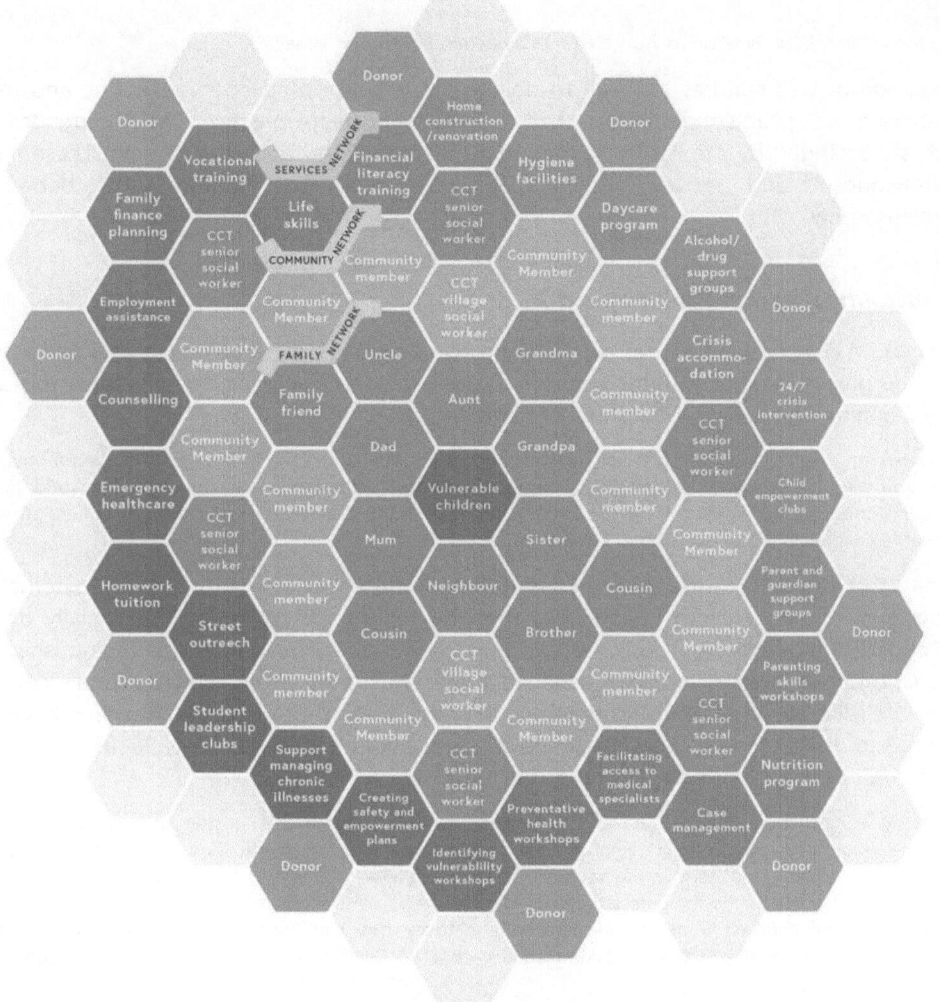

Figure 2. CCT's vision of a self-contained Village Hive approach centred on vulnerable children within empowered family, community and services networks, and strengthened by CCT staff and donor support in its establishment phase (see CCT, 2020).

being separated from their families and reunited hundreds of children in orphanages with their families. Figures 1 and 2 indicate the holistic approach of the Village Hive model, embedding needed services within the community to be integrated under local governance rather than separate NGO authority. Appendix C outlines the Ou Char pilot of the Village Hive model and shows the embedded, inclusive, collaborative approach. Appendix D highlights the main programmes of CCT presented chronologically demonstrating the evolutionary change of its programmes from a "downstream" child rescue approach to an "upstream" community empowerment approach.

CCT follows a partnership approach with families, communities, all levels of government, other NGOs and NGO networks such as Family Care First and large donors such as United States Agency for International Development (USAID). CCT's track record in systems building and sustainable programming has led to CCT being recognised as an innovative leader in the Cambodian child protection sector and becoming the largest implementing partner of the Family Care First/REACT Network (FCF, n.d.).

Findings

> "*I did then what I knew how to do. Now that I know better, I do better*" - Maya Angelou

The evolution of CCT features an urge to do good but built on an open, reflective and responsive approach that enabled significant change from mainstream practice. The findings from this research are arranged by the major themes that emerged from the research data. These include: white saviourism and moving beyond it; when you know better, do better; and Cambodians know.

White saviourism and moving beyond it

In her book, Winkler explained her initial desire to do good and her evolving critical consciousness as she gained greater insights through experience. One incident demonstrates this as she described beginning to teach English to the children:

> The words of Nelson Mandela kept running through my mind: "Education is the most powerful weapon which you can use to change the world". In my own little way, I thought I was changing the world. Insert face palm here... I had a lot to learn about the potential harm that could come from volunteering with vulnerable children (Winkler & Delacey, 2016, pp. 67-8).

Despite this rapidly growing critical consciousness, there was also a dynamic of external pressures designating her a white saviour and pressing for this to continue. This initially occurred through media narratives (e.g., ABC, 2010) and then carried over into the relationships with donors. One donor interviewed for Australian ABC's "Australian Story" stated: "Tara's come and changed their life. What an amazing brave young woman who's just totally inspiring. None of us could hope to do, you know, what she does in one week in a lifetime" (ABC, 2010).

CCT former Operations Manager Erin Kirby stated:

> Everybody wanted to meet with Tara and gave her this white saviour title... that she saved these kids, that's the whole narrative of the original ABC Australian Story. It got put onto her and I think it took her years to see it, to learn that term white saviour and [to] back away from [it]... but at the same time people wouldn't want to give us money unless they got to meet her ... I think it's something that she's had to do a lot of critical reflection on, a lot of work personally... to see how can she still do the work that she wants to do, but in a way that is actively working against that, rather than benefitting from it (pers. comm., 6 August 2021).

However, as Winkler explained, this external tendency to laud her as a white saviour could also be harnessed for positive outcomes:

> Leadership qualities are more often made, not born. The opportunities I've been granted, thanks to my white privilege, has allowed me to cultivate and hone the unique set of skills required to be an effective leader. Truth is, as a young, naive and inexperienced expatriate, I should never have given the opportunity to set up an orphanage in Cambodia. However, I did at least use those privileges to bring Jedtha [CCT Co-Founder] and Sinet [orphanage survivor and Director on the CCT Board] along with me so that they too had the opportunities to develop the leadership qualities that were vital to their ultimate empowerment as they stepped into their rightful roles as CCT's leaders (pers. comm., 10 May 2021).

Sinet Chan described the impact of Winkler's transformation in consciousness:

> If Tara didn't step out of her power, Jedtha wouldn't be able to run this work by himself, as he does today. I think, you can say that [CCT] is not only Jedtha's family, but also Tara's too. It takes courage for a person to give up the one thing that they love the most... for a change that... will make a better future for the Cambodian people. I think that is an incredible role model and incredibly empowering ... we can learn from... that ... It's not easy... you have the power to do anything you want, that is your life, your job, why would you give that away to other people? But I think she inspires us to see what real leadership and a real role model looks like.
>
> So, for me too, as a care leaver facilitator and a care leaver myself, I will never be a care leaver leader for the rest of my life. I am coaching and giving advice to other care leavers to run it by themselves, because

later … I will step aside. This is what new power means: let the local people do it and the right people will come to do the job (pers. comm., 3 September 2021).

The Village Hive model is a repudiation of white saviour practices in development and a step towards capacity sharing. Rather than seeing Global South peoples as incapable and in continuous need of external guidance, CCT Khmer leadership bring needed support and resources to the cooperating communities:

> Handing over a self-contained Village Hive involves restructuring our model, to ensure we are working within public systems and operating out of public facilities, rather than privately run programs in parallel. Handing over to the local council means the commune makes valuable contributions to the Village Hive in the form of human, social and financial capital. This demonstrates the value the community place on the Village Hive and shows their commitment to its long-term success. CCT can then exit and move on to partner with the next community, thereby having an actionable, tangible exit strategy that allows us to scale economically (CCT, 2020, p. 17).

NGOs such as CCT that hold an allyship mindset are able to play helpful roles in supporting communities and governments in creating connections for transformation. Previously, Cambodia lacked a formal child protection system, instead relying on traditional kinship care networks. CCT was well placed to contribute to policy thinking that addressed how Cambodia's formal child protection system could develop. Cambodia features decentralisation from national down to commune levels of governance. UNICEF approached CCT to discuss how child protection operates at the commune level (Winkler, pers. comm, 10 May 2021) because CCT has social workers partner with families and knows how to work at commune level. CCT had been informed by the Signs of Safety programme (see https://www.signsofsafety.net/what-is-sofs/) and introduced new practices to Cambodia that worked according to a strengths-based, non-paternalistic approach that places families in control. "Because CCT has donor rather than grant funding it has allowed for innovative programme design" (Winkler, pers. comm., 10 May 2021). Guiney's earlier research noted a gap between UNICEF and local level leadership (2015, p. 255) which since that time CCT has helped bridge through its embedded practices.

However, Winkler noted that CCT has not fully arrived at a point where empowerment and emancipatory practices are fully enacted because of the ongoing structural context of CCT operations:

> CCT works at a sort of omnishambles of very complicated sectors - the child protection and international development sectors mirror each other in their roots and their historical complexities around colonialism, race, paternalism and power dynamics. It is no wonder we started where we started (pers. comm., 10 May 2021).

When you know better, do better

Currently CCT activities in Cambodia are 100 per cent Khmer operated, a quite radical transformation from the beginnings based on the humanitarian impetus of a young Australian woman. CCT's evolution is built on a dedicated practice of engaged reflection and learning which is the organisation's defining feature. As CCTA Managing Director described it, "… a process of a pilot, reflection and adaption is really that thread that's been through CCT right from the beginning" (Keir Drinnan, pers. comm., 10 July 2021).

The most important example of critical reflection and responsive change is evident in the change in consciousness of Co-Founder Tara Winkler herself. Her ability to reflect on poor practices, to be open to change and to be humble in openly sharing these stories in her book, speeches and media interviews offers others a chance to develop critical consciousness. Winkler has noted:

> I have made all of those mistakes myself. I came to Cambodia as an orphanage tourist, and then I was a voluntourist when I volunteered at the orphanage, and then I set up an orphanage, and then I facilitated

> orphanage tourism. I have learnt from those mistakes... the hard way ... when you make mistakes in development, you're making mistakes with people's lives (cited in Ryan, 2015).

Another example occurred when CCT still ran the orphanage, the children were invited to perform in Singapore. Winkler explained: "The problem is that it is almost impossible to use beneficiaries [the children in care] to raise money without perpetuating 'poor third-world people saved by western NGO' stereotypes" (Winkler & Delacey, 2016, p. 276). However, Winkler began to understand that it harms children's confidence and sense of self and can also harm them in the community (e.g., job prospects). Despite the fact that CCT could raise more funds if it was willing to display the beneficiaries to donors this way, CCT formed a policy that only staff would support CCT fundraising events (Winkler & Delacey, 2016, pp. 275-6).

CCT has piloted programmes and evolved its practices through this dedicated learning approach (See Appendix D). Keir Drinnan explained:

> Two years ago, we looked at the issues and reflected on some of the successful programmes that we'd had and realised that they were all embedded within the local infrastructure. They were partnering with the local stakeholders, [so] ownership and accountability for that program laid with them. That's why they continued on after the funding had stopped. I think that changed our thinking around how we would deliver programmes (pers. comm., 10 July 2021).

This theme of the research suggests that those with a desire to do good for others can be guided through a rigorous process of critical reflection to a more justice-oriented practice in solidarity with communities.

Cambodians know

Development agencies and NGOs have often relied on expatriate consultants to provide expertise. Underpinning the mindset is that the local people lack skills and capabilities. In its early years, CCT also brought in external experts to advise, train and set up processes for managing, for instance, human resources, social work, project management and finance. It now actively avoids hiring foreigner consultants and has transitioned to a 100 per cent Khmer leadership team in Cambodia. As Sinet Chan, a Director on the Board of CCT, stated: "I think that many NGOs in Cambodia don't know about the power of the local people" (pers. comm., 3 September 2021). In her interview, Keir Drinnan explained there was a three-year plan developed in 2019 to transition from foreign involvement to solely Khmer leadership which was expedited due to the COVID pandemic and its border closures.

Jedtha Pon, CCT Co-Founder and Director, stated the value of the shift to 100% Khmer leadership of CCT and the critical discovery of Cambodian voice:

> It's really important for our Cambodian people ... because we know our culture, we speak our language... We know what we want... we live it; we're owning it. Everyone here has accountability for what we are doing.
>
> Previously, the communication team was all foreigners and now it is all Khmers. It has been important for us to recruit and look for local staff to fill such roles. You can imagine that for the social work, Cambodia has only [recently] had university degrees in social work... That has made it hard to find experienced, qualified social workers to do this job. That is why sometimes we needed [foreign experts]. But now... all leadership roles are for Khmer (pers. comm., 20 August 2021).

Chan was one of the original 14 "orphans" that Winkler and Pon removed from the abusive orphanage in 2007. She has unique, valuable knowledge as an orphanage survivor (as she calls herself). As she grew into adulthood, she has taken on leadership roles with CCT, first as a CCT Ambassador and now as one of the Directors on the CCT Board and Founder of its *Khmeng Onka* Care Leavers programme. Chan's voice of lived experience has been important in demanding the banning of orphanage tourism and advocating for governments to regulate their aid funding to orphanages to ensure these do not support trafficking and exploitation of children (see Chan, 2021). She and Winkler both presented to the 2018 Australian Parliamentary Inquiry into establishing a Modern Slavery Act.

Chan's positionality as a child that experienced both abusive and non-abusive forms of institutional care informs her practice in leadership in CCT. She explained in her interview how CCT leadership recognised the harms of institutional care on children and responded by planning the closure of the orphanage and reuniting and reintegrating children with their families. But from her own personal experience, she identified a need to support care leavers[3] (those that leave institutional care to live in the community at adulthood):

> ... back then I was only a CCT Ambassador and only a kid's voice. CCT had changed from an orphanage to reuniting all the children back with their families. But at the same time, I turned 18 ... an adult, so, I moved to living independently ... since then, I became an activist for children's rights, for children survivors from the orphanage. ... now we have a care leavers club which I run. I support care leavers who are living independently after leaving the orphanage. Now we provide support to them as well (Sinet Chan, pers. comm., 3 September 2021).

Another aspect of respect for Cambodian people's knowledge and capabilities is the recognition of the need to work cooperatively with all levels of Cambodian governance. Pon stated: "We are really working in partnership with the government; we cannot work without them. We have very close relationships with the commune, the district, provincial and the government ministries". This has grown particularly true with the Village Hive model which features close working relations from the local to the national level (from the local Commune Council for Women and Children (CCWC) to the national Ministry of Social Affairs, Veterans and Youth Rehabilitation).

Former Operations Manager Erin Kirby explained how CCT staff began working more closely and effectively with commune level governance:

> ... it's not just having an MOU with the government or going to the required meeting, sitting there and getting your name ticked off a list ... When we take on a new family, we get our referral coming from the CCWC, who are responsible for child protection cases at a local level ... [We] then go out with our assessment forms which are government forms that our social workers use ... In the past they would go out and do it themselves and then they would take those forms to be signed off by the CCWC member because that person's really busy ... and not always available. However, now that the social workers understand how ... important it is, they're undertaking collaborative case management in partnership with CCWC and coaching and mentoring government social workers to build their capacity (pers. comm., 6 August 2021).

One key outcome that came with abandoning the orphanage model and evolving to the Village Hive, is an approach that ensures children supported through CCT initiatives remain in the culture and mode of living of the community (as well as in the loving bonds of families). Chan made this clear in her interview when she spoke of life as a child in the CCT orphanage:

> I lived like a queen, because CCT back then was the best; we had tutors ... painting, dancing, music, ... holidays. But it's a small amount of people that received that support. But now there's resources going to different houses, different villages, different communities. Each home then can plan for themselves ... I don't think it is good to empower one person like me or other children, but rather all the donors' support that we get is going to the household ... Helping the ... whole family. I think it's a great [approach], because we know that raising one kid in an orphanage is expensive versus raising them in the family (pers. comm., 3 September 2021).

The Village Hive model was designed to offer "an upstream approach to child protection, delivered sustainably by local communities" (CCT, 2020, p. 7). By "upstream", CCT means working from a preventative approach rather than a reactive approach. This is a holistic way of working with communities, with CCT Khmer teams organised for social work, finance (financial literacy and livelihoods), housing, health and education (See Figure 1).

The most important point of distinction in CCT's Village Hive model from mainstream models of development based on dependency is the embedded exit strategy. CCT described it:

> We are working towards integrating the Village Hive into the public sector where it can be delivered sustainably by communities.
>
> The ultimate aim of the pilot is to create an evidence base for a prevention-focused child protection system that can be scaled across Cambodia.
>
> The pilot will transfer the delivery of the Village Hive multi-sector services from CCT to the local councils, who rightfully hold the responsibility to provide services to their community (CCT, n.d.).

Drinnan explained:

> Ensuring vulnerable families are connected to universal services and support, within their own communities, to enable them to care for and raise their children well. That's working with the local authorities to help embed those services into existing public facilities, such as public schools, within those communities, and providing mechanisms for families to be able to be referred for those supports and services. This includes working with the community to obtain the financial support required through their commune investment plans for social protection mechanisms within their own communities… the Village Hive is utilising the power of networks - family networks, community networks and services networks (pers. comm., 10 July 2021; see also Appendix C).

The issue of "empowerment" is an important one. Used in the development context, there is a connotation that those that bring development empower the communities being developed. Understood this way, an inherent inequality is clear. In contrast, a Freirian approach of mutuality emphasises dialogic learning and action for change (1970, p. 125). CCT demonstrates critical understanding of these issues:

> To achieve systemic change our approach must also be empowering. The term "empowerment" is so overused by charities that it's almost lost its real meaning and has been reduced to a buzzword…
>
> We believe empowerment should be a conversation about power. The people whose lives are affected by programming decisions, need to be the ones making them. They need to have seats at the decision-making table and be the architects of the programs. Our experience has shown us that local people hold all the wisdom required to design and implement programs that safeguard children and empower families in their own communities (CCT, 2020, p. 7).

This case study of CCT demonstrates that a desire to do good is insufficient to build partnerships for sustainable development. Instead, approaches that build local agency, in the interests of local communities and that respect local sovereignty are essential if the crisis of dependency is to be avoided. Empowerment is revealed as enabling change that is valuable to the people in their own context through reflection and action (Kabeer, 2011).

Discussion

This research has offered an in-depth case study to learn from one NGO co-founder who traversed the path from white saviour to ally and who supported her organisation to simultaneously transition from fostering development dependency to fostering local agency through the Village Hive approach. This work continues on from the earlier, in-depth work of Guiney (2015) and reported also in Guiney and Mostafanezad (2015). The latter documented the beginning of the anti-orphanage movement in Cambodia from 2011 and questioned the Cambodian government's ability to fund orphanages and social welfare if private and corrupt orphanages were shut down. CCT's example models an approach where white saviourism and NGO hegemony are abandoned after a committed praxis of transformation. Understanding that Cambodian communities and governments are the foundations for sustainable development, a practice of "working with" has been established based on recognising and supporting their capabilities at the local level through the Village Hive approach.

Freire's work (1970) concerning conscientisation referred to earlier is useful in theorising Winkler's and CCT's work *with* rather than *for* communities in Cambodia. Freire connected conscientisation to action which is based on a trust in the capabilities of communities: "To achieve this praxis, however, it is necessary to trust in the oppressed and in their ability to reason. Whoever lacks this trust will fail to initiate (or will abandon) dialogue, reflection, and communication…" (1970, p. 66). This case study has explained both the evolving conscientisation of Winkler but also of her CCT colleagues that enabled them to co-develop a practice of working with Cambodian families and communities. Instead of the traditional "capacity building" approaches to development, this Freirian practice is best described as "capacity sharing", describing when parties in the development process share experiences and knowledge to strengthen decision making under an agreed development trajectory.

Freire provides a useful way of understanding this transformation:

> Authentic help means that all who are involved help each other mutually, growing together in the common effort to understand the reality which they seek to transform. Only through such praxis—in which those who help and those who are being helped help each other simultaneously—can the act of helping become free from the distortion in which the helper dominates the helped (cited in hooks, 1993, p. 150).

Instead of the "false generosity" of a dependent path to development, Freirian approaches work with and through the receiving community: "True generosity lies in striving so that these hands—whether of individuals or entire peoples—need be extended less and less in supplication, so that more and more they become human hands which work and, working, transform the world" (Freire, 1970, p. 46). The Khmer people involved in the Village Hive pilot embraced the opportunity offered to be more self-determining:

> What I thought would be the biggest challenge was the willingness of the stakeholders to take control of the project and the administrative burden that would come along with it. On reflection, this revealed a personal blind spot. In reality, each and every stakeholder responded with an overwhelming enthusiasm to take on the responsibility (Tara Winkler, pers. comm., 21 December 2021).

A Freirean lens makes us attentive to issues of justice when we analyse the urge to "do good" both in tourism and in development. Power, inequality and injustices are prevalent in this space where the Global North and South meet (Mowforth & Munt, 1998, p. 160). CCT leadership has recognised issues of power throughout its work and through critical reflective praxis has co-developed ways of creating networks of partnerships that are focused on addressing the root causes of not only child endangerment but also continuing underdevelopment in Cambodia.

As Lediard (2016) demonstrated, if justice is placed to the fore, volunteer tourism relations can be transformed "from power-over to power-with" (p. 100). Cammarota argued further that "justice can occur with the oppressed in leadership positions and the oppressor adhering to and following this leadership" (2011, p. 245). This is indeed what CCT's evolution has modelled as CCT has transitioned to a full Khmer leadership team, with CCTA acting as support for donor engagement in Australia.

Freire's conceptualisations of conscientisation and mutuality also suggest that partners from the Global North can and should learn from partners in the Global South. Therefore, the Village Hive model that CCT developed may be considered for adaptability in communities throughout the world. Absent a white supremacy mindset of white saviourism, reciprocal learning and action for development is not only possible but may be imperative as we seek to address multiple, future crises confronting us all.

CCT has exhibited some of the strengths and weaknesses identified by Delvertere and De Bruyn (2009) as a private initiative for development. However, what this research has demonstrated is that a commitment to continual critical reflective praxis has helped it evolve successful programmes built on transitioning to Cambodian leadership, partnership approaches with all stakeholders and a belief in the capacities of families and communities to build their own livelihoods with the right kinds of supports from these partnerships (in contrast to perpetual dependency- see Appendix B).

However, there are some difficulties with the upstream approach of CCT's Village Hive model. Individual donors may prefer an individualised, child sponsorship approach that often pertains to

NGO aid to orphanages (Sinet Chan, pers. comm., 3 September 2021). Drinnan concurred, stating: "it's difficult for donors to understand our model, even institutional donors. I know USAID is adverse [to local authorities managing aid funds given to CCT] … there's a perceived lack of transparency and accountability that such organisations require … " (pers. comm., 10 July 2021). Winkler expanded:

> The biggest challenge encountered in shifting power to local communities turned out to be the Global North's pervasive belief in the corruption and incompetence inherent in Cambodian people, authorities and institutions. This results in a lack of trust that prevents most institutional donors from funding projects in which they don't retain sufficient control. I see this as a glaring impediment to Cambodia's development and the ultimate empowerment of its people (pers. comm., 21 December 2021).

Therefore, efforts at conscientisation must be universal to transform the mindsets of all stakeholders in tourism and development so that we can shift away from notions of "doing good", white saviourism, Western/Eurocentric superiority and perpetual dependency to mutual solidarity for co-building positive futures. As Guiney pointed out, " … a significant reconfiguration of the international system is required" (2015, p. ii) if we are to move beyond addressing the symptoms of poverty to actually address the needs of communities for sustainable development.

Recommendations arising from this research include:

- All children must be protected from the exploitation and commodification that the neoliberal model brings.
- Development must be focused on a commitment to tangible actions for change in the material causes of oppression rather than being the "false generosity" of continual aid.
- Development models should be reconfigured as collective, collaborative and co-created engagements for capacity sharing and empowerment; dependency is anathema.
- Collective dialogue for conscientisation offers a pedagogy of justice to develop co-understanding of the conditions under which projects are undertaken.
- Trust in the capacities of "recipients" is essential to this Freirian approach to development.
- Tertiary studies in tourism and development should include curricula on pedagogies of social justice, white saviourism and applied praxis for working mutually with stakeholders for empowerment and co-created positive change.

Conclusion

With ten per cent of children in Cambodia reportedly not living with their biological parents despite having at least one living parent, children are clearly vulnerable (UNICEF. , 2018, p. 43). Orphanage tourism is a catalyst to child trafficking, exploitation and labour and measures are being implemented to address this. This falls under the SDG agenda as target 8.7 calls for the end of child labour in all of its forms by 2025. Clearly, all stakeholders must support the termination of orphanage tours and associated voluntourism experiences. The case of CCT presented here illustrates these issues but it also points to the imperative to not stop there; it is essential to build a mutual praxis to address the structural injustices at the root of these issues.

This article has argued that we confront a crisis of development which results in poor outcomes for communities. Dynamics exposed here include: white saviourism; a corresponding ideology that the receiving community is inferior and incapable; and a resultant dependency approach to development. When we think about SDG 17 on partnerships for sustainable development in light of the CCT case study presented here, we might think of it in a Freirian sense of mutuality as illustrated in this case study of CCT. This work suggests a guiding framework for more just partnerships: critical consciousness developed through dialogue, co-development of a praxis for change, capacity sharing and trust in the knowledge and capabilities of the people.

Despite the limitations of case study research in terms of non-generalisability, the findings of this research may be used to inform tourism research, education and industry practice. Many university programmes engage students in internationalisation experiences and fieldwork that might involve voluntourism and orphanage visits. While we encourage all tourism advocates to seek to make a positive difference and support those who are disadvantaged in our global community, critical literacy concerning white saviourism, neo-colonial practices and the wider context of structural injustices is essential to inform effective and just praxis (see Scheyvens & Storey, 2014). Additionally, curricula and training for Freirian practices are helpful in guiding efforts to build pedagogies of social justice in the tourism domain.

After the COVID pandemic, development assistance will be even more important as countries of the Global South will have carried a heavier burden, particularly tourism dependent nations. One United Nations Development Program representative placed aid top on the list in addressing the pandemic's impacts: "Governments need to boost foreign aid spending now, and sustain it" (Steiner, 2020, n.p.). This study suggests simultaneously rethinking our approaches to work from a consciousness of mutual solidarity if we are to transition from sustaining poverty to sustainable development.

Notes

1. This corresponds to Elite Development Theory, which Selwyn (2016) described as seeing: "the poor" as human inputs into or, at best, junior partners within elite-led development processes" (p. 781).
2. SKO orphanage is accused of abuse and neglect of the orphans in its care and corruption in the use of donations (see Winkler & Delacey, 2016; Chan, 2017). SKO closed in early 2020.
3. Care leavers are more at risk of exploitation and trafficking (Lumos, 2021, p. 45).

Acknowledgements

The authors would like to thank the Cambodian Children's Trust, and in particular Tara Winkler Keir Drinnan, for their support of this research. We also thank Dr Jacky Desbiolles who provided vital research support. Finally, we appreciate the advice and support provided by the journal's Editors, Guest Editors and anonymous peer reviewers.

Disclosure statement

No potential conflict of interest was reported by the authors.

Funding

This research did not receive any specific grant from funding agencies in the public, commercial, or not-for-profit sectors.

ORCID

Regina A. Scheyvens http://orcid.org/0000-0002-4227-4910

References

ABC. (2010). *Children of a lesser God*. Australian Story. Retrieved October 10, 2021, from https://www.abc.net.au/austory/children-of-a-lesser-god/9172650.

ABC. (2014). *The house of Tara*. Australian Story. Retrieved December 29, 2021, from https://www.youtube.com/watch?v=nRUwbzQipZI.

Bandyopadhyay, R., & Patil, V. (2017). The white woman's burden' – the racialized, gendered politics of volunteer tourism. *Tourism Geographies*, *19*(4), 644–657. https://doi.org/10.1080/14616688.2017.1298150

Banks, G., & Scheyvens, R. (2014). Ethical issues. In R. Scheyvens & D. Storey (Eds.), *Development fieldwork: A practical guide* (pp. 160–187). Sage.

Bhabha, H. K. (1994). *The location of culture*. Routledge.

Bott, E. (2021). My dark heaven': Hidden voices in orphanage tourism. *Annals of Tourism Research*, *87*, 103110. https://doi.org/10.1016/j.annals.2020.103110

Britton, S. (1982). The political economy of tourism in the Third World. *Annals of Tourism Research*, *9*(3), 331–358. https://doi.org/10.1016/0160-7383(82)90018-4

Cammarota, J. (2011). Blindsided by the Avatar. *Review of Education, Pedagogy, and Cultural Studies*, *33*(3), 242–259. https://doi.org/10.1080/10714413.2011.585287

CCT. (2019). *2019 was a year for winners in ICT*. Retrieved October 10, 2021, from https://cambodianchildrenstrust.org/2019-was-a-year-for-winners-in-ict/

CCT. (2020). *2020 Annual Impact Report*. Retrieved October 10, 2021, from https://cambodianchildrenstrust.org/wp-content/uploads/2021/03/CCT_AnnualImpactReport-2020_WEB_FA.pdf

CCT. (n.d.). *Exit strategy*. Retrieved October 17, 2017, from https://cambodianchildrenstrust.org/village-hive/exit-strategy/

Chan, S. (2017). *Testimony to the Australian Parliamentary Inquiry into Modern Slavery*. Retrieved October 3, 2021, from https://cambodianchildrenstrust.org/listen-sinet-chan-and-tara-winkler-give-evidence-at-2017-moden-slavery-inquiry/.

Chan, S. (2021). *Sinet addresses the UN*. Retrieved October 17, 2021, from https://cambodianchildrenstrust.org/sinet-addresses-the-un/.

Cheney, K. E., & Rotabi, K. S. (2014). Addicted to orphans: How the global orphan industrial complex jeopardizes local child protection systems. In C. Harker, K. Hörschelmann, & T. Skelton (Eds.), *Conflict, violence and peace: Geographies of children and young people* (pp. 89–107). Springer.

ChildSafe Movement & G Adventures. (2018). *Child welfare and the travel industry: Global good practice guidelines*. Retrieved October 4, 2021, from https://media.gadventures.com/media-server/dynamic/admin/flatpages/17-502-00-PT-GL_Child_Welfare_Guidelines_v4_03_2018.compressed.pdf.

Cole, T. (2012). The white-savior industrial complex. *The Atlantic*. Retrieved October 21, 2021, from https://www.theatlantic.com/international/archive/2012/03/the-white-savior-industrial-complex/254843/.

Develtere, P., & De Bruyn, T. (2009). The emergence of a fourth pillar in development aid. *Development in Practice*, *19*(7), 912–922. https://doi.org/10.1080/09614520903122378

FCF. (n.d.). *Family Care First-React*. Retrieved October 10, 2021, from www.fcf-react.org

Flanagan, K. (2020). Promising practices – Strengthening families and systems to prevent and reduce the institutional care of children. In J. Cheer, L. Matthews, K.E. van Doore & K. Flanagan (Eds.), *Modern day slavery and orphanage tourism* (pp. 63–81). CABI.

Freidus, A. L. (2017). Unanticipated outcomes of voluntourism among Malawi's orphans. *Journal of Sustainable Tourism*, *25*(9), 1306–1321. https://doi.org/10.1080/09669582.2016.1263308

Freire, P. (1970). *Pedagogy of the oppressed*. Continuum.

Ghosh, B. N. (2021). *Dependency theory revisited*. Routledge.

Giroux, H., & Giroux, S. S. (2006). Challenging neoliberalism's new world order. *Cultural Studies <=> Critical Methodologies*, *6*(1), 21–32. https://doi.org/10.1177/1532708605282810

Global Study of Sexual Exploitation of Children in Travel and Tourism. (2018). *The international summit of child protection in travel and tourism*.

Guiney, T. (2015). *Orphanage tourism in Cambodia: The complexities of "doing good" in popular humanitarianism* [PhD thesis]. University of Otago.

Guiney, T., & Mostafanezhad, M. (2015). The political economy of orphanage tourism in Cambodia. *Tourist Studies*, *15*(2), 132–155. https://doi.org/10.1177/1468797614563387

Guttentag, D. (2011). Volunteer tourism: As good as it seems? *Tourism Recreation Research*, *36*(1), 69–74. https://doi.org/10.1080/02508281.2011.11081661

Hall, C. M., Scott, D., & Gössling, S. (2020). Pandemics, transformations and tourism. *Tourism Geographies*, *22*(3), 577–598. https://doi.org/10.1080/14616688.2020.1759131

Harrison, D. (2012). Pro-poor tourism: Is there value beyond "whose" rhetoric? In T. V. Singh (Ed.), *Critical debates in tourism* (pp. 137–149). Channel View.

Henry, J. (2019). Directions for volunteer tourism and radical pedagogy. *Tourism Geographies*, *21*(4), 561–564. https://doi.org/10.1080/14616688.2019.1567577

Hickel, J., Sullivan, D., & Zoomkawala, H. (2021). Plunder in the post-colonial era. *New Political Economy*, *26*(6), 1030–1047. https://doi.org/10.1080/13563467.2021.1899153

Higgins-Desbiolles, F. (2020). Socialising tourism for social and ecological justice after Covid-19. *Tourism Geographies*, *22*(3), 610–623. https://doi.org/10.1080/14616688.2020.1757748

hooks, b. (1993). bell hooks speaking about Paolo Freire. In P. McLaren & P. Leonard (Eds.), *Paolo Freire: A critical encounter* (pp. 145–152). Routledge.

Illich, I. (1990). To hell with good intentions (1968 speech). In J. C. Kendall (Ed.), *Combining service and learning* (pp. 314–320). National Society for Internships and Experiential Education.

JSCFADT. (2017). *Orphanage trafficking*. Retrieved September 3, 2021, from https://www.aph.gov.au/Parliamentary_Business/Committees/Joint/Foreign_Affairs_Defence_and_Trade/ModernSlavery/Final_report/section.

Kabeer, N. (2011). Between affiliation and autonomy: navigating pathways of women's empowerment and gender justice in rural Bangladesh. *Development and Change*, *42*(2), 499–528. https://doi.org/10.1111/j.1467-7660.2011.01703.x

Kuokkanen, R. (2011). Indigenous economies, theories of Subsistence, and women: Exploring the social economy model for Indigenous governance. *American Indian Quarterly*, *35*(2), 215–240. https://doi.org/10.5250/amerindiquar.35.2.0215

Lediard, D. E. (2016). *Host community narratives of volunteer tourism in Ghana: From developmentalism to social justice* [Theses and Dissertations (Comprehensive). 1862]. https://scholars.wlu.ca/etd/1862.

Lumos. (2021). *Cycles of exploitation*. Retrieved December 19, 2021, from https://lumos.contentfiles.net/media/documents/document/2021/12/LUMOS_Cycles_of_exploitation.pdf.

Manzo, K. (1991). Modernist discourse and the crisis of development theory. *Studies in Comparative International Development*, *26*(2), 3–36. https://doi.org/10.1007/BF02717866

Merriam, S. B. (1998). *Qualitative research and case study applications in education*. Jossey-Bass.

Meyer, D. (2012). Pro-poor tourism: Is there actually much rhetoric? And if so, whose? In T. V. Singh (Ed.), *Critical debates in tourism* (pp. 132–136). Channel View.

Mowforth, M., & Munt, I. (1998). *Tourism and sustainability*. Routledge.

Nee, M. (2020). CCT Program Evaluation Report. Unpublished consultancy report in authors' possession.

Pelling, O. (2019). The secret world of modern-day slavery. *Smith Journal*, 30 (Online). Retrieved May 7, 2021, from https://oliverpelling.com/australia-modern-slavery.

Piekkari, R., & Welch, C. (2018). The case study in management research. In C. Cassell, A. L. Cunliffe & G. Grandy (Eds.), *The SAGE handbook of qualitative business and management research methods* (pp. 345–358). Sage.

Pierce, C. S. (1978). Pragmatism and abduction. In C. Hartshorne & P. Weiss (Eds.), *Collected papers* (pp. 180–212). Harvard University Press.

ReThink Orphanages. (n.d.). *Facts and figures about orphanage tourism*. Retrieved October 7, 2021, from https://rethinkorphanages.org/problem-orphanages/facts-and-figures-about-orphanage-tourism.

Rogerson, C. M., & Saarinen, J. (2018). Tourism for poverty alleviation: Issues and debates in the Global South. In C. Cooper, S. Volo, W. C. Gartner, & N. Scott (Eds.), *The Sage handbook of tourism management* (pp. 22–37). Sage.

Ryan, R. (2015). 'Making mistakes with people's lives': The ethics of orphanages and voluntourism. *ABC Online*. Retrieved October 17, 2021, from https://www.abc.net.au/radionational/programs/latenightlive/tara-winkler-the-ethics-of-orphanages-and-voluntourism/6819486.

Saldaña, J. (2015). *The coding manual for qualitative researchers*. Sage.

Sauer, J. B., & Lyle, R. R. (1997). Narrative, truth, and self. *The Personalist Forum*, *13*(2), 195–222.

Scheyvens, R. (2002). *Tourism for development: Empowering communities*. Prentice-Hall.

Scheyvens, R. (2012). Pro-poor tourism: Is there value beyond the rhetoric? In T. V. Singh (Ed.), *Critical debates in tourism* (pp. 124–132). Channel View.
Scheyvens, R., & Storey, D. (Eds.). (2014). *Development fieldwork: A practical guide*. Sage.
Schwandt, T. A. (1998). Constructivist, interpretivist approaches to human inquiry. In N. K. Denzin & Y. S. Lincoln (Eds.), *The landscape of qualitative research* (pp. 221–259). Sage.
Selwyn, B. (2016). Elite development theory. *Third World Quarterly, 37*(5), 781–799. https://doi.org/10.1080/01436597.2015.1120156
Sharpley, R., & Telfer, D. J. (2015). *Tourism and development: Concepts and issues*. Channel View.
Simpson, K. (2004). 'Doing development': The gap year, volunteer-tourists and a popular practice of development. *Journal of International Development, 16*, 681–692. https://doi.org/10.1002/jid.1120
Sin, H. L. (2010). Who are we responsible to? Locals' tales of volunteer tourism. *Geoforum, 41*(6), 983–992. https://doi.org/10.1016/j.geoforum.2010.08.007
Steiner, A. (2020). How to avert the worst development crisis of this century. *UNDP*. Retrieved October 12, 2021, from https://www.undp.org/blog/how-avert-worst-development-crisis-century.
UNICEF. (2018). A statistical profile of child protection in Cambodia. Retrieved October 11, 2021, from https://www.unicef.org/cambodia/reports/statistical-profile-child-protection-cambodia.
Wearing, S. (2001). *Volunteer tourism: Experiences that make a difference*. CABI.
Wearing, S., & McGehee, N. G. (2013). Volunteer tourism: A review. *Tourism Management, 38*, 120–130. https://doi.org/10.1016/j.tourman.2013.03.002
Wearing, S., Mostafanezhad, M., Nguyen, N., Nguyen, T., & McDonald, M. (2018). Poor children on Tinder'and their Barbie Saviours: towards a feminist political economy of volunteer tourism. *Leisure Studies, 37*(5), 500–514. https://doi.org/10.1080/02614367.2018.1504979
Winkler, T. (n.d.). *Voluntourism & volunteering in Cambodia*. Retrieved July 7, 2021, from https://cambodianchildrenstrust.org/volunteer-in-cambodia/.
Winkler, T., Delacey, L. (2016). *How (not) to start an orphanage … by a woman who did*. Allen & Unwin.
Yin, R. K. (2014). *Case study research: Design and methods*. Sage.

Rethinking the space of tourism, its power-geometries, and spatial justice

Lucia Tomassini and Ian Lamond

ABSTRACT

This article critically reflects on the space of tourism and its intersection with spatial justice. Despite the urgency to embrace a more just, sustainable, tourism the space of tourism has not yet received a sufficiently forceful reflection. Our study contributes to filling that gap by proposing a novel theoretical framework that combines Massey's spatial turn in the social sciences, with the posthuman, affirmative ethics of Braidotti, while drawing on elements of Field Theory/Hodological Space in Lewin. We move from an overview of the sociological space of tourism, articulated by neoliberal capitalism, through a critique built upon the spatial turn in the social sciences, arguing that space is relational, imbued with power geometries, the product of interrelations, and always under construction. We develop a conceptual model that draws on Latour's Actor-Network Theory and aspects of Field Theory, which understands space as an individual lived experience (hodological space). That theorisation is then extended to the space of tourism through the lens of the posthuman, affirmative ethics, and mobile ontologies of both human and non-human actants. In doing so the novel theoretical framework we develop exposes the complexity and nuance of the space of tourism and enables the possibility of spatially just tourism.

Introduction

In the last decades the growth of tourism (UNWTO, 2020) has interweaved with an enduring questioning of such an industry and its role within a neoliberal, globalised, world; underpinned by the intense mobility of capital, goods, and people (Boluk et al., 2019). Tourism has been widely critiqued as an expression of a fluid global society, affected by the "tourist syndrome" (Bauman, 2000; Franklin, 2003). A global society divided by mobility injustice (Sheller, 2018) and social injustice (Bauman, 1996; Ritzer, 2012). While the uncontrolled growth of tourism – understood as overtourism (Milano et al., 2019a, 2019b) – has seen widening economic and social inequalities, and unbalanced power-relations (Bianchi, 2018; Higgins-Desbiolles, 2018; Jamal & Camargo, 2014, 2018), crises, such as the recent COVID-19 pandemic, have exposed new fears and uncertainties for contemporary society and its "tourist syndrome" (Higgins-Desbiolles, 2020; Jamal & Budke, 2020; Jamal & Higham, 2021; Rastegar, 2020). Such uncertainties have been at the centre of a growing academic debate, one that advocates for a critical rethinking of tourism in favour of social and ecological justice (Higgins-Desbiolles, 2020), novel future tourism scenarios (Lew et al., 2020), and the socialising of tourism (Higgins-Desbiolles et al., 2022).

This conceptual article builds on the idea that crises reveal meaning and can facilitate a critique of a given condition (Mbembe & Roitman, 1995; Roitman, 2013; Spracklen & Lamond, 2016). Critique is intimately connected with creation because "critique is not only a sterile opposition but also an active engagement of the conceptual imagination in the task of producing sustainable alternatives" (Braidotti, 2011, p. 6). Hence, our article aims to critically reflect on the mobility and spatial justice within, and throughout, tourism. Bauman, in *Liquid Modernity* (2000), stressed how mobility is the most coveted value and that the freedom to move is a stratifying factor of our times. Urry, in *Sociology beyond societies: mobilities for the twenty-first century* (2000), identified mobilities as a pivotal concept within the emerging spatial social sciences and humanities. However, mobility is not only relational and plural (Caletrío, 2016, 2017; Sheller & Urry, 2006), it also creates winners and losers, and divides as much as it unites (Bauman, 1998). Uneven mobilities and unjust spatialities produce unbalanced power relations because "power is a situation or a process, not an object or an essence" (Braidotti, 2011, p. 4). This urges the adoption of a cartographic approach, where "a cartography is a theoretically based and politically informed reading of the process of power relations" (ibid, p.4). Consequently, a cartographic approach conveys a political understanding of mobility conditions in a globalised context; it does so by identifying "the various power differences between distinct forms, categories, and practices of movement for both human and nonhuman mobile units" (ibid, p.11). This means that an emphasis on complex mobilities is advocated, one that sets out a novel ontological basis for relational spaces and for the understanding of how such relational spaces are produced through social practices, such as tourism (Sheller, 2018).

This article is inspired by the spatial turn in social sciences, stimulated by the works of Lefebvre (1984), Urry (De Gregory & Urry, 1985; Urry, 2000) and Massey (2005). Such pivotal works investigated the complex movements of capital, together with the resulting sociological space. More specifically our work draws on the three propositions on space elaborated by Massey (2005), conceiving space as the product of interrelations, as the sphere of a contemporaneuos plurality and multiplicity, and always under construction. Massey's work influenced the development of a mobilities turn in sociology by stimulating ideas of mobile sociology (Sheller, 2017; Urry, 2000), theorisations of spaces as flow and network (Sheller, 2018), and a growing interest in studies of places as the product of socio-spatial relations imbued with power-geometries (Bélanger & Silvey, 2020; Sheller, 2017, 2018). Hence, Massey's ruminations on space (2005) encourage us to rethink the space of tourism and the role justice plays in such spaces.

The sociological space of tourism, together with its practices, are largely the product of modern western culture and the production systems of neoliberal capitalism (Bianchi, 2018; Higgins-Desbiolles, 2018). In such neoliberal space tourism has been discussed as disregarding the needs and well-being of the local community (Higgins-Desbiolles, 2019a, 2019b; Higgins-Desbiolles & Bigby, 2022; Tomassini et al., 2021) urging a call to action to socialise tourism (Higgins-Desbiolles et al., 2022). Conversely, we understand and investigate tourism as mobility *praxis* in an unjust spatiality and as a lack of affective relationships. To interpret such unjust mobility *praxis,* we draw on aspects of Field Theory and hodological space (Janni, 1984; Lewin, 1951, 2015) arguing that behaviours - here meant as "mobility actions" - are impinged by the totality of coexisting and interdependent forces within a space that - by becoming an individual lived experienced - results in a hodological space. The hodological space (ὁδός in ancient Greek means "way") is neither a Euclidean space nor a space of knowledge but the space of the itinerary as an aspect of Lefebvrian lived and perceived space (Lefebvre, 1984), understood, and experienced by a subject in movement (Janni, 1984). Though the notion of the *subject,* as we shall discuss later, also needs some re-thinking. Consequently, our work examines how neoliberal spatiality, with its unjust power-geometries, result in uneven mobility actions that are the expression of a mono-dimensional hodological space made of given, privileged, uneven, preferred itineraries, and relations.

From Lewin (1951, 2015) we take the idea that in order to rethink the spatiality of tourism we need to understand spatiality as an affective field of complex and nuanced relationships, rich in itineraries and trajectories. We argue that such richness and heterogeneity present a landscape of multiplicity that cannot be grasped through a singular conceptual narrative. So that a spatially just Tourism can emerge, we suggest, a posthuman affirmative ethics is needed to contrast the commodification and depoliticisation of justice in tourism (Guia, 2021). This is possible because such an ethical approach "rests on an enlarged sense of interconnection between self and others, including the non-human or 'earth' others" (Braidotti, 2013, p. 190) and "bases the ethical relation on positive grounds of joint projects and activities" (ibid. p.190).

This approach enables us to rethink a novel posthuman spatial/relational ontology for a just tourism space, and to interweave the notions of space and justice; where justice not only concerns "people's lives, their well-being, homes, communities, and work; the places they travel to, and the journeys they undertake" (Jamal, 2019, p. 28). Through a networked nomadic ontology (Braidotti, 2011; Latour, 2007) and posthuman affirmative ethics (Braidotti, 2013; Guia, 2021), we argue that justice also concerns a human and non-human otherness that has, so far, largely been racialised, naturalised, and sexualised, in negative terms, as a diminished otherness. Numerous are the initiatives and projects trying to overcome an unfair spatiality, ground in a diminished otherness, while stressing the urgency of a posthuman, affirmative, ethics that emphasise political responsibility and an ethical commitment to tourism space. Projects like "Migrantour" (www.migrantour.org) and "Roots Guide" (www.rootsguide.org) aim to unpack the potential of both migrant heritage and connections with different places and people in urban spaces (Ormond & Vietti, 2022). Similarly, initiatives like "Marry an Amsterdammer for a day" (Nicholls-Lee, 2019) propose temporary affective relationships, as a means of addressing a tourist/local dualism. whilst challenging overtourism as a phenomenon. A gendered approach to spatial justice is largely discussed by Leslie Kern (2020) in "Feminist City: Claiming Space in a Made-Man World" as well as being part of the work of Jane Jacobs (Adhya, 2012; Brandes Gratz, 2011). The urgency of socialising animal-based tourism via a posthuman lens is discussed by Carol Kline (2022) and emerges also in projects like "How is Water" (www.cyanplanet.org/how-is-water) where people are virtually connected with the ocean and marine environment conservation via empathic emotional experiences created through immersive media. Hence, in this article, we set the ground for novel conceptualisations of a future, just, tourism, where "for the future to be open, space must be open too" (Massey, 2005, p. 12). Such an approach draws together the mapping of such spaces (cartography) with an understanding of how the map's features are associated and connected (topography). It means exploring how the notion of relational space, imbued with multiplicity, heterogeneity, and always under construction, can intersect tourism and disrupt its neoliberal, anthropocentric, power geometries and forces, while prompting affective encounters and actions of affirmative ethics (Braidotti, 2011, 2013) and deep justice (Guia, 2021). Facilitating a perspective on tourism that no longer needs to be synonymous with a commodification of space (Gainsforth, 2019).

The structure of our article is as follows. First, it provides a critical overview of the tourism space produced by modern, neoliberal, capitalism, with uneven spatiality, mobility injustice, and neo-colonial practices. Then, it proceeds to present a perspective on Field Theory (Lewin, 1951, 2015) and hodological space (Janni, 1984), through a discussion of conceptualisations of open spatiality (Latour, 2007; Massey, 2005), posthuman Nomadic Theory, and affirmative ethics (Braidotti, 2011, 2013). In doing so our article offers a novel theoretical framework, one that is grounded in an open, posthuman, relational ontology, that can facilitate a spatially just tourism.

Modern and neoliberal spatiality

This section draws on the idea that the search for wealth and individual self-interest (see Adam Smith, 1999: "Wealth of Nations" originally published in 1776) has been construed as a rational rather than affective enterprise, one that, together with its constant seeking for profit and capital accumulation (Granovetter, 2018; Power et al., 2017), as stressed by Weber (1930) in "Protestant Ethics and the Spirit of Capitalism," constitutes the foundations of the modern cultural political economy (Sum & Jessop, 2015). Hence, the current debate on the fragilities of advanced global capitalism is grounded in the idea that advanced capitalism has been underpinned by misleading values dating back to the Modern age and classical political economy (Bruni, 2020). Such misleading values have resulted in theorisations of the wealth of nations, economic growth, capital accumulation, and distribution, that largely disregards social collective wellbeing (Bruni, 2020; Bruni & Porta, 2005). By conceiving space "as knowledge and action, in the existing mode of production" (Lefebvre, 1984, p. 11), the modern profit-oriented economic vision led to a construal of space that largely comprehended it in terms of a flat surface for transit and trading, with centres and peripheries. Eriksen (2010, p. xi) in his foreword of the 2010 edition of Wolf's *Europe and the people without history* stresses how "the centrifugal forces of modernity are continuously counteracted by the centripetal forces of community, and the human need to make sense of the world together in collectives ensures a certain degree of cohesion" (p. xi). Moreover, he highlights, "the networked capitalist world [has become] ... a framework, or scaffolding, for almost any serious inquiry into contemporary cultural and social dynamics" (p. xiii).

In this study we elaborate on how modern capitalism, in its current neoliberal manifestation, has produced a tourism space that is imbued with social inequalities, uneven spatial and mobility justice, and unbalanced power-geometries (Bianchi, 2009; Boluk et al., 2019; Higgins-Desbiolles et al., 2019; Jamal, 2019). Such spatial injustice has been largely understood as "directly attributable to the global expansion of European mercantilism and then capitalism" (Britton, 1982, p.333) as "the most important cause of structural distortions" (Ibid. p.348). Hence, the debate on tourism justice in the neoliberal capitalism appears deeply intertwined with the evolution of colonial space-economy (Britton, 1980, 1982, 1991) as well as with neocolonialism (Everingham & Motta, 2022; Mbaria & Ogada, 2016; Mkono, 2019; Wijesinghe et al., 2019). We have become used to imagining space as a surface, continuous, and given, and as a land stretching out around us (Massey, 2005). A surface in which people, capital, and products transit, and tourism mobility appears as an assertion of privilege and power imbalance (Tomassini & Cavagnaro, 2020). "So, easily this way of imagining space can lead us to conceive of other places, peoples, cultures simply as phenomena on 'this' surface" (Massey, 2005, p. 4). Space itself has consequently become a commodified entity (Lefebvre, 1976, 1984; Massey, 1994, 1999), where justice is pushed to the side as a secondary or irrelevant consideration. As Braidotti (2011, p. 25) argues, "[a]dvanced capitalism is a different engine in that it promotes the marketing of pluralistic differences and the commodification of the existence, the culture, the discourses of 'others', for the purpose of consumerism. Consequently, the global system of the post-industrial world produces scattered and polycentered, profit-oriented power relations."

The tourism phenomenon, together with its practices, have been largely understood as organically intertwined with the development of industrial capitalism, the advent of novel technologies, infrastructures, systems of production, and novel topological and cartographic competencies (Borocz, 1992). The tourism industry has grown within a Euclidean space, scientifically measured, and mapped by the Modern Western World. The Enlightenment's humanistic tradition has identified and legitimised leisure travelling as an interest towards an "otherness," one that is made of geographical distances, ethnic, cultural, and species diversities, that has created centres and peripheries as well as a juxtaposition between an ideal moral, cognitive, aesthetic Western model, and those of the other (Braidotti, 2011, 2013). Moreover, D'Eramo (2017) highlights how the seriality, anonymity, and automatisation of the touristic procedure and touristic space resonate with those of the seriality, anonymity, and automatisation of activities inside a traditional industrial plant. Both serialities are shaped by the 20th century concept of urbanity and the principle of urban zoning, assigning to each urban space a specific mono-functional destination like housing, industrial activities, trade, and leisure activities (D'Eramo, 2017; Moskowitz,

1998). This conveyed not only a fragmentation of the space and its functions, but also a flattening of the multiplicity and richness of the relations taking place in such spaces (Massey, 2005, 2009). Hence, the contemporary space of tourism is closely interlaced with the way space has been constructed. A construction that is understood via the misleading profit-oriented values of the Modern age (Bruni, 2020), unbalanced power-relations, uneven mobility practices of modern and neoliberal capitalism (Massey, 2005; Wolf, 2010), the commodification of "others" (Braidotti, 2011, 2013), the 20th century concept of functional urban spatiality (D'Eramo, 2017), and the neoliberal paradigm of cities whose aim is to create economic value (Gainsforth, 2019). Spatiality, composed in this way, becomes a domain of market prominence and resource management for profit. The mobility of people becomes one framed by commercial payoff and, at a personal level, leisure becomes something that requires negotiation. Within such a frame of reference, the orientation of an organisation's environmental agenda becomes construed as a route to obtain a competitive advantage. From an individual perspective, touristic mobility becomes a way of suggesting/displaying status and the accruing of social and cultural capital. The relational affectivity and social justice of such mobility becomes obscured, often implicit and overlooked, and frequently side-lined, or deliberately ignored. To both radically unpack such spatiality and propose an alternative framework an alternative construal of space is required; it is that to which we will now turn.

Field theory and hodological space: space as a field of forces and relationships

As an initial step towards rethinking the space of tourism and the affective associations of tourist mobilities and the environment needed for a more just, ethical, and sustainable tourism (Boluk et al., 2019; Jamal & Higham, 2021), we begin by drawing out ideas associated with those of lived-space, or hodological space, developed by Kurt Lewin (2015).

Hodological space is a behavioural/interactional space; one that is lived. As such it is neither homogeneous, isotropic, nor multi-dimensional. It is a field of forces with privileged spots, directions, and itineraries (Janni, 1984). Within neo-liberalism/late capitalism this becomes expressed as an uneven flat functional spatiality, made of unbalanced forces and geometries of power. Whilst it can be argued that such a spatiality resonates with Lewin's ontology in his formulation of field theory, with its implicit acceptance of an ontological binary of subject and environment; it is not his ontology that interests us. Articulated in this way, lived-space within neo-liberalism/late capitalism manifests as uneven mono-dimensional space that, once operationalised, can sustain unjust tourism and mobility praxis. Whilst our article goes back to Lewin's work (1951, 2015) and draws on concepts within Field Theory, and hodological space, we can develop a novel theoretical framework for an open mobile spatial justice in tourism by interweaving those ideas with a Posthuman ontology and an affirmative ethics.

Concerns within tourism studies, such as those discussed in much of the debate around over tourism or the negative environmental and cultural impact of visitor mobilities, tend towards a distinction between the action/behaviour of the tourist and the environment (Higgins-Desbiolles, 2019a) stressing the urgency to socialise tourism (Higgins-Desbiolles, 2020; Higgins-Desbiolles et al., 2022) as well as to reconnect the space of tourism and citizenship (Tomassini et al., 2021). Such arguments are commonly framed around the tourist's impact on the setting, either directly through their bodily presence in the visited space, or indirectly through such factors as transculturation (Ortiz, 2019), cultural commodification (Macleod, 2006), cultural appropriation (Said, 1979), or a consequence of various technologies of mobility (Kaplan, 2002; Vannini et al., 2021). We can interpret Lewin's conceptualisation of field and hodological space - the latter elaborated further by Janni (1984) – to move the focus from the individual's actions to the relationship between the exhibited behaviour and the context in which that behaviour occurs.

For Lewin, behaviours are impinged by the totality of *co*existing and *inter*dependent forces within a space, understood as lived-space. The core idea of a spatiality conceptualised as the lived-space of a route, or journey, that is manifest through the field of relations - and forces - between the subject in movement and the space it lives in and travel to, is what we take from Lewin (2015) and hodological space (Janni, 1984). Whilst we do not embrace his ontology, we draw on his idea of understanding space as a field of forces shaping behaviours, here referred to as a mobility praxis. Hodos (ὁδός) translates as the road, route, or the way (as in direction and form of travel) taken. Lived-space (hodological space) is not a straightforward, Euclidean, path, but the space of the itinerary, as lived, understood, and experienced by a subject in movement (Janni, 1984). A Euclidean trajectory would resonate with the uneven flattened plain of the neo-liberalist framework we discussed earlier. One where the individual traverses a space whilst the two remain two disentangled entities. Consequently, the regime of power resides in the former and the latter becomes little more than that on which it acts. While the geometries of power may be challenged and confronted, they remain in place and, for the most part, intact. Such a neoliberal uneven flat spatiality, imbued with unjust power-geometries and forces, results in uneven mobility actions and (lack of) affective relations that are expressions of a mono-dimensional hodological spaces made of given, preferred, itineraries and movements on a flat surface. Hence, a space understood via an ontology rooted in the dualism subject/object seems not to allow for a deep justice (Guia, 2021; Guia & Jamal, 2020) nor spatial justice.

Lewin's first thorough exposition of hodological (lived) space appears in *The Principles of Topological Psychology* (Lewin, 2015). Lewin locates behaviour as the function of a relationship between the person and their environment and represents that relationship through a simple equation:

$$B = f\ (P,\ E)$$

Where B stands for behaviour, *f* – is a function of the person (P), and their environment (E). The focus is on the relationship, not the P or the E, or the impact one has on the other. For Lewin the person could never be fully grasped without a consideration of their relationship to the environment, and vice versa. What we are proposing differs from Lewin but is strongly connected back to this formulation.

First, our attention is on the mobility praxis of tourism, here understood as the affective relations, taking place in tourism space. Such affective relations reach beyond the behaviour of individuals or groups to one that is revealed through human and non-human associations as they are made apparent through the geometries of power. Relationally, this might include tangible and intangible aspects of culture/heritage, imaginaries of space, place, and time, as well as behaviour of visitors, hosts, residents, and diasporic communities. Borrowing from Doreen Massey's three propositions on space (2005), which we will discuss in depth shortly, Lewin leads us to recognise space is always under construction and the product of interrelations and interactions (albeit one, we suggest, of human and non-human participants). Consequently, our formulation replaces the E of Lewin with an s (spatiality), using the lower case to indicate that this is both neither a thing-in-itself nor a preferred perspective.

Second, as we seek to establish an affective tourism relation that is just, ethical, and environmentally sustainable, it is important to value the active part non-human actors play in such a framework. To use the language of Bruno Latour, the focus is not simply on human actors but on human and non-human *actants* (Latour, 2007). For Latour (2004, 2018) actants emerge from a complex network of associations, where, in place privileged positions, there are perspectives on the network. Such *perspectives* encompass those of human and non-human actors (hence his preference for the more neutral term *actant*). We thus replace the "P" (Person in Lewin) with an "a," for actant. Maintaining our stylistic choice, where a lower-case s was adopted for spatiality (s) to indicate the lack of a privileged perspective, we use a lower-case "a" in this position.

Because we are seeking to emphasise the mobility praxis of tourism in terms of affective relations taking place in tourism space, we have chosen to use a capitalised "A" in place of Lewin's "B" for *behaviour*. We are now able to re-articulate, for our purpose, the basic formula espoused in Topological Psychology (Lewin, 2015) as:

$$A = f(a,s)$$

Expressed in words: The affective relation of tourism (A), is a function (f) of the relationship between the actants (a) and the principles of spatiality (s) adopted, where the focus of attention is on the relationship rather than the factors in that relationship, as things-in-themselves. As a base-line statement this goes some way to rethinking the space of tourism, however, if we are to grasp this in a way that enables us to challenge the power geometries manifest within an identified area it is important to unpack further what an open relational space looks like as well as the needed ontological base and ethical approach required to prompt spatial justice. It is towards that which we now turn.

For an open relational space

We argue that despite the urgency to embrace a more just, ethical, and sustainable tourism (Boluk et al., 2019; Jamal & Higham, 2021) in an enduring state of crisis (Roitman, 2013), the theoretical and political implications of tourism space and its rethinking have not received a thorough reflection. To address that gap we built on Massey's (1994, 2005, 2009) theorisations of space to re-think the spatiality of tourism and its relationship with justice. By using this approach, justice not only concerns human beings and their lives and wellbeing at home, at work, at leisure, and on holiday (Jamal, 2019) but it also embraces a posthuman stance. Such a posthuman stance deeply questions the anthropocentric and neo-colonially dominant approaches to the "otherness" that inhabits the world, which becomes understood as a naturalised, racialised, sexualised "otherness" (Braidotti, 2011, 2013). Posthumanism, and the Nomadic Theory associated with it (Braidotti, 2011), advocates a justice rooted into an affirmative ethics, one that "rests on an enlarged sense of inter-connection between self and others, including the non-human or 'earth' others, by removing the obstacles of self-centred individualism on the one hand, and the barriers of negativity on the other" (Braidotti, 2013, p. 190).

Beside chiming with the networked ontology of Bruno Latour (2007), this approach conveys an affirmative ethics based on collective bonds and proactive commitment, where a global affective community - made of networked locally situated communities - is kept together by the positive ground of shared projects and activities (Braidotti, 2011, 2013). In Braidotti's words (2013, p. 49): "I define the critical posthuman subject within an eco-philosophy of multiple belongings, as a relational subject constituted in and by multiplicity, that is to say a subject that works across differences and is also internally differentiated, but still grounded and accountable." Hence the posthuman affirmative ethics - and its implications for spatial justice - is grounded in a rhizomatic subjectivity, as discussed by Deleuze and Guattari (1987) in terms of a non-hierarchical subjectivity, that lacks centre and peripheries, that misses a beginning, a middle, and an end, as well as a privileged position or point of view. According to this approach, spatial justice in tourism can emerge, as the result, from the intersection of a posthuman affirmative ethics with a rhizomatic relational subjectivity in an open space made of mobile ontologies.

In our reasoning we draw heavily of the spatial theory of Doreen Massy, as expressed - particularly – in *For Space* (Massey, 2005) and the collection of essays in *Space, Place, and Gender* (Massey, 1994). Massey's critical geography conceives spatiality through three intertwined modalities, all of which need to be in place for the emergence of spatiality. Like a carefully structured architectural form, the removal of one of these modalities will result in an absence of spatiality. Those intertwined propositions are, first: "we recognise space as the product of interrelations, as constituted through interactions, from the immensity of the global to the intimately tiny" (2005,

p. 9). Second: "we understand space as the sphere of the possibility of the existence of multiplicity in the sense of contemporaneous plurality" (ibid), and finally: "we recognise space as always under construction" (ibid). Anderson (2008, p. 228) argues: "the three propositions [...] aim to enable us to ponder the challenges and delights of spatiality and subsequently open up the political to the challenge of space [...] disrupting how political questions are formulated, perhaps intervening in current arguments, and perhaps contributing to alternative imaginations that enable different spaces to be." Thus, the approach we propose is grounded in a conceptualisation of space as a lively, heterogeneous, and political response to global capitalism, social inequalities, and the power geometries it generates (i.e. space is imbued with and a product of relations of power, which have their own geography). Spatiality and mobility are both shaped by and reproduce power differentials in society (Bélanger & Silvey, 2020).

In Massey's spatial theory (2005), the first two modalities of Massey (2005) also raise questions about who/what is involved in such multiple and heterogeneous interrelations; how space is constructed at those points where such interrelations take place, and through what process(es) that construction occurs. It is here we are drawn to bring in the actor network theory of Latour (2007). It is important to note, at this stage, that Massey argues in *For Space* (2005) place is not a location, it is an event. That event of place draws in what Castoriadis (1997), and Anderson (1983) refer to as the imaginary of place. However, event as process takes on an *evental* aspect, one that sits at the foundation of all three of Massey's modalities. It also resonates with the removal of a privileged position when it comes to Latour's *actant(s)* because - in his terms (Latour, 2007) - *actant(s)* and spatiality are little more than different perspective in a network of associations.

In *Reassembling the Social* (2007) Latour identifies non-human active agents and the social as emergent from diverse perspectives in a network of associations. In seeking to overcome the linguistic assumption of actors as human beings, Latour replaces the term actor with *actant* – depersonalising it in the process. A central aspect of *actants* is the recognition that they are not relationally privileged. *Actants* are assembled through a network of associations. In this sense *actant* differs from Lewin's (1951, 2015) person as Latour (2007) suggests that the actant lacks a privileged position being grounded in a relational, rhizomatic ontology. Following Latour (2007), since *actants* are removed from a privileged position, there can no longer be any claim to a hierarchy in how *actants* interact with the environment. It is this theme that Latour explores in *Politics of Nature* (2004) and *Down to Earth* (2018). In those texts he considers hierarchical ontologies, referred to in *Down to Earth* as a view from Sirius, as central to sustaining a divide between ourselves, errantly construed as a primary actant, and our planet. A divide that has led to the existing climate emergency, and exploitative practices throughout human history. He argues that it is only by flattening the ontology (i.e. adopting a view that is *down to earth*) that real progress and justice can occur.

Hence, such theorisations of space (Massey, 1994, 2005) - together with their interplay with Latour's (2007) notion of *actant* and posthuman Nomadic Theory/Affirmative Ethics (Braidotti, 2011, 2013) - set the ground for a novel theoretical framework that can result in a rethinking of spatial justice in tourism. As Massey (2005, p. 11) stresses: "only if the future is open is there any ground for a politics which can make a difference. [...] For the future to be open, space must be open too" (ibid, p. 11-12).

A posthuman theoretical framework of mobile ontologies for spatial justice in tourism

In this study a critical reflection on spatial justice in tourism is intertwined with posthuman conceptualisations questioning the humanist paradigm, with its presumed *civilising* ideal of universal commodification (Braidotti, 2006, 2013, 2019). That approach is broken as it has resulted in mass explotation of people and place. Massey's ruminations on space (2005), the Posthumanism mobile ontology with its relational rhizomatic subjectivity and affirmative ethics (Braidotti, 2011, 2013; Latour, 2007) constitute a theoretical perspective that can offer us guidance on a novel understanding of tourism, and spatial justice in tourism studies (Cohen, 2019; Guia, 2021; Guia & Jamal, 2020). We have critically reflected on a novel ontic foundation for a sustainable just tourism space that provides a conceptually rich picture of space, tourism, and mobility from a perspective of field theory, hodological space, posthuman ontology, and affirmative ethics.

In this study we have assumed that tourism has been intrinsically, historically, ontologically, epistemologically embedded in the idea of "otherness" (e. g. in terms of geographical distance, cultural difference, encounters with other human or non-human subjects and species) and in the existence of power-relationships and power-geometries (e. g. locals-visitors, host-guest, etc.). Moreover, tourism - in neoliberal capitalism and global mobility practices - is highly representative of a space conceived as a flat "transit" surface; namely a commodified space largely understood as *tourism destination* and imbued with *mono-dimensional uneven and privileged itineraries* (Janni, 1984; Lewin, 2015). The tourists' relation and interplay with a naturalised, racialised, and sexualised otherness in the commodified neoliberal space, where tourism takes place, exposes privileged uneven itineraries, relations, and mobilities practices that have led to unbalanced power relations and uneven power geometries, resulting in ever widening problems of spatial justice.

Our novel theoretical framework for spatial justice in tourism assumes an open spatiality made of multiple connections, relations, and networks. Such open spatiality is multiple, always under construction, and never given (Massey, 2005). A space where human and non-human *actants* constitute such spatiality through a networked non-hierarchical ontology (Latour, 2007). This means that tourism, if it is to be sustainable, just, and future-proof, should not happen in a flat, commodified, space, continuous and given. Neither is it appropriate for tourists to transit through a surface made of advantaged itineraries and mobility practices. Spatial justice via tourism cannot occur through such assertions of privilege, power imbalance, and mobility injustice. Using Guia's words (2021, p. 517): "taking Posthumanism seriously means actively resisting the co-optation of tourism by the market, that is, learning to contest neoliberism [...] it paves the way for reintroducing political responsibility, solidarity, and advocacy as positive world-making practices with which to subvert the current commodification and de-politicisation of all forms of tourism."

Hence, we assume that spatial justice in tourism can occur by conceiving tourists as *co-actants,* entangled in a relational spatiality made of multiple relations with multiple others (Latour, 2007; Massey, 2005). As such, tourists can create spatial justice by conceiving the space in which they move as open, posthuman, mobile, and relational. In doing so, we take a critical stance on neoliberal tourist mobilities, actions, and practices, that currently express an uneven hodological space that is neither isotropic nor multidimensional; it is a field of forces with privileged spots, directions, and itineraries (Janni, 1984). In place of that, our article draws on the work of Lewin (1951, 2015), by extending his conceptualisation of Field Theory and hodological space to tourism studies, interweaving that perspective with a Posthuman axiology and ontology, to offer a novel theoretical framework for an open, multidimensional, mobile spatial justice in tourism.

We have revisited Lewin's equation $B = f\,(P, E)$ (Lewin, 2015), rearticulating it as $A = f\,(a, s)$.

Moreover, we understood those terms relationally, through an open mobile spatiality, made of posthuman *actants*. This has meant exploring the interplay of ideas in the work of Latour (2007), Massey (1994, 2005, 2009), and Braidotti (2011, 2013). What emerges is a theoretical framework where spatial justice in tourism *can* occur. It conceives tourists not as moving through linear mono-dimensional itineraries and mobility practices that are rooted in privileged power-geometries and relations. Instead, tourists are conceived as *actants* within a relational network of associations, actively co-constituting a spatiality that is not a field of forces with privileged perspectives, directions, and itineraries; instead, it is an open relational space "exploded" into a multiplicity of dimensions, networks, and relations. In such an open, posthuman, spatiality, human and non-human *actants* (e.g. those articulated as *tourists*) lack a privileged position in relational networks, whilst acknowledging they are active components of ethically driven affirmative relations and actions.

Hence, we assume that justice in such an open, mobile, relational, spatiality can happen through a posthuman affirmative ethics. Posthumanism with the aligned Nomadic Theory (Braidotti, 2011, 2013) offers a promising ethical approach to rethink spatial justice in tourism. Such a promising sustainable, ethical, affective approach consists in the positive affirmation of a monadic interconnected subjectivity and relies on a broader feeling of interconnection and belonging among humans and non-human others, beyond a self-centred individualism (Braidotti, 2013). Hence, "to be posthuman [...] implies a new way of combining ethical values with the well-being of an enlarged sense of community, which includes one's territorial or environmental interconnections" (ibid, p. 190). This means: "Posthuman theory also bases the ethical relation on positive grounds of joint projects and activities, not on the negative or reactive grounds of shared vulnerability" (ibid, p.190). Moreover, "the stated criteria for this new ethics include non-profit; emphasis on the collective; acceptance of relationality and of viral contaminations; concerted efforts at experimenting with and actualizing potential or virtual options; and a new link between theory and practice, including a central role for creativity" (ibid, p. 191). Therefore, "the pursuit of collective projects aimed at the affirmation of hope, rooted in the ordinary micro-practices of everyday life, is a strategy to set up, sustain and map out sustainable transformations" (ibid, p. 192).

Casting a posthuman affirmative ethics into an open mobile relational spatiality means opening tourists up to various transformative relationships rooted into the affirmation of hope and belonging to a global community, with its localised situations, by being engaged with the positive affirmation of a sustainable, just, tourism.

Conclusion

This article offers a theoretical reconsideration of the tourism space, rethinking its connection to spatial justice. We argue that as spatially just touristic relations have, to date, been argued from within a framework that ultimately undermines spatially just tourism the *space* of *tourism* has only received limited attention. Consequently, its theoretical, practical, and political implications require a radical rethinking.

Our theoretical article aims at casting a novel light onto spatial justice in tourism; it does this by offering a novel theoretical ground for its reconceptualisation. In doing so we have interweaved a critical reflection on modern and neoliberal spatiality through Massey's (1994, 2005) theorisations on space, Latour's monadology of actants in his actor-network theory, and the Posthuman Nomadic Theory and affirmative ethics of Braidotti (2011, 2013). For this we have also drawn on Lewin's developments within (1951, 2015) Field Theory and hodological space in the social sciences (Janni, 1984).

Our novel approach has led us to unpack the base relationships of tourism's uneven mobility and spatiality as well as its ontic foundation. We did that by reshaping Lewin's (2015) original conceptual equation on the articulation of the relationship between behaviour and the environment. Our iteration presents the mobility praxis of tourism as affective relations that take place in tourism space as a function of a posthuman actant and an open relational spatiality. For us, this does not mean offering a sterile critique of tourism space with its unbalanced power relations, uneven mobilities, and unjust spatiality but, on the contrary, it means offering a critique conceived as creation and active engagement with possible sustainable alternatives (Braidotti, 2011).

The proposed theoretical lens, together with our reframing of Lewin's formula, is not a conclusion, meant as a universal principle that postulates general rules or seeks universalisation. Instead, its aim is to reposition and reorientate critical thinking in tourism studies from a perspective that breaks those associations which, in previous perspectives, have inadvertently rested on exploitative and exclusionary frameworks. We have done this by challenging practices of spatial injustice and uneven mobility that are rooted in unbalanced power relationship as well as in privileged itineraries and relations.

Our contribution aims at supporting the building of further knowledge, opening novel lines of investigation in tourism studies. It does so by exploring novel disruptive (evental), and ground-breaking approaches to spatial justice in tourism. We understand this study as a contribution to rethinking tourism, one that goes beyond unfair and misleading dualisms and juxtapositions grounded in an alterity that alienates the *other*. It represents a call to action to investigate how power-geometries work in touristic spatial relations and ask which affective relations can prompt spatial justice. This means that future research should take the proposed conceptualisation of $A = f$ (a, s) and investigate it further, testing it in different contexts, situations, and places as - for instance - in the aforementioned projects and initiatives like "Migrantour," "Roots Guide," "Marry an Amsterdammer for a day," and "How is water." This would convey a novel understanding of spatial justice in tourism by exploring the mobility praxis and affective relations as actions of posthuman affirmative ethics taking place within tourism space.

Disclosure statement

No potential conflict of interest was reported by the authors.

References

Adhya, A. (2012). Jane Jacobs and the theory of placemaking in debates on sustainable urbanism. In S. Hirt & D. Zahm (Eds.), *The urban wisdom of Jane Jacobs* (pp. 215–228). Routledge.

Anderson, B. (1983). *Imagined Communities. Reflections on the origin and spread of nationalism*. Verso.

Anderson, B. (2008). For Space (2005): Doreen Massey. In P. Hubbard, R. Kitchin, & G. Valentine (Eds.), *Key texts in human geography* (pp. 227–235). Sage.

Bauman, Z. (1996). *Tourists and vagabonds: Heroes and victims of postmodernity*. Institut für Höhere Studien.

Bauman, Z. (1998). *Globalisation: The human consequences*. Polity Press.

Bauman, Z. (2000). *Liquid modernity*. Polity Press.

Bélanger, D., & Silvey, R. (2020). An Im/mobility turn: Power geometries of care and migration. *Journal of Ethnic and Migration Studies*, *46*(16), 3423–3440. https://doi.org/10.1080/1369183X.2019.1592396

Bianchi, R. (2009). The 'critical turn' in tourism studies: A radical critique. *Tourism Geographies*, *11*(4), 484–504. https://doi.org/10.1080/14616680903262653

Bianchi, R. (2018). The political economy of tourism development: A critical review. *Annals of Tourism Research*, *70*, 88–102. https://doi.org/10.1016/j.annals.2017.08.005

Boluk, K. A., Cavaliere, C. T., & Higgins-Desbiolles, F. (2019). A critical framework for interrogating the United Nations Sustainable Development Goals 2030 Agenda in tourism. *Journal of Sustainable Tourism*, *27*(7), 847–864. https://doi.org/10.1080/09669582.2019.1619748

Borocz, J. (1992). Travel-capitalism: The structure of Europe and the advent of the tourist. *Comparative Studies in Society and History*, *34*(4), 708–741.

Braidotti, R. (2006). Posthuman, all too human: Towards a new process ontology. *Theory, Culture & Society, 23*(7–8), 197–208. https://doi.org/10.1177/0263276406069232

Braidotti, R. (2011). *Nomadic theory: The portable Rosi Braidotti*. Columbia University Press.

Braidotti, R. (2013). *The Posthuman*. John Wiley & Sons.

Braidotti, R. (2019). A theoretical framework for the critical posthumanities. *Theory, Culture & Society, 36*(6), 31–61. https://doi.org/10.1177/0263276418771486

Brandes Gratz, R. (2011, November 16). Jane Jacobs and the power of women planners. *Bloomberg CityLab*. https://www.bloomberg.com/news/articles/2011-11-16/jane-jacobs-and-the-power-of-women-planners?utm_medium=social&utm_content=citylab&utm_source=facebook&utm_campaign=socialflow-organic&fbclid=IwAR3qvihQxCJGOK6HAaqveeHpR1XltVLM0mVa7bzpQn9ibaploguY2PG_KXU

Britton, S. G. (1980). The evolution of a colonial space-economy: The case of Fiji. *Journal of Historical Geography, 6*(3), 251–274. https://doi.org/10.1016/0305-7488(80)90081-X

Britton, S. G. (1982). The political economy of tourism in the third world. *Annals of Tourism Research, 9*(3), 331–358. https://doi.org/10.1016/0160-7383(82)90018-4

Britton, S. G. (1991). Tourism, capital, and place: Towards a critical geography of tourism. *Environment and Planning D: Society and Space, 9*(4), 451–478. https://doi.org/10.1068/d090451

Bruni, L. (2020). *Economics, wellbeing, and happiness: A historical perspective*. Edward Elgar Publishing.

Bruni, L., & Porta, P. (2005). *Economics and happiness: Framing the analysis*. OUP Oxford.

Caletrío, J. (2016, September 2). Mobilities paradigm. *Mobile Lives Forum*. https://en.forumviesmobiles.org/marks/mobilities-paradigm-3293

Caletrío, J. (2017, February 28). Bauman (Zygmunt). *Mobile Lives Forum*. https://en.forumviesmobiles.org/marks/bauman-zygmunt-3538

Castoriadis, C. (1997). *The imaginary institution of society*. MIT Press.

Cohen, E. (2019). Posthumanism and tourism. *Tourism Review, 74*(3), 416–427. https://doi.org/10.1108/TR-06-2018-0089

D'Eramo, M. (2017). *Il selfie del mondo. Indagine sull'età del turismo*. Feltrinelli.

De Gregory, D., & Urry, J. (1985). *Social relations and spatial structures*. Springer.

Deleuze, G., & Guattari, F. (1987). *A thousand plateaus: Capitalism and schizophrenia*. University of Minnesota Press.

Eriksen, T. H. (2010). Foreword to the 2010 edition: A book for the twenty-first century. In E. R. Wolf (Ed.), *Europe and the people without history* (pp. x–xvii). University of California Press.

Everingham, P., & Motta, S. C. (2022). Decolonising the 'autonomy of affect' in volunteer tourism encounters. *Tourism Geographies, 24*(2–3), 223–243. https://doi.org/10.1080/14616688.2020.1713879

Franklin, A. (2003). The tourist syndrome: An interview with Zygmunt Bauman. *Tourist Studies, 3*(2), 205–217. https://doi.org/10.1177/1468797603041632

Gainsforth, S. (2019). *Airbnb città merce. Storie di resistenza alla gentrificazione digitale*. Derive Approdi.

Granovetter, M. (2018). *The sociology of economic life*. Routledge.

Guia, J. (2021). Conceptualizing justice tourism and the promise of posthumanism. *Journal of Sustainable Tourism, 29*(2–3), 503–519. https://doi.org/10.1080/09669582.2020.1771347

Guia, J., & Jamal, T. (2020). A (Deleuzian) posthumanist paradigm for tourism research. *Annals of Tourism Research, 84* 102982. https://doi.org/10.1016/j.annals.2020.

Higgins-Desbiolles, F. (2018). Sustainable tourism: Sustaining tourism or something more? *Tourism Management Perspectives, 25*, 157–160. https://doi.org/10.1016/j.tmp.2017.11.017

Higgins-Desbiolles, F. (2019a, April 8). Tourists behaving badly are a threat to global tourism, and the industry is partly to blame. *The Conversation*. https://theconversation.com/tourists-behaving-badly-are-a-threat-to-global-tourism-and-the-industry-is-partly-to-blame-112398

Higgins-Desbiolles, F. (2019b, May 22). Rethinking tourism so the locals actually benefit from hosting visitors. *The Conversation*. https://theconversation.com/rethinking-tourism-so-the-locals-actually-benefit-from-hosting-visitors-116066

Higgins-Desbiolles, F. (2020). Socialising tourism for social and ecological justice after COVID-19. *Tourism Geographies, 22*(3), 610–623. https://doi.org/10.1080/14616688.2020.1757748

Higgins-Desbiolles, F., & Bigby, B. C. (2022). A local turn in tourism studies. *Annals of Tourism Research, 92*, 103291. https://doi.org/10.1016/j.annals.2021.103291

Higgins-Desbiolles, F., Carnicelli, S., Krolikowski, C., Wijesinghe, G., & Boluk, C. (2019). Degrowing tourism: Rethinking tourism. *Journal of Sustainable Tourism, 27*(12), 1926–1944. https://doi.org/10.1080/09669582.2019.1601732

Higgins-Desbiolles, F., Doering, A., & Bigby, B. C. (2022). *Socialising tourism for social and ecological justice*. Routledge.

Jamal, T. (2019). *Justice and ethics in tourism*. Routledge.

Jamal, T., & Budke, C. (2020). Tourism in a world with pandemics: Local-global responsibility and action. *Journal of Tourism Futures, 6*(2), 181–188. https://doi.org/10.1108/JTF-02-2020-0014

Jamal, T., & Camargo, B. A. (2014). Sustainable tourism, justice, and an ethic of care: Toward the Just Destination. *Journal of Sustainable Tourism*, *22*(1), 11–30. https://doi.org/10.1080/09669582.2013.786084

Jamal, T., & Camargo, B. A. (2018). Tourism governance and policy: Whither justice? *Tourism Management Perspectives*, *25*, 205–208. https://doi.org/10.1016/j.tmp.2017.11.009

Jamal, T., & Higham, J. (2021). Justice and ethics: Towards a new platform for tourism and sustainability. *Journal of Sustainable Tourism*, *29*(2–3), 143–157. https://doi.org/10.1080/09669582.2020.1835933

Janni, P. (1984). *La mappa e il periplo. Cartografia antica e spazio odologico*. Bretschneider Editore.

Kaplan, C. (2002). Transporting the subject: Technologies of mobility and location in an era of globalisation. *Publications of the Modern Language Association of America*, *117*(1), 32–42. https://doi.org/10.1632/003081202X63492

Kern, L. (2020). *Feminist city: Claiming space in man-made world*. Verso Books.

Kline, C. (2022). Socialising animal-based tourism. In F. Higgins-Desbiolles, A. Doering, & B. C. Bigby (Eds.), *Socialising tourism: Rethinking tourism for social and ecological justice* (pp. 195–213). Routledge.

Latour, B. (2004). *The politics of nature*. Harvard University Press.

Latour, B. (2007). *Reassembling the social: An introduction to actor-network-theory*. Oxford University Press.

Latour, B. (2018). *Down to earth*. Polity Press.

Lefebvre, H. (1976). *Survival of capitalism*. Allison and Busby.

Lefebvre, H. (1984). *The production of space*. Blackwell Pub.

Lew, A., Cheer, J. M., Haywood, M., Brouder, P., & Salazar, N. B. (2020). Visions of travel and tourism after the global COVID-19 transformation of 2020. *Tourism Geographies*, *22*(3), 455–466. https://doi.org/10.1080/14616688.2020.1770326

Lewin, K. (1951). *Field theory in social sciences: Selected theoretical papers*. Harper & Brothers Publishers.

Lewin, K. (2015). *The principles of topological psychology*. Martino Fine Books.

Macleod, D. (2006). Cultural commodification and tourism: A very special relationship. *Tourism Culture and Communication*, *6*(2), 71–84. https://doi.org/10.3727/109830406777410580

Massey, D. (1994). *Space, place, and gender*. Polity Cambridge.

Massey, D. (1999). Imagining globalisation: Power-geometries of time-space. In A. Brah, M. J. Hickman, & M. M. Ghaill (Eds.), *Global futures: Migration, environment and globalisation*. Explorations in Sociology (pp. 27–44). Palgrave Macmillan.

Massey, D. (2005). *For space*. Sage.

Massey, D. (2009). Concepts of space and power in theory and in political practice. *Documents D'anàlisi Geogràfica*, 55, 15–26.

Mbaria, J., & Ogada, M. (2016). *The big conservation lies*. Lens & Pens Publishing.

Mbembe, A., & Roitman, J. (1995). Figures of the subject in times of crisis. *Public Culture*, *7*(2), 323–352. https://doi.org/10.1215/08992363-7-2-323

Milano, C., Cheer, J. M., & Novelli, M. (2019a). *Overtourism: Excesses, discontents and measures in travel and tourism*. Cabi.

Milano, C., Novelli, M., & Cheer, J. M. (2019b). Overtourism and degrowth: A social movements perspective. *Journal of Sustainable Tourism*, *27*(12), 1857–1875. https://doi.org/10.1080/09669582.2019.1650054

Mkono, M. (2019). Neo-colonialism and greed: Africans' views on trophy hunting in social media. *Journal of Sustainable Tourism*, *27*(5), 689–704. https://doi.org/10.1080/09669582.2019.1604719

Moskowitz, M. (1998). Zoning the industrial city: Planners, commissioners, and boosters in the 1920s. *Business and Economic History*, *27*(2), 307–317.

Nicholls-Lee, D. (2019, June 5). Amsterdam invites tourists to "marry" a local for a day. *The Guardian: International Edition*. https://www.theguardian.com/travel/2019/jun/05/amsterdam-fake-wedding-tourists-local-untouristy

Ormond, M., & Vietti, F. (2022). Beyond multicultural 'tolerance': Guided tours and guidebooks as transformative tools for civic learning. *Journal of Sustainable Tourism*, *30*(2–3), 533–549. https://doi.org/10.1080/09669582.2021.1901908

Ortiz, F. (2019). "Transculturation" and Cuba. In A. Chomsky, B. Carr, & P. M. Smorkaloff (Eds.), *The Cuba reader: History, culture, politics* (pp. 25–26).Duke University Press.

Power, S., Di Domenico, M., & Miller, G. (2017). The nature of ethical entrepreneurship in tourism. *Annals of Tourism Research*, *65*, 36–48. https://doi.org/10.1016/j.annals.2017.05.001

Rastegar, R. (2020). Tourism and justice: Rethinking the role of governments. *Annals of Tourism Research*, *85*, 102884. https://doi.org/10.1016/j.annals.2020.102884

Ritzer, G. (2012). Vagabonds and tourists. In G. Ritzer (Ed.), *The Wiley-Blackwell encyclopedia of globalisation*. DOI: 10.1002/9780470670590.wbeog940

Roitman, J. (2013). *Anti-crisis*. Duke University Press.

Said, E. (1979). *Orientalism*. Vintage.

Sheller, M. (2017). From spatial turn to mobilities turn. *Current Sociology*, *65*(4), 623–639. https://doi.org/10.1177/0011392117697463

Sheller, M. (2018). *Mobility justice: The politics of movement in an age of extremes*. Verso Books.

Sheller, M., & Urry, J. (2006). The new mobilities paradigm. *Environment and Planning A: Economy and Space, 38*(2), 207–226. https://doi.org/10.1068/a37268

Smith, A. (1999). *Wealth of nations*. Wiley Online Library.

Spracklen, K., & Lamond, I. R. (2016). *Critical event studies*. Routledge.

Sum, N., & Jessop, B. (2015). *Towards a cultural political economy: Putting culture in its place in political economy*. Edward Elgar.

Tomassini, L., & Cavagnaro, E. (2020). The novel spaces and power-geometries in tourism and hospitality after 2020 will belong to the 'local. *Tourism Geographies, 22*(3), 713–719. https://doi.org/10.1080/14616688.2020.1757747

Tomassini, L., Schreurs, L., & Cavagnaro, E. (2021). Reconnecting the space of tourism and citizenship: The case of tourists' hubris. *Journal of Tourism Futures, 7*(3), 337–349. https://doi.org/10.1108/JTF-10-2020-0176

UNWTO. (2020, January 20). *International tourism growth continues to outplace the global economy*. https://unwto.org/international-tourism-growth-continues-to-outpace-the-economy

Urry, J. (2000). *Sociology beyond societies: Mobilities for the twenty-first century*. Routledge.

Vannini, P., Budd, L., Jensen, O., Fisker, C., & Jirón, P. (2021). *Technologies of mobility in the Americas*. Peter Lang Verlag.

Weber, M. (1930). *The Protestant ethic and the spirit of capitalism*. Allen and Unwin.

Wijesinghe, S. N. R., Mura, P., & Bouchon, F. (2019). Tourism knowledge and neocolonialism - A systematic critical review of the literature. *Current Issues in Tourism, 22*(11), 1263–1279. https://doi.org/10.1080/13683500.2017.1402871

Wolf, E. (2010). *Europe and the people without history*. University of California Press.

Index

Note: **Bold** page numbers refer to tables; *italic* page numbers refer to figures.

African educational system 170
African national judicial systems 172
Agyeman, J. 5, 190
air contamination 41, 45
Alderman, D. 22
Alfred, G. T. 41
All Our Relations: Native Struggles for Land and Life (LaDuke) 40
alternative tourism: in China 205; markets 211–215, *214*; participation mechanism 208, 211, 217, 218
American sanctions 155
American voluntary service missions 243
Anderson, B. 268
anemia 45
Anitha, S. 68
anti-orphanage tourism 101
Asian financial crisis 209
Astleithner, F. 190
Ateljevic, I. 19
Australia: migrant food delivery workers 128, 130; platform workers 133; precarious workers 133; temporary migrants 130, 133, 138
Australia's Modern Slavery Act 2018 241
Azcárate, M. C. 157

backpackers 207, 208, 213, 217
backpacking 207, 213, 217
Baidu Tieba 210
Bakas, F. 68
Baldwin, J. 29
Bandyopadhyay, R. 245
Barragán-Paladines, M. J. 10
Bauman, Z. 112, 113, 115; *Liquid Modernity* 262
Baum, T. 64
Beaumont, N. 195
Bell, D. 20, 22, 25, 32
Benjamin, S. 8, 22, 26
Berlin *Refugee Voices Tours* (2021) 111
Bhabha, H. K. 245
Bhatia, B. 10
Bhutan's initial tourism programme 7
Bianchi, R. V. 59, 157, 158
Bigby, B. C. 8, 10, 47, 49
biodiversity 3, 4, 178, 179, 228
biological diversity 3–4
BIPOC *see* Black, Indigenous and People of Colour (BIPOC)
Black feminist literature 67
Black, Indigenous and People of Colour (BIPOC) 8, 19, 20, 22, 24, 25, 31
Black interests converge 22
The Black Paris Project 29
Black people: civil rights victories 22; storytelling practices 31
Black storytelling traditions 30
Black travelers 25
Black Travel Movement (BTM) 25–27
Boluk, K. A. 42
Borowiak, C. 63
borromean connections: ethical responsibility 115–116; hostipitality and deconstruction 116–117; strangerhood 114–115
Bott, E. 241, 243
Boukhris, L. 29
Braidotti, R. 264, 267, 269, 270
Braun, V. 156, 192
Brayboy, J. 23
Bright, C. 25, 26, 31
Britton, S. G. 242
BTM *see* Black Travel Movement (BTM)
Buda, D.-M. 9
Buil, I. 82
Burawoy, M. 129
Burrai, E. 9
Buy Kenya Build Kenya 176
Buzinde, C. 10, 11, 19

Camargo, B. A. 59, 62, 189
Cambodia: anti-orphanage movement 254; CCT activities 251–252; child labour 256;

child protection system 251; communities 255; dependency theory 242; family-based care approach 248; Modern Slavery Act 247; orphanage tourism 243–244; tourism and development 241–243; volunteer tourism 243–244; White saviourism 244–245
Cambodian Children's Trust (CCT): ambassador 252; background 247–249, *248, 249*; co-founder and director 251; evolution 250; holistic approach *248,* 249; operations manager 250, 253; sister organisation 247; Village Hive model 248–249, 251; working in mutuality 245–246
Campbell, I. 129
Cancer Alley 38
Canosa, A. 8, 11
capitalism: advanced 264; characterise broader 193; direct/indirect products 3; dynamics and fundamental contradictions 168; free policy and governance processes 188; neoliberal consumer 193, 197; revolutionary change 188
Carpenter, K. 101
Carr, A. 19
cartographic approach 262
Castoriadis, C. 268
Caterina-Knorr, T. 10, 11
Caton, K. 23
CCT *see* Cambodian Children's Trust (CCT)
CCT Australia (CCTA) 247, 248, 251, 255
Chambers, D. 19, 25, 26
Chan, S. 250, 252, 253
chat piles 44, 45, *45, 49*
child labour: global challenges 99; labour rights 100; and orphanage tourism 8, 100–101; participation rights 102; policy and planning 101–102; poverty-stricken countries 99; protection rights 102; provision rights 102; tourism 99–100
children's human rights 102, 103
child rights: children's participation in tourism 102, *103*; facilitating relational spaces 105; Lundy model 103; participatory approach 105; safe relational spaces 104; tourism 98, 102–105, *104*; tourism industry 98
child safe practices 103
"child safe" tourism 101
child's education 99
child trafficking crises 100–101, 104
Chilisa, B. 170, 180
China: alternative tourism 205; economic crisis 206, 210, 212, 217, 218; mass tourism in 206, 209, 210, 212; political economy 208; qiongyou 205–208 *see also* qiongyou, in China; state-led tourism industries 207–208, 217–218
Chinese rural tourism destinations *see* rural tourism destinations
Chin, J. 23
the chronic conservation crisis 227–228, **228**
Clarke, V. 156, 192
clean energy technologies 5
climate change crisis 37, 69; challenges 2; clean energy technologies 5; climate justice 4–6; disproportionately affects 3; ethical implications 5
Clinton, Bill 28
Cole, S. 9, 68
Cole, T. 245
colonial and imperial theory 60
colonialism: direct/indirect products 3; tourism literature 67
community-based tourism enterprises (CBTEs) 179
community citizenship behavior for the environment (CCBE): perceived justice 81; and perceived justice 80; perceived justice effects 81–82; place identity 81–83; small tourism business entrepreneurs 78–79; subjective well-being 81–83
composite counter-story (CCS) 24–27
conceiving space 262, 264
consumption-oriented tourism 209, 210, 215
The contemporary (anthropocentric) tourism system 3
Convention on the Elimination of all Forms of Discrimination against Women (CEDAW) 64
Cook, D. A. 25, 27
Corntassel, J. 41
corporate social responsibility 78
Coulthard, G. 41, 54; *Red Skin, White Masks: Rejecting the Colonial Politics of Recognition* 41
COVID-19 pandemic: ecological health 196; economic impacts 99; Global South 77; health crisis 99; international travel restrictions 97; justice issue 131; negative conditions 128; platform workers 128; precarious workers 133; socio-cultural life 196; tourist-historic city 196–197; vaccines 9
Cresswell, T. 207
critical race theory (CRT): application 20; backlash against 27–29; composite counter-story methodology 24–25; counter-storytelling 22–25; development 21; interest convergence 22; path toward resistance 29–31; race and racism 21; racial discrimination and inequity 20; sustainable tourism studies 25–27; tenets 21, **21**; thoughts 31; unpacking 20–22
critical race tourism: advocates 21; development 20; and *endarkened storywork* challenge 32; framework 20, 31, 32; story *vs.* counter-story 22; theory 32
critical rethinking, of tourism 261–262, 271

critical tourism scholarship 20, 25, 97
Cuevas-Parra, P. 102
cultural identity crises 232–234, *233*

Damrosch, L. F. 153
dark tourism 38
Darwin, C. 226, 227
Dashper 9
De Bruyn, T. 241, 242, 255
decolonisation: anti-racist approach 60; assertion 11; power asymmetries 9–10; transformative actions 11; truth-telling narratives 10–11
Deen, A. 63
De Groot, R. S. 229
Deleuze, G. 267
Delgado Bernal, D. 24
Delgado, R. 21
Deliveroo 128, 133
Deloria, V., Jr. 41
Denzin, N. K. 19, 20
D'Eramo, M. 264
Derrida, J. 112, 114, 116
Dery, I. 30
DeSantis, R. 32
designers, qiongyou 212–213, 217
destination justice 207
destination management organisation (DMO) 195, 196
Develtere, P. 241, 242, 255
Diaz-Parra, I. 189
digital and financial rights 159
digital platforms: Deliveroo 128, 133; DoorDash 133; Facebook 136; Instagram 28; Menulog 133; Twitter 25; Uber Eats 128, 133, 136
Dillard, C. B. 30
Dillette, A. 25, 26
Dillette, S. 26
Dixson, A. D. 20, 25, 27
Donald, K. 65
donkey friends 208
Donnor, J. K. 29
DoorDash 133
van Doore, K. E. 100, 244
van Doorn, N. 141
Douban 210
Dredge, D. 190, 195
Drinnan, K. 252, 254, 256
Drury, A. C. 152
Dudley, K. 25
Duffy, R. 62, 153

East Africa: safari tourism 10; tourism policies 168
eco-feminist paradigm 59, 62
ecological inclusiveness: development principles 169, **169**; in Kenya 175; in Rwanda 178–179
ecological justice: human and biological diversity 3–4; tourism's role 3
economic crisis 206, 210, 212, 217, 218
Ellis, E. 153
employing sovereignty theory 10
employment 20, 51, 59, 68, 120, 129, 155, 161, 187, 227
endarkened storywork framework 20, 30–32
environmental harms: with human rights 65; management 37; settler-colonial state 42
environmentalism 4, 98
environmental justice: conservation and sustainability 4; human and biological diversity 3–4; tourism's role 3
Environmental Justice Policy 4
environmental movements 4
environmental protection 4, 76, 87, 178, 216, 234
Environmental Protection Agency (EPA) 46, 47, 52
equity theory 80, 87–89
Eriksen, T. H. 264
Escobar, A. 7
Estes, N. 42
The Ethical Research Involved Children (ERIC) 104
ethical responsibility: Emmanuel Levinas's ideas 112; Levinas' ethics 113; in tourism 115–116, 122–123
ethinking tourism 39, 271
Eurocentric modernity 7
Europe: tourists 157, 158; worst humanitarian crisis 110
Europe and the people without history (Wolf) 264
European Union training funds 112

Facebook 136, 138
Fainstein, D. R. 192
Fair Work Commission 135
Fannon, F. 41
Farzin, F. 151
female sex tourism 26
feminism 59, 61, 62, 66, 98
feminist alternative economics (FAE): core tenets 68, **69**; feminist ethics of care 60–62; gender and tourism sustainability 59; human rights based economy 64–66; labour process 60; social solidarity economy 62–64
feminist alternative tourism economics (FATE) 66, *67*; analysis 59; feminist perspectives 69; implementation 69–70
feminist ethic of care: disability and tourism provision 62; just destination 62; Kohlberg's model 60–61; Maori philosophy 61; post-COVID-19 recovery 62; Rawls' theory 60–61; sex tourism 62; social entrepreneurship 62; tourism ethical field research 62

Fennell, D. A. 62, 153
Ferguson, L. 61
field theory 265–267
financial crisis 99, 209, 210, 212
Flanagan, K. 241
floating hotel tourism model 229, 234
Floyd, G. 22
Font, X. 82
food delivery workers: challenges 134; complex injustices 138; global research 130; GPS tracking displays 133; rating system 139; safety procedures 139; shadowing 132; social media interactions 133; target migrant workers 130; temporary migrants 135–136; worker profiles 132, **132**
food sovereignty 224–225; broader application 225; ideas 225; theory and practice 225
forcibly displaced people: climate change crisis 110; geopolitical changes 110; interviews 117–120, **118**
For Space (Massey) 267, 268
Fraser, N. 66, 131, 141
Freedom of Information Act 28
Freire, P. 245–246, 254, 255
"The Frugal Chinese" 215
frugality: qiongyou, in China 215–217; travel strategies 211

Galápagos Biosecurity Agency (ABG) 228
Galápagos islands: the chronic conservation crisis 227–228, **228**; colonial theory 233; COVID pandemic 231–232, *232*, 234; cruise-based visitation model 230; cultural identity crises 232–234, *233*; Ecuadorian *vs.* International visitor arrivals 232, *232*; "floating hotel" tourism model 229, 234; food security 231; human-environment relations 226; special laws 227, 228, 233; tourism crisis evolution 228–231
gender and race lines 60
gender-based violence 59
gender equality 58, 59, 63, 64, 67, 68, 161
gender justice 59, 64, 69, 159–161
gender pay gap 60, 65
Ghosh, B. N. 242
Gilligan, C.: anti-patriarchal stance 61; ethic of care 61
Glasgow Declaration 5
Gleeson, B. 4
global crises 62; child trafficking 100–101; Covid-19 pandemic and climate change 69; environmental conservation 4; impacts 3; society 60; sociocultural wellbeing 4; "9/11" terrorist attacks 99; tourism and justice 3–6; tourism social injustices 98–101
global financial crisis 99
globalization 38, 171, 193
Global North: climate change-driven impacts 5; economic relations 242; social and environmental problems 5; voluntourism opportunities 241
global pandemic 1, 3, 8, 9, 19, 31, 241
global political–economic system 99
global solidarity 8
Global South: climate change 9; global crises 168; hospitality workers 59; tourism-dependent communities 5; tourism development policies 168, 170
global tourism system 60, 103
Gouws, A. 61
Graham, A. 8, 11
grassroots traveler 206–207, 218
Grimwood, B. 19
Guattari, F. 267
Guba, E. G. 156
Guia, J. 269
Guiney, T. 254
Guinier, L. 28
Gupta, J. 168, 170, 172
Gympaki, M. 68

Hai, N. T. T. 64
Hales, R. 153
Hallak, R. 82
Hall, C. M. 9–11, 99, 151, 158, 191
Hamedinger, A. 190
Haraway, D. J. 11
Harman, S. 9
Harris, A. P. 29
Hart, R. A. 103
Harvey, D. 188
Hatley, E. L. 8, 10, 37, 42–44, 46, 47–53
health crisis 3, 99
Henry, J. 245
heritage tourism 22, 31, 77, 208
He, X. 81, 83, 87
Hickel, J. 11
Higgins-Desbiolles, F. 4, 7, 10, 19, 59, 62, 153, 171, 217, 225
Higham, J. 59
Hillenkamp, I. 63
hodological space 262, 263, 265–267, 269, 270
Hoffstaedter, G. 9
Hopkins, J. 49
hostipitality: city-tours guide 116–117; Jacques Derrida's theorisations 112; in tourism 112–114
Houston, D. 160
"How is Water" 263, 271
human dignity 131, 224
human diversity 3–4
human physiology 36

human rights: certain stakeholders 99; impacts 152; issues 99; violations 5, 101, 110, 206; and women 64
human rights based economy (HRBE): capitalist neo-liberal economic system 64; and girl child 64; hotel business 65; Indigenous values systems 66; principles and laws 64; and women 64
Human Rights Based Economy (HRBE) 64–66, 68, 69
Human Rights Council 65
Hunt, C. A. 10

Illich, I. 243
inbound tourism 209
inclusive development framework: The African Union's Agenda 2063 182; deductive codes 172, **173**; ecological inclusiveness 169, **169,** 175, 178–179; economic aspects 171; Kenya and Rwanda 167–183; principles 169, **169**; relational inclusiveness **169,** 169–170, 175–177, 179–180; social inclusiveness 169, **169,** 174–175, 177–178; three pillars 174, 180
Indian Removal Act of 1830 43
Indian Territory 43
India tourism 123
Indigenous communities: environmental movements 4; in Hawaii 225; lifeways 7; mobilization 224
Indigenous environmental justice 4
Indigenous intelligence systems 41
Indigenous people: Eurocentric perspective 23; in North America 42
Indigenous perspectives 39–42, 242
Indigenous resurgence 39–42
industrialization 38, 40, 54, 209
Ingold, T. 132
Instagram 28
International Labour Organisation (ILO) 99, 129
International Monetary Fund (IMF) 7, 168
international sanctions: nuclear agreement 154; politico-economic crises 150
International Summit on Child Protection in Travel and Tourism 244
Iranian tourism 150; civil aviation 158; interviews 156; 2015 nuclear agreement 154; oil revenues 154; US law travelers 158
Iran–Iraq war 154
Isaac, R. K. 206
Izurieta, J. C. 10

Jaan Bai 248
Jacobs, J. 263
Jamal, T. 59, 62, 152, 153, 189
Janni, P. 266
Jarosz, L. 225
Jim, R. 8, 10, 37, 42, 43, 46, 47, 49–53
Joppe, M. 186, 187
Josefsson, J. 98, 100
Journal of Sustainable Tourism 26, 222
Jover, J. 189
Judd, S. S. 192
justice: and dignity 131; global crises 3–6; human and biological diversity 3–4; migrant food delivery workers 131; *see also specific entries*
justice theory 222, 223, 234
justice tourism 205–207, 217, 218

Kalisch, A. 9
Kant, I. 152
Kenya tourism policy documents: *Buy Kenya Build Kenya* 176; ecological inclusiveness 175; policy documents 173–174; relational inclusiveness 175–177; social inclusiveness 174–175; sustainable tourism development 176
Kern, L. 263
Kirby, E. 253
Kirkwood, N. G. 49, 52
Kline, C. 263
Kloutsiniotis, P. V. 80
Kozinets, R. V. 210

labour market 60, 128, 130, 134, 136, 138, 142
Ladson-Billings, G. 29
LaDuke, W. 38, 40; *All Our Relations: Native Struggles for Land and Life* 40
Lamond, I. R. 10
Latour, B. 266–270; *Down to Earth* 268; *Politics of Nature* 268; *Reassembling the Social* 268
Laughter, J. 8
Laydjiev, I. 98
LEAD Agency's Toxic Tours 37, 38
Lediard, D. E. 246, 255
Lee, B. 28
Leeds, England: the context 119–120; cultural walking tours 112; socio-cultural fabric 118
Lefebvre, H. 187–188, 262; theory 188
Lefebvre's The Right to the City 187–188, 190, 197; geographical and spatial treatment 187; vision of city life 188
Levinas, E. 112–116
Lewin, K. 263, 265, 267, 268–271; *The Principles of Topological Psychology* 266
Lincoln, Y. S. 20, 156
Lindberg Falk, M. 68
Li, Q. 10, 87
Liquid Modernity (Bauman) 262
Liu, Y. 80, 83, 87, 88
lived-space (hodological space) 266
Living Planet Report 2022 3
local community-centered tourism 205, 207
Local Environmental Action Demanded (LEAD): autoethnographic dialogue with 46–50,

47, 49; co-founders 43; founders 42; leaders 46–50
local judicial systems 172
local support workers: interviews 117–120, **118**; public and non-governmental sectors 117, **118**
Lovelock, B. 62
Lovelock, K. 62
Low, N. 4
Lumumba, P. L. O. 170, 171, 180, 181
Lundy, L. 103
Luo, X. 208
Lyle, R. R. 246

mainstream tourism 206, 210, 215, 217
de Man, F. 59
Maori philosophy 61
"Marry an Amsterdammer for a day" 263
Massey, D. 262, 266–270; *For Space* 267, 268; *Space, Place, and Gender* 267
mass tourism 206, 208–210, 212
McGehee, N. G. 243
McLaren, D. 190
Menulog 133
Messing, K. 65
micro, small and medium enterprises (MSMEs) 177
migrant food delivery workers: agency 136–137; challenges 133; collaboration 137; COVID crisis 135; experiences 131; food and financial support 138; *vs.* independent workers 135–136; injustices ecosystem framework 140, *140*; navigate injustices 140, *140*; resistance 137; *see also* food delivery workers
"Migrantour" 263
migration: definition 111; residency rules 228
Mihail, D. M. 80
Mika, J. P. 7
Mika, R. A. 7
Miller, H. 158
Miller, R. 23
mobile ontologies, spatial justice 268–270
mobile sociology 262
mobilities 115, 157, 158, 207, 262, 265
mobility praxis 262, 265–267, 270, 271
modern capitalism 264
Modern Slavery Act 65, 247, 253
modern spatiality 263–265
Morgan, N. 19
Mostafanezhad, M. 254
Mottiar, Z. 63
Mowatt, R. 19
Mowforth, M. 242
multilateral trading system 151
Munt, I. 242
Myanmar tourism 154
Myths, Legends and Old Sayings (Shade) 40

Nairn, R. W. 49
narcissistic emotion 119
"National Tourism Work Conference" 209
navigating injustices: agency 136–137; collaboration 137; resistance 137
neocolonialism 242
neo-liberal globalisation 62
neoliberalism 205, 207, 218
neo-liberalism/late capitalism 265
neoliberal space tourism 262
neoliberal spatiality 263–265
netnography 210–211
Neuenkirch, M. 151
Neumeier, F. 151
New Public Management (NPM) policies 195
The New York Times 22
NGOs *see* non-government organisations (NGOs)
Nketsian, N. K. 181
Nobel, R. 60, 66
Nomadic Theory 267, 268, 270
non-BIPOC scholars 19
non-government organisations (NGOs) 10, 241
North American feminism 61
Novelli, M. 173, 182
Nunkoo, R. 223
Nyíri, P. 206, 208

Okello, M. 173, 182
online communities 210, 213
open relational space 267–268, 270
Ordoñez L. A. 10
Organ, D. W. 82
organizational citizenship behavior (OCB) 78; engagement 80; exchange relationships 80; *vs.* organizational identification 82; and perceived justice 80
organizational citizenship behavior for the environment (OCBE) 77–79
Ormond, M. 117
orphanage tourism: anti-slavery policy 243; Cambodia 243–244; child rights 100–101; Global North *vs.* South 241
orphanage trafficking 100
orphanage volunteering programmes 101
otherness 263, 264, 267, 269

Paddison, B. 10, 11
Palestine tourism 206
Pan-African: framework 172; movement 172
Pan African Vision 182
participation rights: child labour 102; children and young people's 102
Patil, V. 245
Peksen, D. 152
Perkins, E. E. 11
Pezzullo, P. C. 38, 39, 50, 51

place attachment theory 81
place identity and subjective well-being: effects 81–82; mediation 82–83; perceived justice effects 81; positive impacts 87; role 82
platform economy: cumulative impacts 136; external support 138; independent workers and temporary migrants 135–136; materials and methods 131–133; navigating injustices 136–137; precarious platform work 134–135; precarious workers 133–134; traditional ethnographic methods 131
Plumwood, V. 6
policymaking and governance structures: economic policy and local development 193–195; global policymakers 190; governance and stakeholder engagement 190–191, 195–196; Lefebvre's the right to the city 188–189; semi-structured interviews 192, **192**; spatial (in)justice 188–191, 193–197; stakeholder groups 195; thematic analysis framework 192, **193**; unjust geographies 188–190
Politics of Nature (Latour) 268
Pon, Jedtha 247, 252
postcolonialism 98
poverty 5, 58, 61, 63, 69, 168, 186, 241
precarious platform work 134–135
precarious workers 133–134
precarious works: cumulative impacts 136; global North impacts 129; justice and dignity 131; platforms and migrant workers 129–131
Price, R. 129
The Principles of Topological Psychology (Lewin) 266
Pritchard, A. 19
protection rights 102
provision rights 102
public policy 59, 187, 189, 191
Pulido, L. 160
Purcell, M. 190
Pürckhauer, A. 66

qiongyouers 206, 208, 211, 217, 218; alternative travel agents 212–213; female 215, 216; frugality as justice 215–217; local embeddedness of 213–215, *214*; travel practices of 212–215, *214*
qiongyou, in China 205–208; collective social action of poor 212, *212*; cultural legitimacy of 216–217; designers 212–213, 217; frugality as justice 215–217; limitations 211, 218; netnography 210–211; semi-structured interviews 211; state-led tourism commodification and crisis 209–210; ticket evasion and services 213–215, *214*, 217
"qiongyou tianxia" 210
Quiñones, B. R. 63

Rarick, C. A. 154
Rastegar, R. 4, 39, 52, 97, 131, 153
Rawls, J. 152, 153, 170
Rawls' theory 60, 153
Reassembling the Social (Latour) 268
Recovering Black Stories in Qualitative Research (Toliver) 30
Red Skin, White Masks: Rejecting the Colonial Politics of Recognition (Coulthard) 41
refugee-crisis 110; and asylum-seekers 111; co-creative hospitable tourist experiences 123–124; creating and delivering tourism services 111; ethical responsibility 122–123; individual and collective identities 121; novel theoretical approach 111; the "strangerhood" concept 120–122; territorial identity 120; and tour-guiding 111, 116–117
Refugee Voices Tours 111
Regulations on the Administration of Travel Agencies 209
relational inclusiveness **169,** 169–170; in Kenya 175–177; in Rwanda 179–180
reliability and validity test: measurements 84; rural tourism destinations 85
restricted mobility rights 157–159
Riddell, S. 102
The Right to the City: re-theorisation 189–190; Soja's spatial treatment 188; theorisation 188
Riordan, T. 9
Robbins, S. P. 79
Robinson, R. N. S. 9, 129
Roche, S. 101
Rogerson, C. M. 242
Roitman, J. 130
"Roots Guide" 117, 263, 271
Rose, D. B. 10
Rosenberg, E. 151
Rousseau, C. K. 20
rural tourism destinations: closer and longer-lasting relationships 79; community-based small businesses 76; critical stakeholders 76; data analysis 84, **85**; environmental management systems 77; equality/inequality emerges 77; future research 89; geographical and psychological boundaries 79; hypotheses development 78–83; limitations 89; literature review 78–83; measurements 84; methods 83–85; perceived justice 80; quality of life 78; questionnaire-based survey 83; reliability and validity test 85; research findings 86–88; resources and environment 79; results 85–86; sampling 83–84, **84**; small tourism business entrepreneurs 78–80; social fabric 82; stronger familial and cultural ties 79; structural equation modeling 85–86, **86,**

87; study sites 83–84, **84**; theoretical and managerial implications 88–89
Rustler, A. 152
Rwanda tourism policy documents: challenges 177; ecological inclusiveness 178–179; Global South nations 168; inclusive development framework 169, **169**; policy documents 173–174; relational inclusiveness 179–180; social inclusiveness 177–178; training system 177
Rydzik, A. 68

Saarinen, J. 9, 242
sanctions: application 153; data collection and analysis 155–157; digital rights/communication rights 159; ethical perspectives 152–154; financial restrictions 159; foreign policy measures 150; foreign policy positions 152; gender justice 159–161; implications 161–162; legitimacy and effectiveness 152; multilateral trading system 151; political goals 151; post-Cold War era 149; restricted mobility rights 157–159; study context 154–155; tourism and SDGs 151–152
Sauer, J. B. 246
Scheyvens, R. A. 10, 39, 51, 52
Schwartz, J. M. 8
SDGs *see* Sustainable Development Goals (SDGs)
Seifert, A. M. 65
self-determination theory 81
sex tourism 26, 62
Seyfi, S. 9, 151, 152
Shade, H. 38, 40; *Myths, Legends and Old Sayings* 40
shadowing 132
Simpson, K. 245
Simpson, L. B. 41
small tourism business entrepreneurs (STBEs): business characteristics 77; characteristics 78; community citizenship behavior for the environment 78–79; discretionary and informal behaviors 78; discretionary environmental behavior 76, 77; distinctiveness 78; environmental practices 77; general environmental attitudes 77; importance 79; lifestyle motivations 78; with local communities 79; perceived justice 80; place identity 82; rural destinations 78–80; social interactional factors 77; sustainable behavior 78; systematic mechanisms 77
smart sanctions 150
Smith, M. 62, 153
Social and Solidarity Economy Enterprises and Organisations (SSEEOs) 63
social exchange theory 80, 82, 87–89
social identity theory 81, 82
social inclusiveness 169, **169**; in Kenya 174–175; in Rwanda 177–178
socialising tourism 4
social solidarity economy (SSE) 62–64
The Social Solidarity Economy 66
social tourism 207
socio-cultural action 205, 206, 215–217
socio-ecological crises 3
Sociology beyond societies: mobilities for the twenty-first century (Urry) 262
soil contamination 45
Soja, E. W. 187–190, 194
Soler-i-Martí, R. 63
solidarity: cold and disembodied rules 9; importance 8; justice principles 9
Solidarity Economy movement 63
sovereignty theory 222; contemporary research 224; food sovereignty 224–225; hallmark archival research methods 226; and tourism 225–226
space: field of forces and relationships 265–267; open relational 267–268, 270; as surface 264
space of tourism: field theory 265–267; hodological space 262; modern spatiality 263–265; neoliberal spatiality 263–265; open relational space 267–268; posthumanism mobile ontology 268–270
Space, Place, and Gender (Massey) 267
spatial injustice 264
spatiality: modern 263–265; neoliberal 263–265
spatial justice 267; economic policy and local development 193–195; governance and stakeholder engagement 190–191, 195–196; mobile ontologies posthuman theoretical framework 268–270; posthumanism mobile ontology 268–270; tourist-historic city 196–197; the tourist-historic city 196–197; and unjust geographies 188–190
spatial theory 268
state-led tourism commodification: qiongyou, in China 209–210
state-led tourism industries, China 207–208, 217–218
Stefancic, J. 21
Stephenson, M. L. 158
Stevenson, N. 9
strangerhood: people's identities 120–122; Zygmunt Bauman's notion 112
Structural equation modeling 85–86, **86,** *87*
study sites 83–84, **84**
Su, L. 81, 82
Superfund program 52
Superfund Site program 46
Sustainable Development Goals (SDGs) 150, 151; ethical challenge 151; tourism destinations 151; women rights and empowerment 160

sustainable tourism 24, 25–27, 175, 178, 198, 234
Swanson, S. R. 82
Syrian civil war 110

Tales of a City Tours 112, 117
Tar Creek and Grand Lake watershed systems: background 43–46; discussion 50–54; hazard ranking model 46; heavy metal sediment contamination 37; LEAD Agency's Toxic Tours 37; methodology 42–43; results 46–50; toxic algae blooms 37; tribal nations and crisis 43–46, *44, 45*
temporary migrants 130, 133, 135–136, 141
Teresa, A. Q. 82
"9/11" terrorist attacks 99
Terry, W. C. 65
ticket evasion, qiongyou, in China 213–215, *214,* 217
Tilly, C. 226
Tisdall, E. K. M. 102
Toliver, S. R. 20, 29–31; *Recovering Black Stories in Qualitative Research* 30
Tomassini, L. 10
Topological Psychology 267
Torabian, P. 158
tour-guides: cultural creativity 116; new configurations 116
tourism: B-Corp business model 68; element 61; ethical responsibility 115–116; *vs.* forced migration 111; Green's legacy 22; "Gross National Happiness" approach 7; intercultural contact theory 8; preliminary framework 2, *2,* 10–11; public policy 187; recognition 7; sanctions-ridden destinations 150; social-natural sciences 4; solidarity 8–9; strangerhood 114–115; sustainable futures 2, 6–10
Tourism Advisory Board (TAB) 196
tourism employment 65
tourism participation: and justice issues 206–207; low-budget options 215; promotion 211
tourism policy documents: African-informed approach 171; anti-environmental behaviors 175; deductive or concept-driven coding 172; Global South nations 168; inclusive development framework 169, **169**; investor tailored approaches 181; Kenya and Rwanda 167–183; knowledge production systems 171; pro-environmental behaviors 175; sustainable tourism 175; textual analysis 172–173
tourism scholars 19
tourism space 262–265, 269, 270
tourism studies: child labour 99–100; feminist and postcolonial approaches 103; fundamental concepts 20; post-pandemic 105, 106; redirecting 97; rethinking 97, 102; scholarship type 20
tourist-historic cities 187
tourist syndrome 261
toxic tourism: emerging field 38–39; empowerment framework 39, 51–53; Indigenous perspectives 39–42; Indigenous resurgence 39–42; knowledgeable tour guides 38; people between their environments 50–51; search for justice 39; traditional perspectives 53–54
toxic tours 8, 10
travel agencies 208, 209, 212–213
#TravelingWhileBlack 25
Trump, Donald 28
Tuan, L. T. 80
Twitter 25

Uber Eats 133, 136
Ukrainian crisis 110
UNCRC *see* United Nations Convention on the Rights of the Child (UNCRC)
unemployment 59, 64, 210, 217
unilateral sanctions 151
United Nations Convention on the Rights of the Child (UNCRC) 99, 102
unmasking and unmaking racism 20
UN Sustainable Development Goal 8.7 100
Unwin, T. 188
Urban, J. 66
Urry, J.: *Sociology beyond societies: mobilities for the twenty-first century* 262

Van Zyl, M. 61
Vegelin, C. 168–170, 172
Vergunst, J. L. 132
Verschuur, C. 63
Vieira, T. 141
Vietti, F. 117
Vijay, D. 141
The Village Hive model 248–249, 251, 253, 255
volunteer tourism 243–244
voluntourism 100
Vo-Thanh, T. 9

Wall, J. 98, 100
Wang, J. 10
Wang, S. 83
Wang, Y. 10
Watene, K. 7
water crises 37
water pollution 5, 37
van der Watt, H. 51, 52
Wearing, S. 243
Weaver, F. 22
Weber, M. 264
web ethnography 210

White, R. 186
White saviourism 244–245, 250–251
Whyte, K. P. 6
Williams, P. 24
Wilson 9
Winkler, T. 246, 247, 250, 251, 253, 254
Wolf, E.: *Europe and the people without history* 264
women rights 161
Women's motives 63
The 2021 World Economic Forum Insight
  Report 59
World Travel and Tourism Council (WTTC) 155
World Wars I 44
World Wars II 44
Wu, M. 87
Wu, M.-Y. 10
Wu, X. 10

Xie, J. 208
Xu, H. 82, 84

Yang, Y. 10
Yin, R. K. 192

Zhang, H. 82, 84
Zhang, Y. 82
Zhihu 210
Zhu, X. 208